Quiet Daily Geomagnetic Fields

Edited by
Wallace H. Campbell

1989

Birkhäuser Verlag
Basel · Boston · Berlin

Reprint from Pure and Applied Geophysics
(PAGEOPH), Volume 131 (1989), No. 3

Editor's address:

Wallace H. Campbell
United States Department of the
Interior Geological Survey
Mailstop 968
Federal Center Box 25046
Denver, CO 80225
USA

Library of Congress Cataloging in Publication Data

Quiet daily geometric fields / edited by Wallace H. Campbell.
  p. cm.
  »Reprint from Pure and applied geophysics (PAGEOPH), volume 131
(1989), no. 3« – – T. p. verso.
    ISBN-13: 978-3-7643-2338-7   e-ISBN-13: 978-3-0348-9280-3
    DOI: 10.1007/978-3-0348-9280-3
    1. Magnetism, Terrestrial – – Diurnal variation.
I. Campbell, Wallace H. (Wallace Hall), 1926 –.   II. Pure and applied geophysics.
QC831.Q54   1989   538'.742--dc20   89-7045

CIP-Titelaufnahme der Deutschen Bibliothek

Quiet daily geomagnetic fields / ed. by Wallace H. Campbell. –
Reprint. – Basel ; Boston ; Berlin : Birkhäuser, 1989
  Aus: Pure and applied geophysics ; Vol. 131, No. 3

NE: Campbell, Wallace H. [Hrsg.]

Contents

PAGEOPH, Vol. 131, No. 3 (1989)

0033–4553/89/030315–17$1.50 + 0.20/0

An Introduction to Quiet Daily Geomagnetic Fields

Wallace H. Campbell[1]

Abstract—On days that are quiet with respect to solar-terrestrial activity phenomena, the geomagnetic field has variations, tens of gamma in size, with major spectral components at about 24, 12, 8, and 6 hr in period. These quiet daily field variations are primarily due to the dynamo currents flowing in the *E* region of the earth's ionosphere, are driven by the global thermotidal wind systems, and are dependent upon the local tensor conductivity and main geomagnetic field vector. The highlights of the behavior and interpretation of these quiet field changes, from their discovery in 1634 until the present, are discussed as an introduction to the special journal issue on Quiet Daily Geomagnetic Fields.

Key words: Quiet daily geomagnetic field variations, *Sq*, *L*, lunar variations, ionospheric dynamo currents, thermotidal currents.

1. The Sq Field Variations

The daily record of geomagnetic variations at any world location typically shows a multitude of irregular changes in the field that represent the superposition of many spectral components whose amplitudes generally increase with increasing period. Unique current sources in the upper atmosphere and magnetosphere have been identified as origins of many of these spectral field variations. On occasion there are days when the magnetic records are smoothly changing with primarily 24-, 12-, 8-, and 6-hour period spectral components dominating the field composition with few of the irregularly appearing, shorter or longer period changes present. On these days, the oscillations of the three orthogonal field components produce records that are predictably similar to others recorded many days earlier. Such records describe the "quiet daily geomagnetic field variations," and when the small but persistent effects ascribed to the lunar (*L*) tidal forces are removed, the changes are referred to as "*Sq*" for "solar quiet" fields. Some authors, particularly those interested in earth crustal conductivity studies, prefer to call the *Sq* "Daily Variations" or "Diurnal Variations." More correctly, "Diurnal Variation" should

[1] U.S. Geological Survey, Denver Federal Center, P.O. Box 25046, Mail Stop 968, Denver, CO 80225, U.S.A.

be restricted to meaning only the 24-hr spectral component of the quiet field variation.

Solar activity, identified with the sunspot number, controls the percentage of magnetically quiet (Sq producing) days in a year as an inverse relationship. The quietest geomagnetic levels usually occur on, or a year after, the minimum in sunspot number. Because of the 10.6-year cycle in solar activity, a similar cycle of geomagnetically quiet years occurs. The annual percentage of the 3-hr magnetic activity indices, Kp, that are below a value of 3 can be used as a measure of the quietness of any year. In 1934, 1945, 1955, 1965, and 1976, these percentages were 86, 78, 85, and 80. In contrast, the peak activity years had about 40 to 60 percent of the Kp indices less than 3. For quiet days, some authors select the five that are published each month as an item of "Geomagnetic and Solar Data" in the *Journal of Geophysical Research*. Unfortunately, such "quietest days" may encompass considerable activity in a fully disturbed month; part of the reported solar cycle amplitude change in Sq is likely due to the inclusion of such disturbed data. However, with a more careful selection of quiet days, an increase of Sq amplitude and a change of the global appearance of Sq amplitude contours do occur in active years.

The Sq field changes slowly in amplitude and phase through the months of the year. Outside the polar regions, the quiet-day variations about a daily mean level are modulated by the daytime solar zenith angle; the records show a relatively flat appearance in the night hours. There is a clear annual increase in amplitude during the summer months and a seasonal phase shift of the maximum (earlier in summer and later in winter) for Sq at the midlatitude stations. The data from equatorial stations display a clear semiannual, equinoctial enhancement of amplitudes. There seems to be a North/South Hemisphere asymmetry in the quiet variations that may be associated with the greater distance between the magnetic dip pole location and the spin axis pole in the Southern Hemisphere than in the Northern Hemisphere. Inside the polar cap, an enhanced E-region ionization and local electric fields associated with the magnetospheric processes that introduce field-aligned currents from the magnetosphere are superposed upon the local dynamo processes, which are the dominant Sq source at the lower latitudes.

Away from the polar cap, the major driving force for the quiet day field changes seems to arise from the dynamo-current process in the ionospheric E region between 90 and 130 km. At that altitude, the interplay of the Sun-energized thermotidal motions, thermospheric winds of global scale, and lunar tidal forces move the local ionization that has been generated, in large measure, by the solar radiation. Electrons in this dynamo region are relatively collison free. In contrast, the positive ions there are rather strongly subjected to collisions with the neutral atmosphere. As a result, a current of electrons cannot be neutralized by an opposite current of positive ions. This electron current is responsible for the Sq and L fields.

Depending upon the direction of the earth's main field, with respect to a local driving electric field direction, three directional conductivities coexist in this interesting dynamo region of the ionosphere. At most latitudes the Hall conductivity (where the electric field is perpendicular to the component of the earth's magnetic field, and the current direction is perpendicular to the electric field) dominates. There is some evidence that at night hours, when the solar production of electrons ceases and recombination processes have essentially removed the D- and E-region ionization, the residual lower F-region may be sufficiently conducting to support a low amplitude Sq effect.

The dynamo current that flows as the ionospheric plasma is moved through the earth's field is similar, in a way, to the current that is generated in a hydroelectric plant as conducting wires are moved, by the water turbines, through the field of a large magnet. The major difficulties with this analogy are that the ionosphere is a plasma varying in time with a tensor conductivity which is dependent upon the directions of the forces and fields in the region, and that the magnetic field from the earth differs in direction and intensity with location about the earth.

A special electrojet current condition is established at the magnetic dip equator where the earth's field, at E-region heights, is horizontally directed south to north. There, the Hall current flow, inhibited by the ionospheric boundaries perpendicular to the flow, causes a polarization field to be established in opposition to the flow. Under those unique equatorial conditions, the effective conductivity parallel to the boundary (perpendicular to the earth's magnetic field) is increased. The net result is a considerable enhancement of the east-west ionospheric conductivity that, for quiet daily variations, allows the flow of an intense eastward current called the "equatorial electrojet."

2. History of Understanding Sq

The first reports of a regular quiet-day variation of field came from careful observations of the motion of the end point of the long magnetic compass needle by Hellibrand in 1634 at London (VON HUMBOLDT, 1863). The quiet variation was also independently discovered later by Tachard in 1685 in Siam (AAB, 1925). However, the variation was not truly established until GRAHAM (1724) made the first calibrated routine determinations of quiet changes in declination, $Sq(D)$. By the latter part of 18th Century, measurements by CANTON (1759) and by Celcius, Hiorter, Cassini, Gilpin, and Beaufoy (VON HUMBOLDT, 1863) had established that the seasonal change of the diurnal variation amplitudes was more than a local phenomenon.

Three major contributions to the study of Sq can be attributed to GAUSS (1841). He increased the sensitivity of the magnetic observations by a new design that affixed a mirror to the compass magnet allowing its angular change to be more easily determined at a distance. He organized a Magnetic Union of simultaneous observations about Europe (truly the first international geophysical cooperation) to separate the local and regional field changes. The everlasting gift to geomagnetism was his development of the mathematical method for separating the external and internal contributions to the surface field by a unique global analysis of the earth's main field. However, the application of this new "Spherical Harmonic Analysis" (SHA) to the Sq fields had to wait another 40 years until reliable worldwide observations of daily variations were available.

Baron VON HUMBOLDT (1863), an explorer and universal scientist, has been called the father of geomagnetism in the Americas because of his measurements there (1799–1804) and his encouragement of further observations. His publications, influential letters, and pressures on the Royal Society in Britain (VON HUMBOLDT, 1938) were responsible for SABINE'S (1847, 1857a,b) establishment of standardized observatories at Toronto (Canada), St. Helena (South Atlantic Ocean), Cape of Good Hope (South Africa), and Hobarton (Tasmania, Australia). SABINE (op. cit.) subsequently discovered that the magnetic field intensity varied in parallel with the sunspot changes and also determined the Sq and lunar field changes at his observatories.

In 1846, Charles Brooke (AIRY, 1896) constructed the first self-recording photographic magnetometer using a gas-flame light source, a mirror arrangement similar to that of Gauss, and a rotating cylinder with photographic paper (not much different from the recording variometers of today!). The instrument was subsequently installed in the Kew Observatory (near London) then under the supervision of Balfour Stewart (Report of the British Association for 1859, cf. WALKER, 1866). The resulting continuous, sensitive recordings were to make possible Stewart's later sensational discovery of the existence of a conducting ionized layer in the atmosphere above the earth.

The first modern textbook on geomagnetism was published by WALKER in 1866. That book went beyond the simple cataloging of current knowledge to a level of skillful interpretation of the global geomagnetic observations. For example, WALKER (1866) noticed a geomagnetic ordering of the Sq behavior (Figure 1) and proposed the existence of a unique magnetic equator (pg. 126): "Such a line would divide the globe into parts which might be called the North and South Magnetic Hemispheres. In leaving this line either northward or southward the characteristics of the mean solar diurnal variation present themselves and this variation gradually increases in amount as we recede further from the dividing line."

Although the discovery of the ionosphere has been formally credited to the later radiowave studies in the mid-1920's, the region was, in fact, described 43 years earlier by STEWART (1882) who reasoned that "conductive currents established by the Sun's heating influence in the upper regions of the atmosphere are to be regarded as conductors moving across lines of magnetic force and are, thus, the vehicle of electric currents which act upon the magnet[-ometer]." Relying on his supporting laboratory studies of ionized rarefied gases, he concluded that the

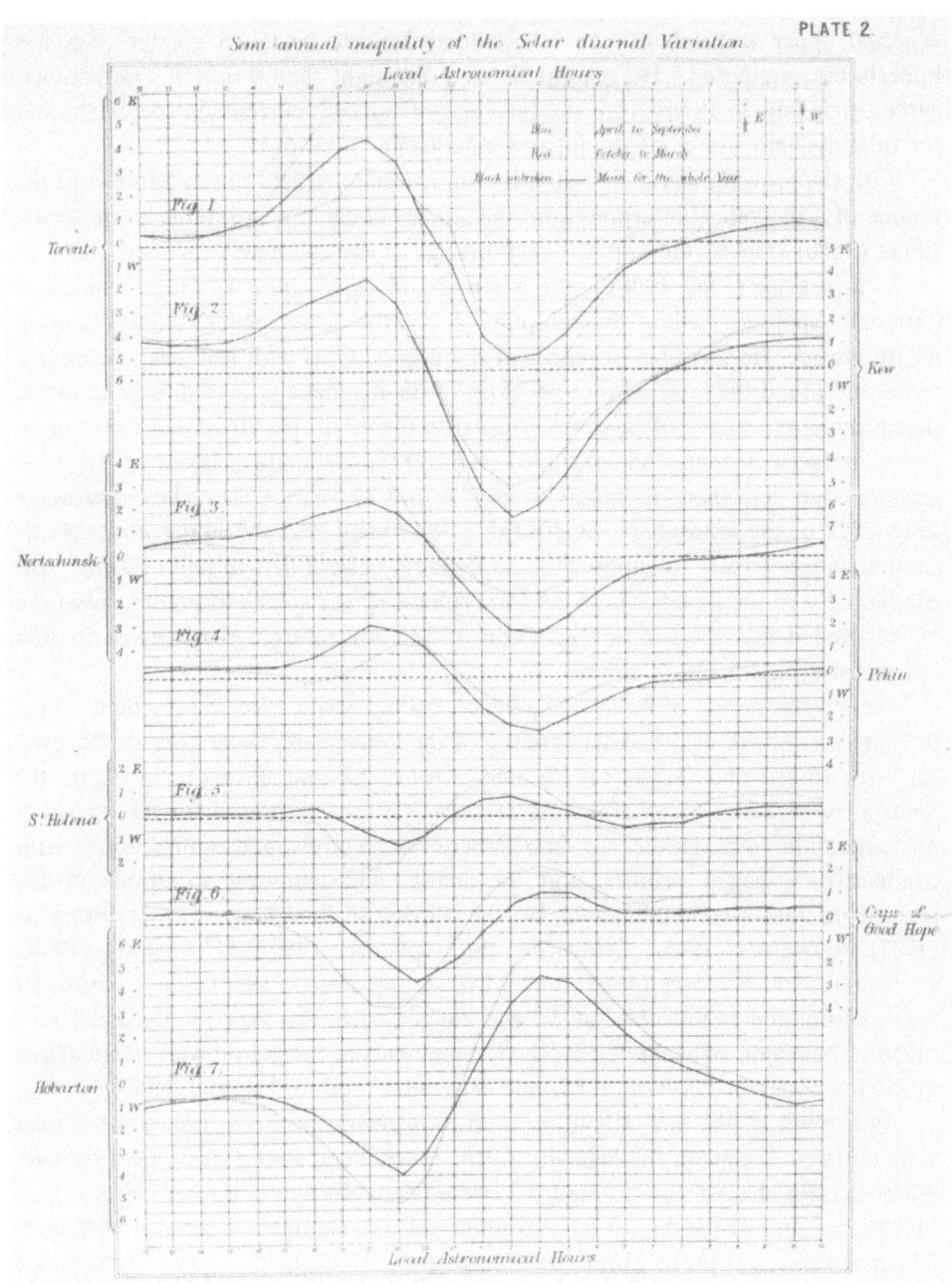

Figure 1

Daily variation of $Sq(D)$ at Toronto (56°N.; eastern Canada), Kew (53°N.; near London, England), Nertschinsk (42°N.; Nertchinsk, Siberia), Peking (28°N.; Beijing, China); St. Helena (−12°S.; Island, South Atlantic Ocean), Cape of Good Hope (−30°S.; Capetown, South Africa), and Hobarton (−53°S.; Hobart, Tasmania) observatories (at approximate geomagnetic latitude). Plate 2 of E. Walker *Terrestrial and Cosmical Magnetism* (Deighton, Bell, and Co., Cambridge 1866) 386 pp.

expected upper atmosphere region "conductivity may be much greater than has hitherto been supposed." He also predicted a day/night change in that conductivity, surface field contributions from the secondary (induced) currents in the earth, and the tidal-dynamo source of the lunar geomagnetic variations.

With the growing worldwide distribution of accurate field observations and the outline of a theoretical interpretation, the studies of Sq moved rapidly to maturity. ELLIS (1880) tracked the sunspot relationship of Sq at Greenwich from 1841 to 1877. SCHUSTER (1889, 1908) made a remarkable innovation to Gauss' spherical harmonic analysis method by adapting it to the geomagnetic daily variation measurements. He was able to represent the observations with just a few harmonic terms and found the external/internal ratio of the magnetic potential to be about 4, thus proving the source of Sq to be external to the earth. He attributed the internal part to induced currents and concluded (rightly) that "the upper layers of the earth must conduct less than the inner layers." A full 17 years before the "radiowave discovery" of the ionosphere, SCHUSTER (1908) computed the upper atmospheric conductivity necessary to support the Sq dynamo current system and ascribed this conductivity to the ionization of the atmosphere by the solar ultraviolet radiation. He concluded that the sunspot variation of Sq resulted from the increase in that ionization at high sunspot times.

The 40-year period of radiowave science, ending about 1966, represented a time of explosive growth in the understanding of the ionosphere. Soon after its "discovery" by BREIT and TUVE (1925) and APPLETON and BARNETT (1925), the ionosphere became a focus of intense research. Of special interest for the theory of Sq generation were studies of atmospheric oscillations and ionization, tensor conductivities, height profiles, and the spatial and temporal variations of the ionosphere that were highlighted by the works of PEDERSEN (1927), PEKERIS (1937), COWLING (1945), CHAPMAN and COWLING (1952), CHAPMAN (1956), RATCLIFFE and WEEKES (1960), and MAEDA and KATO (1966). The ionospheric wind distribution responsible for Sq was calculated by MAEDA (1955) and KATO (1956). The major ionospheric physics developments of the period were summarized in the textbooks of DAVIES (1965) and RISHBETH and GARRIOTT (1969).

As a result of the new attention upon ionospheric research, numerous studies were initiated to explain the behavior of the ionospheric source of Sq field changes whose surface characteristics (Figure 2) were being clearly established. HASEGAWA (1936a,b,c) studied the day-to-day variability of Sq. The intense equatorial electro-jet current was brought to world attention by BARTELS and JOHNSON (1940a,b) and EGEDAL (1947, 1948). Theoretical explanations of this large increase in $Sq(H)$ variation near the dip equator locations were proposed by CHAPMAN (1951) and HIRONO (1952, 1953). Major refinements of the dynamo theory of Sq were contributed by MAEDA (1952), FEJER (1953, 1964), BAKER (1953), BAKER and MARTYN (1953), LUCAS (1954), MAEDA (1955), and KATO (1956). The International Geophysical Year, of solar active year 1958, spawned a growth of magnetic

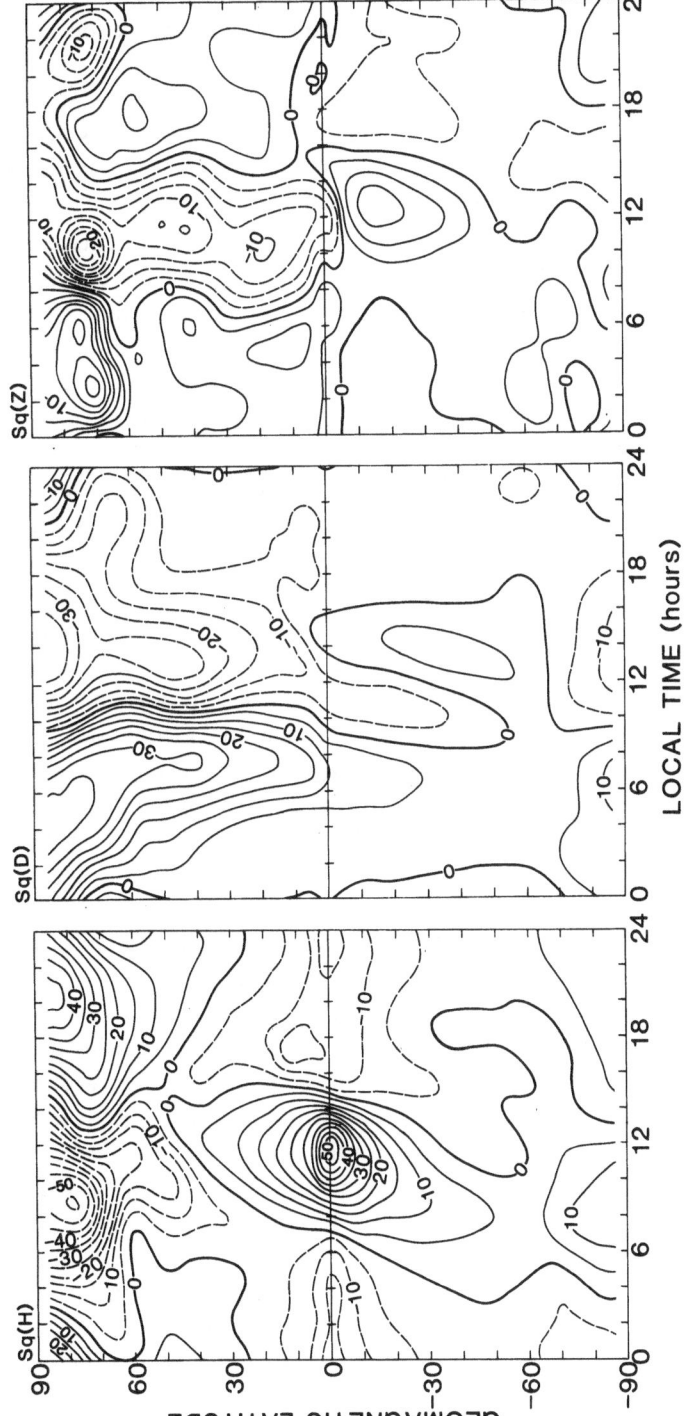

Figure 2

Amplitude contours (in gammas) for surface measurements of Sq in local time versus geomagnetic latitude. H, D, and Z components are represented at left, center, and right, respectively, on 21 June 1965 for variations at 1700 UT with local noon at 75°W. longitude. Contours are shown for 5–gamma increments in H and D but 2.5–gamma increments in Z. Positive and negative field values are indicated by solid and dashed lines, respectively.

observatories and cooperative studies that continued through the 1966 sunspot minimum year; many reports of Sq behavior came from this period (PRICE, 1969).

Research during the period beginning in the mid-1960's elucidated the magnetosphere/ionosphere interactions (PUDOVKIN, 1974); the discoveries forced a complete reevaluation of the ionospheric dynamo theory. However, RICHMOND *et al.* (1976) and RICHMOND (1979) concluded that winds in the 80 to 200 km region arising from solar heating were still the most likely cause of Sq currents and that magnetospheric sources are of only secondary importance at middle and low latitudes. Some researchers investigated the possibility of conjugate effects in midlatitude Sq current systems that could arise from the assymmetry of heating and conductivity because of the difference in times of sunrise and sunset and seasonal inequality between hemispheres (VAN SABEN, 1970; WAGNER *et al.*, 1980; RICHARDS and TORR, 1986).

Although a great number of Sq spherical harmonic analyses have been performed over the years, only a few are of sufficient detail to be noted here. CHAPMAN (1919), using 21 observatories at latitudes between $+60°$ and $-60°$, analyzed the 1902 (sunspot minimum) and 1905 (sunspot maximum) years separating the solstitial and equinoctial months. HASEGAWA and OTA (1937) studied the September 1933 and June 1934 sunspot minimum data. BENKOVA (1940) analyzed 46 global (including polar) stations for the May-August 1933. PRICE and WILKINS (1963) studied UT samples of 1932/33. The 1958 International Geophysical Year of sunspot maximum, with about 70 quality world magnetic observatories, produced a wealth of geomagnetic records analyzed with different SHA techniques by MATSUSHITA and MAEDA (1965a), MISHIN *et al.* (1966), PARKINSON (1971), SUZUKI (1973), MALIN (1973), and MALIN and GUPTA (1977). Studies of the subsequent 1965 sunspot minimum period with the greatest ever world distribution of observatories have been analyzed by WINCH (1981) as well as CAMPBELL and SCHIFFMACHER (1985, 1988). Figure 3 is an example of such analyses.

Early progress in the studies of the equatorial electrojet currents of Sq were summarized by CHAPMAN and RAO (1965) with their analysis of the field variations at nine equatorial observations. They described the location of the jet with respect to the geographic equator, the Sq range increased with increased sunspot numbers, and the loss of equatorial field enhancement with the depletion of the local E-region ionization. The electrojet extends about 600 km in latitude; but it is wider over Peru than at other locations. SCHIELDGE *et al.* (1973) modeled the ionospheric dynamo for the equatorial region and found that the winds driving the dynamo were sensitive functions of location as well as local and universal time. They also investigated the contribution of field-aligned components to the equatorial observations. RASTOGI and IYER (1976) found that the amplitudes of the diurnal and semidiurnal components increase linearly with sunspot number. Negative depressions in the equatorial H variation sometimes occur around sunrise and sunset

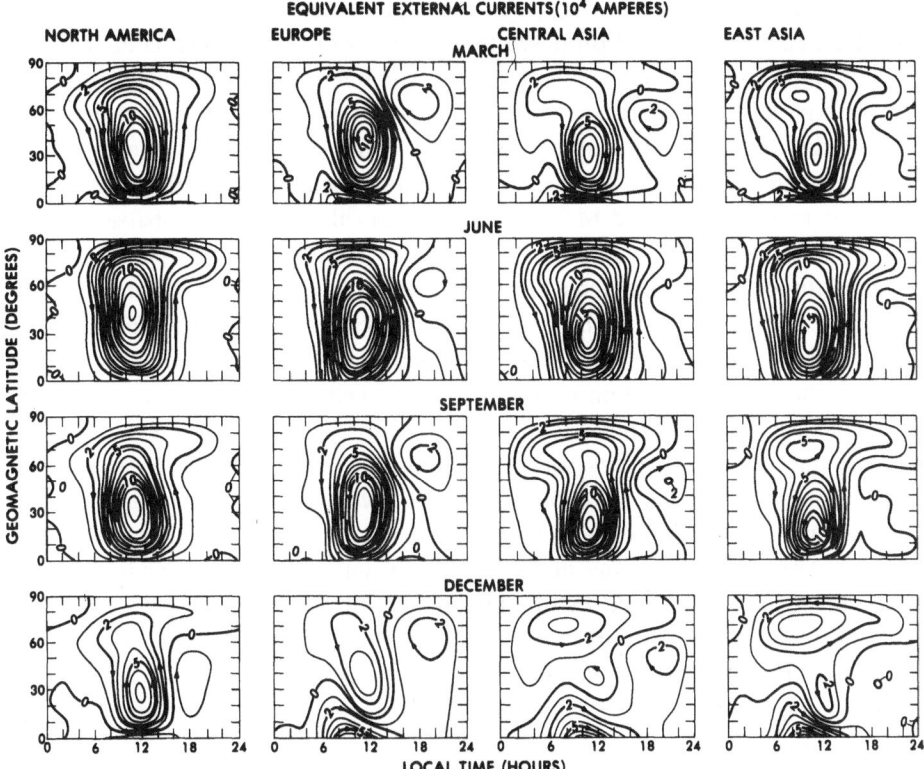

Figure 3

Equivalent ionospheric source current for *Sq* daily variations of field in continental regions of North America (first column), Europe (second column), central Asia (third column), and east Asia (fourth column). Examples for the four selected months of March, June, September, and December 1965 are given in the top to bottom rows, respectively. Each pattern in local time versus latitude coordinates shows the equivalent current contours in 10^4- A step with arrows for the required flow direction. A midnight zero current level was assumed.

(MAYAUD, 1977). Studies of this "Counter Electrojet" behavior (e.g., RASTOGI, 1974; FORBES and LINDZEN, 1976; MARIOTT *et al.*, 1979; BHARGAVA *et al.*, 1980) established the basic properties of this phenomenon and its relationship to unique lunar and *Sq* vortex current effects. REDDY and DEVASIA (1981) investigated the height and latitude of the electrojet currents and fields due to local winds. FORBES (1981) published the first comprehensive review of the equatorial electrojet current system covering the physical process, the field observations, and the supporting *in situ* measurements. However, recent publications (MACDOUGALL, 1979; MAEDA *et al.*, 1982; MAEDA, 1986; ANANDARAO and RAGHAVARAO, 1987; RAGHAVARAO *et al.*, 1988) indicate that the equatorial region behavior of *Sq* still is not fully understood.

The need to consider a special polar cap contribution to Sq different from the usual dynamo effect was first proposed by HASEGAWA (1940). Much later NAGATA and KOKUBUN (1962) confirmed that the intimate interaction of the magnetosphere with the high latitude ionosphere produces a special quiet daily variation field Sq^p. Other studies (e.g., KAWASAKI and AKASOFU, 1967; IIJIMA, 1973) also supported the need to separate out the unique part of the quiet field variation that occurs only in the polar cap. ROBLE and MATSUSHITA (1975) calculated the global-scale joule heating of Sq and found it greatest at high latitudes. An early viewpoint was that the high-latitude behavior differed from the low latitude ionospheric dynamo Sq in two ways. There seemed to be a variation dependent upon the interplanetary magnetic field direction of the solar wind upon its encounter with the earth's magnetosphere and a variation resulting from the field-aligned currents connecting the magnetosphere to the polar ionosphere even in the quietest times.

With the advent of *in situ* solar-terrestrial field measurements by satellite, it became apparent that the direction of the solar wind field at the magnetospheric boundary was important to the response of particles and fields measured at high latitudes. The toward or away directions of the solar field often divide the solar ecliptic plane into identifiable sectors that rotate with the sun. SVALGAARD (1973), MATSUSHITA et al. (1973), and CAMPBELL (1976) traced this sector effect in the polar cap surface magnetic field. It would subsequently develop that the Y (east/west) component of the IMF (interplanetary magnetic field) was more closely tied to the "sector" behavior of the high-latitude fields and that the quiet conditions prevailed when the IMF Z (north/south) component was zero or only slightly positive (FRIIS-CHRISTENSEN and WILHJELM, 1975; MATSUSHITA and XU, 1982; ZANETTI et al., 1982; FRIIS-CHRISTENSEN et al., 1985; HEPPNER and MAYNARD, 1987; CLAUER and FRIIS-CHRISTENSEN, 1988).

Field-aligned currents flowing from the magnetosphere to the auroral ionosphere were first proposed by BIRKELAND (1908), but observational verification was not possible until the appearance of research satellites (ARMSTRONG and ZUNDA, 1973). IIJIMA and POTEMRA (1976a, b) studied the general behavior of these currents at quiet and active times, relating the quiet system to the Sq^p. Part of the field-aligned current system was found to persist in the quiet times (AKASOFU and AHN, 1981; LEVITIN et al., 1982; ZANETTI et al., 1983; RASMUSSEN and SCHUNK, 1988). In recent years, data from several special polar chains of surface observatories have fueled the modeling of high-latitude magnetosphere/ionosphere processes (KISABETH, 1979; AKASOFU et al., 1980; KAMIDE et al., 1981; AKASOFU et al., 1981; AKASOFU and AHN, 1981; RICHMOND and KAMIDE, 1988). A much clearer picture of the polar fields and thermospheric tides and winds has evolved (MATSUSHITA and XU, 1982; ZANETTI et al., 1983; AFANASEVA, 1984; FRIIS-CHRISTENSEN et al., 1985; KILEEN and ROBLE, 1986; FESEN et al., 1986; REES and FULLER-ROWELL, 1987; RASMUSSEN and SCHUNK, 1988).

There are many aspects of Sq changes that are still not sufficiently understood to permit a prediction of quiet-field variation levels at all world locations. Such subjects as the contribution to Sq by the nighttime currents (TAKEDA and ARAKI, 1985), the ocean influence on Sq (TAKEDA, 1985), the variability of Sq (BUTCHER and BROWN, 1981; BRIGGS, 1984), and the dependence of Sq on magnetospheric parameters (KUZNETSOV and PTITSYNA, 1987) clearly need further study.

3. Lunar Effects in the Quiet Field Variations

The gravitational pull of the moon on the rotating earth raises the atmosphere and ocean levels both on the side facing the moon and on the side away from the moon. This semidiurnal tidal effect occurs because the moon's attraction overbalances the centrifugal force on the facing side of the earth; whereas, the centrifugal force overbalances attraction on the away side. Within the E and lower F regions of the ionosphere, the resulting transport of dayside ionization through the earth's main field causes dynamo currents to arise that may be detected at the earth's surface as a lunar geomagnetic variation of about 1 to 10 gammas. The moon circles about the earth in the same direction as the earth's spin, taking 24 hr 50.5 min per rotation. Lunar time is computed from a midnight when the moon is 180° away from its zenith position. Because the ionization is in step with the 24-hr solar day and the twice daily tides are in step with the lunar day, there is, consequently, not only a semidiurnal lunar tidal component but also a solar component to the lunar field variations. The lunar siderial period is 27.3 days, so there is some difficulty in distinguishing monthly lunar changes from solar activity (27-day period) changes.

The presence of lunar variations in the quiet field records at Prague was announced to the Bohemian Society of Sciences in 1841 by M. Kreil (WALKER, 1866, pg. 109). SABINE (1853) used the Toronto, St. Helena, and Hobarton records to confirm the lunar effect. Early progress in detection of lunar changes on a global basis was summarized in CHAPMAN and BARTELS (1940) and MATSUSHITA (1977). The definitive textbook on atmospheric tides was published by CHAPMAN and LINDZEN (1970). Basic documentation of the almost universally used Chapman-Miller lunar analysis method has been provided by MALIN and CHAPMAN (1970). A slightly modified, fast lunar analysis method for high resolution, short data series was introduced by MATSUSHITA and CAMPBELL (1972). SCHLAPP and WEEKES (1973) have developed a solar-hour technique of analysis as an alternative to the Chapman-Miller method.

Although a great many publications on the lunar effects upon the geomagnetic field have appeared over the years, only a few of the most significant need be mentioned in this brief review. VAN BEMMELEN (1912, 1913) was the first to apply spherical harmonic techniques to the lunar field study. The wind distribution at ionospheric heights responsible for the lunar geomagnetic effect was calculated by

MAEDA (1955) and KATO (1956). CHAPMAN (1919) followed a few years later with a more detailed analysis. MATSUSHITA and MAEDA (1965b), GUPTA and CHAPMAN (1969), TARPLEY (1970), GUPTA and MALIN (1972), and MALIN (1973) provided the global picture of lunar field changes from data of the IGY period. A review of lunar analysis techniques and a thorough study of lunar fields during the 1964/65 quiet year period was carried out by WINCH (1981). In recent years, MATSUSHITA and XU (1983, 1984) reevaluated lunar field data of earlier IGY studies.

4. Special PAGEOPH Issue

This special issue of *Pure and Applied Geophysics* grew from a collection of papers that was presented at the International Association of Geomagnetism and Aeronomy Assembly at Vancouver, Canada, on 12 August 1987. Most of those original presentations that were selected for this issue were considerably revised and expanded. To complete the topical coverage on this subject, several other authors who had not attended the Vancouver meeting were invited to contribute.

The collection that follows has been arranged to start with a description of methods for a selection of geomagnetically quiet days. Next, there is a discussion of what constitutes Sq variations and some of the present attitudes regarding the physical processes. Following the two papers on the equatorial Sq variations, the collection closes with reviews of the lunar field changes. The authors of this special *PAGEOPH* issue hope that this focus upon recent progress in the study of quiet daily geomagnetic fields will be as rewarding to the readers as it has been to the authors in preparation.

REFERENCES

AAB, P. S. (1925), *Report on Terrestrial Magnetism in Siam*, Int. Assoc. Geomag. Aeron., *Bul 5*, 84–86.
AFANASEVA, V. I. (1984), *Diurnal Variations of the Magnetic Field of the Earth and the Structure of the Terrestrial Magnetosphere*, Geomag. Aeron. (English ed.) *24*, 379–384.
AIRY, G. B., *Autobiography* (Cambridge Univ. Press 1896) 414 pp.
AKASOFU, S. I., and AHN, B. H. (1981), *Distribution of the Field-aligned Currents, Ionospheric Currents, and Electric Fields in the Polar Region on a Very Quiet Day and a Moderately Disturbed Day*, J. Geophys. Res. *86*, 753–760.
AKASOFU, S. I., KAMIDE, Y., and KISABETH, J. (1981), *Comparison of Two Modelling Methods for Three-dimensional Current Systems*, J. Geophys. Res. *86*, 3389–3396.
AKASOFU, S. I., KISABETH, J., AHN, B. H., and Romick, G. J. (1980), *The S^p_q Magnetic Variation, Equivalent Current, and Field-aligned Current Distribution Obtained from the IMS Alaska Meridian Chain of Magnetometers*, J. Geophys. Res. *85*, 2085–2091.
ANANADARAO, B. G., and RAGHAVARAO, R. (1987), *Structural Changes in the Currents and Fields of the Equatorial Electrojet Due to Zonal and Meridional Winds*, J. Geophys. Res. *92*, 2514–2526.
APPLETON, E. V., and BARNETT, M. A. F. (1925), *Local Reflections of Wireless Waves from the Upper Atmosphere*, Nature *115*, 333–334.

ARMSTRONG, J. C., and ZUNDA, A. J. (1973), *Three-axis Magnetic Measurements and Field-aligned Currents in the Auroral Region: Initial Results*, J. Geophys. Res. *78*, 6802–6807.

BAKER, W. G., (1953), *Electric Currents in the Ionosphere, Part II. The Atmosphere Dynamo*, Phil. Trans. Roy. Soc. London, *A246*, 295–305.

BAKER, W. G., and MARTYN, D. F. (1953), *Electric Currents in the Ionosphere, Part I. The Conductivity*, Phil. Trans. Roy. Soc. London *A246*, 281–294.

BARTELS, J., and JOHNSON, H. F. (1940a), *Geomagnetic Tides in the Horizontal Intensity at Huancayo, 1:* Terr. Mag. Atmos. Electr. (J. Geophys. Res.) *45*, 269–308. *2:* (1940b) Terr. Mag. Atmos. Electr. (J. Geophys. Res.) *45*, 485–592.

BENKOVA, N. P. (1940), *Spherical Harmonic Analysis of the Sq-variations May-August 1933*, Terr. Mag. Atmos. Electr. (J. Geophys. Res.) *4*, 425–432.

BHARGAVA, B. N., SASTRI, N. S., ARORA, B. R., and RAJARAM, R. (1980), The Afternoon Counter-electrojet Phenomenon, Ann. Geophys. *36*, 231–240.

BIRKELAND, K., On the cause of magnetic storms and the origin of terrestrial magnetism, in *The Norwegian Aurora Polaris Expedition 1902–3*, Vol. 1 (H. Ascheoug, Christiania, Norway 1908).

BREIT, G., and TUVE, M. A. (1925), *A Radio Method of Estimating the Height of the Conducting Layer*, Nature *116, 357*,

BRIGGS, B. H. (1984), *The variability of the Ionospheric Dynamo Currents*, J. Atmos. Terr. Phys. *26*, 419–429.

BUTCHER, E. C., and BROWN, G. M. (1981), *On the Nature of Abnormal Quiet Days in the Sq (H)*, Geophys. J. Roy. Astr. Soc. *64*, 513–526.

CAMPBELL, W. H. (1976), *Polar Cap Geomagnetic Field Responses to Solar Sector Changes*, J. Geophys. Res. *81*, 4731–4744.

CAMPBELL, W. H., and SCHIFFMACHER, E. R. (1985), *Quiet Ionospheric Currents of the Northern Hemisphere Derived from Quiet Field Records*, J. Geophys. Res. *90*, 6475–6486. (Correction, J. Geophys. Res. *91*, 9023–9024, 1986.)

CAMPBELL, W. H., and SCHIFFMACHER, E. R. (1988), *Quiet Ionospheric Currents of the Southern Hemisphere, Derived from Geomagnetic Records*, J. Geophys. Res. *93*, 933–944.

CANTON, C. (1759), *An Attempt to Account for the Regular Diurnal Variation of the Magnetic Needle*, Phil. Trans. Soc. London *A49*, 398–445.

CHAPMAN, S. (1919), *The Solar and Lunar Diurnal Variation of the Earth's Magnetism*, Phil. Trans. Roy. Soc. London. *A218*, 1–118.

CHAPMAN, S. (1951), *The Equatorial Electrojet as Detected from the Abnormal Current Distribution above Hunacayo, Peru and Elsewhere*, Arch. Met. Geophys. Biolaim. Ser. *A4*, 368–390.

CHAPMAN, S. (1956), *The Electric Conductivity of the Ionosphere: A Review*, Nuovo Cime. *4*, Suppl., 1385–1412.

CHAPMAN, S., and BARTELS, J. *Geomagnetism* (Oxford University Press, London 1940) Chap. XX, pp. 684–698.

CHAPMAN, S., and W. G. COWLING, *Mathematical Theory of Non-Uniform Gases*, 2nd ed. (Cambridge Univ. Press, London 1952).

CHAPMAN, S., and LINDZEN, R. S., *Atmospheric Tides* (D. Reidel Co., Dordrecht, Holland 1970) 200 pp.

CHAPMAN, S., and RAO, K.S.R. (1965), *The H and Z Variations Along and Near the Equatorial Electrojet in India, Africa and the Pacific*, J. Atmos. Terr. Phys. *27*, 559–581.

CLAUER, C. R., and FRIIS-CHRISTENSEN, E. (1988), *High-latitude Fields and Currents During Strong Northward Interplanetary Magnetic Field Observations and Model Simulation*, J. Geophys. Res. *93*, 2749–2757.

COWLING, T. G. (1945), *The Electrical Conductivity of an Ionized Gas in a Magnetic Field with Applications to the Solar Atmosphere and the Ionosphere*, Proc. Roy. Soc. *A183*, 453–479.

DAVIES, K., *Ionospheric Radio Propagation*, Nat. Bur. Std. Monograph *80* (U. S. Government Printing Office, Washington, D.C. 1965) 300 pp.

EGEDAL, J. (1947), *The Magnetic Diurnal Variation of the Horizontal Force Near the Magnetic Equator*, Terr. Mag. Atmos. Electr. (J. Geophys. Res.) *52*, 449–451.

EGEDAL, J. (1948), *Daily Variation of the Horizontal Force at the Magnetic Equator*, Nature *161*, 443–444.

ELLIS, W. (1880), *On the Relation Between the Diurnal Range of Magnetic Declination and Horizontal Force at Greenwich, 1841 to 1877, and the Period of Solar Spot Frequency*, Phil. Trans. Roy. Soc. London *A171*, 541–560.

FEJER, J. A. (1953), *Semidiurnal Currents and Electron Drifts in the Ionosphere*, J. Atmos. Terr. Phys. *4*, 184–203.

FEJER, J. A. (1964), *Atmospheric Tides and Associated Magnetic Effects*, Rev. Geophys. *2*, 275–309.

FESEN, G. G., DICKINSON, R. E., and ROBLE, R. G. (1986), *Simulation of the Thermospheric Tides at Equinox with the National Center for Atmospheric Research Thermospheric General Circulation Model*, J. Geophys. Res. *91*, 4471–4489.

FORBES, J. M., (1981), *The Equatorial Electrojet*, Rev. Geophys. *19*, 469–504.

FORBES, J. M., and LINDZEN, R. S. (1976), *Atmospheric Solar Tides and their Electrodynamic Effects, II. The Equatoiral Electrojet*, J. Atmos. Terr. Phys. *38*, 991–920.

FRIIS-CHRISTENSEN, E., KAMIDE, Y., RICHMOND, A. D., and MATSUSHITA, S. (1985), *Interplanetary Magnetic Field Control of High-latitude Electric Fields and Currents Determined from Greenland Magnetometer Data*, J. Geophys. Res. *90*, 1325–1338.

FRIIS-CHRISTENSEN, E., and WILHJELM, J. (1975), *Polar Cap Currents for Different Directions of the Interplanetary Magnetic Field in the Y-Z Plane*, J. Geophys. Res. *80*, 1248–1260.

GAUSS, C. F., *Allgemeine Theorie des Erdmagnetismus, Resultate aus den Beobachtungen des Magnetischen Vereins im Jahre 1838*, In eds. C. F. Gauss and W. Weber. English translation by E. Sabine and R. Taylor in *Scientific Memoirs Selected from the Transactions of Foreign Academies and Learned Societies and from Foreign Journals* (J. and R. E. Taylor Pub., London 1841) pp. 184–251.

GRAHAM, G. (1724), *An Account of Observations Made of the Variation of the Horizontal Needle at London in the Latter Part of the Year 1722 and Beginning 1723*, Phil. Trans. Roy. Soc. London *383*, 96–107.

GUPTA, J. C., and CHAPMAN, S. (1969), *Lunar Daily Harmonic Geomagnetic Variation as indicated by Spectral Analysis*, J. Atmos. Terr. Physics. *31*, 233–252.

GUPTA, J. C. and MALIN, S. R. C. (1972), *Seasonal Variations in the Solar and Lunar Daily Geomagnetic Variations*, Geophys. J. Roy. Astr. Soc. *30*, 11–18 .

HASAGAWA, M. (1936a), *On the Type of Diurnal Variations of Terrestrial Magnetism on Quiet Days*, Proc. Imp. Acad. Tokyo *12*, 88–90.

HASAGAWA, M. (1936b), *Representation of the Field of Diurnal Variations of Terrestrial Magnetism by the Method of Graphic Integration*, Proc. Imp. Acad. Tokyo *12*, 225–228.

HASAGAWA, M. (1936c), *On the Progressive Change of the Field of Diurnal Variations of Terrestrial Magnetism*, Proc. Imp. Acad. Tokyo *12*, 277–280.

HASAGAWA, M., and OTA, M. (1937), *An Analysis of the Field of Diurnal Variations of Terrestrial Magnetism of Differnet Types*. Proc. Imp. Acad. of Japan *13*, 65–73.

HASAGAWA, M., *Provisional Report of the Statistical Study on the Diurnal Variations of Terrestrial Magnetism in the North Polar Region*, I.U.G.G.–A.T.M.E Bull. No. 11 (ed. Goldie, A.H.R.) (Edinburgh, 1940) pp. 311–318.

HEPPNER, J. P., and MAYNARD, N. C. (1987), *Empirical High-latitude Electric Field Models*, J. Geophys. Res. *92*, 4467–4489.

HIRONO, M. (1952; 1953), *A Theory of Diurnal Magnetic Variations in Equatorial Regions and of the Ionosphere E Region*, J. Geomag. Geoelectr. *4*, 7–21; *5*, 22–38.

IIJIMA, T. (1973), *Enhancement of the Sq^p Field as the Basic Component of Polar Magnetic Disturbance*, Rep. Ionos. Res. Space Res. Japan *27*, 199–203.

IIJIMA, T., and POTEMRA, T. A. (1976a), *The Amplitude Distribution of Field-aligned Currents at Northern High Latitudes Observed by Triad*, J. Geophys. Res. *81*, 2165–2174.

IIJIMA, T., and POTEMRA, T. A. (1976b), *Field-aligned Currents in the Dayside Cusp Observed by Triad*, J. Geophys. Res. *81*, 5971–5979.

KAMIDE, Y., RICHMOND, A. D., and MATSUSHITA, S. (1981), *Estimation of Ionospheric Electric Fields, Ionospheric Currents, and Field-aligned Currents from Ground Magnetic Records*, J. Geophys. Res. *86*, 801–813.

KATO, S. (1956), *Horizontal Wind Systems in the Ionospheric E Region Deduced from the Dynamo Theory of Geomagnetic Sq Variation, Part II. Rotating Earth*, J. Geomag. Geolectr. *8*, 24–37.

KAWASAKI, K., and AKASOFU, S. I. (1967), *Polar Solar Daily Geomagnetic Variations on Exceptionally Quiet Days*, J. Geophys. Res. *72*, 5363–5371.

KILLEEN, T. L., and ROBLE, R. G. (1986), *An Analysis of High Latitude Thermospheric Wind Pattern Calculated by a Thermospheric General Circulation Model, 2. Neutral Particle Transport*, J. Geophys. Res. *91*, 11291–11307.

KISABETH, J. L., *On calculating magnetic and vector potential fields due to large-scale magnetospheric current systems and induced currents in an infinitely conducting earth*, In *Quantitative Modeling of Magnetospheric Processes*, Geophys. Monogr. Ser. *21* (ed. Olsen, W. P.) (Amer. Geophys. Union, Washington, D. C. 1979) 473 pp.

KUZNETSOV, B. M., and PTITSYNA, N. G. (1987), *Variation of the Quiet Level of the Geomagnetic Field at Middle Latitudes in the Solar Activity Cycle and as a Function of the Intensity of Ionospheric and Magnetospheric Current Systems*, Geomag. Aeron. (English Ed.) *27*, 535–539.

LEVITIN, A. E., AFONINA, R. G., BELOV, B. A., and FELDSTEIN, Y. I. (1982), *Geomagnetic Variation and Field-aligned Currents at Northern High Latitudes, and their relations to the Solar Wind Parameters*, Phil. Trans. Roy. Soc. London *A304*, 253–301.

LUCAS, I. (1954), *The Dynamo Theory of Geomagnetic Tides*, Archiv. Elek. Übertr. *8*, 123.

MACDOUGALL, J. W. (1979), *Equatorial Electrojet and Sq Current Systems, I and II*, J. Geomag. Geoelectr. *31*, 341–372.

MAEDA, H. (1955), *Horizontal Wind Systems in the Ionosphere E Region Deduced from the Dynamo Theory of Geomagnetic Sq Variation*, Part I, J. Geomag. Geoelectr. *7*, 121–132.

MAEDA, H., IYEMORI, T., ARAKI, T., and KAMEI, T. (1982), *New Evidence of a Meridional Current System in the Equatorial Ionosphere*, Geophys. Rev. Lett. *9*, 337–340.

MAEDA, K. (1952), *Dynamo-theoretical Conductivity and Current in the Ionosphere*, J. Geomag. Geoelectr. *4*, 63–82.

MAEDA, K. (1986), *A New Transport Process in the Equatorial E Region and its Effect on the Electron Density Profile*, J. Geomag, Geoelectr. *38*, 759–769.

MAEDA, K., and KATO, S. (1966), *Electrodynamics of the Ionosphere*, Space Sci. Rev. *5*, 57–79.

MALIN, S. R. C. (1973), *Worldwide Distribution of Geomagnetic Tides*, Phil. Trans. Roy. Soc. London *A274*, 551–594.

MALIN, S. R. C., and CHAPMAN, S. (1970), *The Determination of Lunar Daily Geophysical Variations by the Chapman-Miller Method*. Geophys. J. Roy. Astr. Soc. *19*, 15–35.

MALIN, S. R. C., and GUPTA, J. C. (1977), *The Sq Current System During the International Geophysical Year*, Geophys. J. Roy. Astr. Soc. *49*, 515–529.

MARRIOTT, R. T., RICHMOND, A. D., and VANKATESWARAN, S. V. (1979), *The Quiet-time Equatorial Electrojet and Counter Electrojet*, J. Geomag. Geoelectr. *31*, 311–340.

MATSUSHITA, S. (1977), *Upper-atmospheric Tidal-interaction Effects on Geomagnetic and Ionospheric Variations—A Review*, Ann. Geophys. *33*, 115–125.

MATSUSHITA, S., and CAMPBELL, W. H. (1972), *Lunar Semidiurnal Variations of the Geomagnetic Field Determined from the 2.5-min Data Scalings*, J. Atmos. Terr. Phys. *34*, 1187–1200.

MATSUSHITA, S., and MAEDA, H. (1965a), *On the Geomagnetic Quiet Solar Daily Variation Field During the IGY*, J. Geophys. Res. *70*, 2535–2558.

MATSUSHITA, S., and MAEDA, H. (1965b), *On the Geomagnetic Lunar Daily Variation Field*, J. Geophys. Res. *70*, 2559–2578.

MATSUSHITA, S., TARPLEY, J. D., and CAMPBELL, W. H. (1973), *IMF Sector Structure Effects upon the Quiet Geomagnetic Field*, Radio Sci. *8*, 963–972.

MATSUSHITA, S., and XU, W. Y. (1982), *Equivalent Ionospheric Current Systems Representing Solar Daily Variations of the Polar Geomagnetic Field*, J. Geophys. Res. *87*, 8241–8254.

MATSUSHITA, S., and XU, W. Y. (1983), *Equivalent Ionospheric Current Systems Representing Lunar Daily Variations of the Polar Geomagnetic Field*, J. Geophys. Res. *88*, 7143–7154.

MATSUSHITA, S., and XU, W. Y. (1984), *Seasonal Variations of L Equivalent Current Systems*, J. Geophys. Res. *89*, 285–294.

MAYAUD, P. N. (1977), *The Equatorial Counter Electrojet—A Review of its Geomagnetic Aspects*, J. Atmos. Terr. Phys. *39*, 1055–1070.

MISHIN, V. M., BAZARZHAPOV, A. D., NEMTSOVA, E. I., and PLATANOV, M. L. (1966), *The Method of Analytical Representation of "Instantaneous" Fields of Magnetic Variations* (in Russian), Geomagn. Issled. *8*, 5–22.

NAGATA, T., and KOKUBUN, S. (1962), *An Additional Geomagnetic Daily Variation Field (S_q^p Field) in the Polar Region on Geomagnetically Quiet Days*, Rep. Ionosph. Space. Res. Japan. *16*, 256–274.

PARKINSON, W. D. (1971), *An Analysis of the Geomagnetic Diurnal Variation During the International Geophysical Year*, Gerlands Beitr. Geophys. *80*, 199–232.

PEDERSEN, P. O., *Propagation of Radio Waves, etc.*, Danmarks Naturvidenskabelige Samfund., Nr. 15 a/b (Copenhagen, 1927) 244 pp.

PEKERIS, C. L. (1937), *Atmospheric Oscillations*, Proc. Roy. Soc. *A158*, 650–671.

PRICE, A. T. (1969), *Daily Variations of the Geomagnetic Field*, Space Sci. Rev. *9*, 151–197.

PRICE, A. T., and WILKINS, G. A. (1963), *New Methods for the Analysis of Geomagnetic Fields and their Applications to the Sq Fields of 1923/33*, Phil. Trans. Roy. Soc. London *A256*, 31–98.

PUDOVKIN, M. I. (1974), *Electric Fields and Currents in the Ionosphere*, Space Sci. Rev. *16*, 727–770.

RAGHAVARAO, R., SRIDHARAN, R., SASTRI, J. H., AGASHE, V. V., RAO, B. C. N., RAO, P. B., and SOMAYAJULU, V. V., *The equatorial ionosphere*, In *World Ionosphere/Thermosphere Study*, WITS Handbook, Vol. 1 (eds. Liu, C. H., and Edwards, B.) (SCOSTEP Secretariat, Univ. Ullinois, Urbana, Ill. 1988) pp. 48–93.

RASMUSSEN, C. E., and SCHUNK, R. W. (1988), *Ionospheric Convection Inferred from Interplanetary Magnetic Field-dependent Birkeland Currents*, J. Geophys. Res. *93*, 1909–1921.

RASTOGI, R. G. (1974), *Lunar Effects in the Counterelectrojet near the Magnetic Equator*, J. Atmos. Terr. Phys. *36*, 167–170.

RASTOGI, R. G., and IYER, K. N. (1976), *Quiet Day Variation of Geomagnetic H-field at Low Latitudes*, J. Geomag. Geolectr. *28*, 461–479.

RATCLIFFE, J. A., and WEEKES, K., *The ionosphere*, Chap. 9 in *Physics of the Upper Atmosphere* (ed. Ratcliffe, J. A.) (Academic Press, New York 1960) pp. 377–470.

REDDY, C. A., and DEVASIA, C. V. (1981), *Height and Latitude Structure of Electric Fields and Currents due to Local East-West Winds in the Equatorial Electrojet*, J. Geophys. Res. *86*, 5751–5767.

REES, D., and FULLER-ROWELL, T. J. (1987), *Hemispheric Asymmetries in Thermospheric Structure and Dynamics*, Mem. Natl. Inst. Polar Res. *48*, Spec. Issue, 134–160.

RICHARDS, P. G., and TORR, D. G. (1986), *Thermal Coupling of Conjugate Ionospheres and Tilt of the Earth's Magnetic Field*, J. Geophys. Res. *91*, 9017–9021.

RICHMOND, A. D. (1979), *Ionospheric Wind Dynamo Theory: A Review*, J. Geophys. Res. *31*, 287–310.

RICHMOND, A. D., MATSUSHITA, S., and TARPLEY, J. D. (1976), *On the Production Mechanism of Electric Currents and Fields in the Ionosphere*, J. Geophys. Res. 547–555.

RICHMOND, A. D., and KAMIDE, Y. (1988), *Mapping Electrodynamic Features of the High Latitude Ionosphere from Localized Observations: Technique*, J. Geophys. Res. *93*, 5741–5759.

RISHBETH, H., and GARRIOTT, O. K. *Introduction to Ionospheric Physics* (Academic Press, New York 1969) 331 pp.

ROBLE, R. G., and MATSUSHITA, S. (1975), *An Estimate of the Global-Scale Joule Heating Rates in the Thermosphere due to Time Mean Currents*, Radio Science *10*, 389–399.

SABINE, E. (1847), *The Very Variable Direction of Magnetic Declination in One Half of the Year at Longwood House, St. Helena*, Phil. Trans. Roy. Soc. London, Pt. 1, 54.

SABINE, E. (1853), *On the Influence of the Moon on the Magnetic Declination at Toronto, St. Helena and Hobarton*, Phil. Trans. Roy. Soc. London *A143*, 549–560.

SABINE, E. (1857a), *On the Evidence of the Existence of the Decennial Inequaltiy in the Solar-diurnal Magnetic Variations, and its Non-existence in the Lunar-Dirunal Variation, of the Declination at Hobarton*, Phil. Trans. Roy. Soc. London *147*, 1–8, .

SABINE, E. (1857b), *On What the External Magnetic Observatories Have Accomplished*, Proc. Roy. Soc. London *35*, 1–19.

SCHIELDREG, J. P., VANKATESWARAN, S. V., and RICHMOND, A. D. (1973), *The Ionospheric Dynamo and Equatorial Magnetic Variations*, J. Atmos. Terr. Phys. *35*, 1045–1061.

SCHLAPP, D. M., and WEEKES, K. (1973), *The Determination of the Lunar Tides, I. Methods of Analysis*, J. Atmos. Terr. Phys. *35*, 1811–1831.

SCHUSTER, A. (1889), *The Diurnal Variation of Terrestrial Magnetism*, Phil. Trans. Roy. Soc. London *A180*, 467–518.

SCHUSTER, A. (1908), *The Diurnal Variation of Terrestrial Magnetism*, Phil. Trans. Roy. Soc. London *A208*, 163–204.

STEWART, B., *Hypothetical Views Regarding the Connection between the State of the Sun and Terrestrial Magnetism*, In *Encyclopedia Britannica*, 9th ed., Vol. *16*, 181–184, 1882.

SUZUKI, A. (1973), *A New Analysis of the Geomagnetic Sq Field*, J. Geophys. Res. *25*, 259–280.

SVALGAARD, L. (1973), *Polar Cap Magnetic Variations and their Relationship with the Interplanetary Magnetic Sector Structure*, J. Geophys. Res. *78*, 2064–2078.

TAKEDA, M. (1985), *UT Variation of Internal Sq Currents and the Oceanic Effect During 1980 March 1–18*, Geophys. J. Roy. Astr. Soc. *80*, 649–659.

TAKEDA, M., and ARAKI, T. (1985), *Electric Conductivity of the Ionosphere and Nocturnal Currents*, J. Atmos. Terres. Phys. *47*, 601–609.

TARPLEY, J. D. (1970), *The Ionospheric Wind Dynamo-I, Lunar Tide*, Planet. Space Sci. *18*, 1075–1090.

VAN BEMMELEN, W. (1912), *Die lunare Variation des Erdmagnetismus*, Meteorol. Z. *29*, 218–225.

VAN BEMMELEN, W. (1913), *Berichtigung zu meiner Abhandlung über die lunare Variation des Erdmagnetismus*, Meteorol. Z. *30*, 589–594.

VAN SABEN, D. (1970), *Solstitial Sq Currents Through the Magnetosphere*, J. Atmos. Terr. Phys. *32*, 1331–1336.

VON HUMBOLDT, A., Cosmos, Vol. V (English translation by E. C. Otté and W. S. Dallas) (Harper and Bros, Pub., New York 1863).

VON HUMBOLDT, A. (1938), *Reprint of 1838 letter to the (U.K.) Royal Geographical Society*, Nature *141* 299.

WAGNER, C. V., MÖHLMANN, D., SCHÄFER, K., MISHIN, V. M., and MATVEEV, M. I. (1980), *Large-scale Electric Fields and Currents and Related Geomagnetic Variations in the Quiet Plasmasphere*, Space Sci. Rev. *26*, 391–446.

WALKER, E., *Terrestrial and Cosmical Magnetism* (Deighton, Bell, and Co., Cambridge 1866) 336 pp.

WINCH, D. E. (1981), *Spherical Harmonic Analysis of Geomagnetic Tides, 1964–1965*, Phil. Trans. Roy. Soc. London *303*, 1–104.

ZANETTI, L. J., POTEMRA, T. A. DOERING, J. P. LEE, J. S. FENNELL, J. F. and HOFFMAN R. A. (1982) *Interplanetary Magnetic Field Control of High-latitude Activity on July 29, 1977*, J. Geophys. Res. *87*, 5963–5975.

ZANETTI, L. J., BAUMJOHANN, W., and POTEMRA, T. A. (1983), *Ionospheric and Birkland Current Distributions Inferred from the MAGSAT Magnetometer Data*, J. Geophys. Res. *88*, 4875–4884.

(Received/accepted June 13, 1988)

PAGEOPH, Vol. 131, No. 3 (1989)

0033–4553/89/030333–09$1.50 + 0.20/0

Geomagnetic Quiet Day Selection

Jo Ann Joselyn[1]

Abstract—Based on published literature and the response to a questionnaire sent to geomagnetic field, ionospheric and magnetospheric researchers, several methods of choosing periods of quiet conditions based on geomagnetic records, as well as other observed parameters, have been identified. Caveats with respect to using geomagnetic indices to select quiet periods include the following:

1. Geomagnetic disturbances are strongly local. Even if the data from all available observatories indicate quiet behavior, there is the distinct possibility that some other location, not sampled, may be disturbed.

2. Geomagnetic indices are convenient but imperfect indicators of geomagnetic activity. Indices based on a quiet-day reference level have uncertainties comparable to the threshold value for quiet conditions. Indices representing average conditions during a 24-hr UT day may not be appropriate.

3. Geomagnetic activity does not fully reflect the range of possible factors that influence the ionosphere or magnetosphere.

Key words: Quiet day, geomagnetic indices.

Introduction

The purpose of this brief overview is to consider the options and accompanying risks inherent in choosing various geomagnetic indices or even "uneventful" geomagnetic observations to select quiet data intervals for geomagnetic, ionospheric or magnetospheric studies.

Any extended time series of a geophysical quantity contains periods of agitation and periods of calm. It is the task of the physicist to understand the basic underlying behavior of the parameter under study, presumably the periods of calm, and then the factors that incite and quench the agitation. Periods of calm are especially requisite for studies of the earth's intrinsic geomagnetic field. For studies of the ionosphere and magnetosphere the level of disturbance of the geomagnetic field serves as a surrogate for the level of disturbance of nearly all other ionospheric/magnetospheric variables. For example, a study of basic ionospheric

[1] NOAA Space Environment Laboratory, Space Environment Services Division, R/E/SE2, 325 Broadway, Boulder, Colorado 80303, USA.

behavior might begin by collecting data for geomagnetically quiet days, while data for a disturbed ionosphere is virtually guaranteed on disturbed geomagnetic days. However, at the outset, it must be cautioned that quiet geomagnetic conditions are not a sufficient condition for a quiet ionosphere and magnetosphere, and may not even be a necessary condition. The chief difficulties for sufficiency lie in the time constants for complete decay from disturbed conditions (e.g., the ionosphere "remembers" yesterday's active conditions even though today may be quiet), and the lack of global geomagnetic coverage. Because geomagnetic disturbances can be quite local, there can be doubt with ground-based data that the grid of magnetic observatories is tight enough. And while it is probably true that a quiet magnetic field is necessary for a quiet magnetosphere, more research is needed with regard to the details of the coupling between the solar wind and the magnetosphere to rule out the possibility of a quiet magnetic field but a disturbed magnetosphere, especially within the polar caps. Conversely, especially for the ionosphere, disturbances are possible in the absence of a geomagnetic signature. The X-ray emission from a significant solar flare will disturb the sun-lit ionosphere without significantly affecting the global geomagnetic field, although a short-lived excursion known as a crochet or solar flare effect can sometimes be seen (e.g., RICHMOND and VENKATESWARAN, 1971). Energetic proton showers from some solar flare events can also strongly ionize the lower ionosphere over the entire polar cap without generating geomagnetic activity (REID, 1965). But for most geophysical research and commercial operations and the amateur radio and other private applications, a quiet magnetic field is enough to assure that factors associated with disturbances are inconsequential.

A particular advantage of geomagnetic data is that it is conveniently indexed and can be quickly scanned to choose periods of interest. This is not true of most other geophysical data that are typically plotted on strip charts or presented on other analog outputs (even if the data were originally digital) and then examined visually. The pervasiveness of geomagnetic effects and the convenience of geomagnetic indices led to the routine selection and publication of "international quiet days" (and also "international disturbed days") in 1906 (JOHNSTON, 1943), a practice that continues, and continues to be valuable, today.

There have been several preceding reviews and discussions of the procedures (and hazards) of selecting quiet geomagnetic days. In particular, the reader is referred to ROSTOKER (1972), CAMPBELL (1979), and MAYAUD (1980). These papers stress the absence of a consensus international standard of geomagnetic quiet, and that "quietness" is itself strongly application dependent. Therefore, in preparation for this contribution I mailed a survey questionnaire on geomagnetic quiet day selection to a cross-section of scientists known to have interest in research of the geomagnetic field, the ionosphere and/or the magnetosphere. The replies, which were very illuminating and represent a considerable effort on the part of the responders that was much appreciated, are incorporated, often directly, into the paragraphs below.

Quiet-Day Indicators

The most common expectation for a "list of geomagnetically quiet days," is the list of five international quiet days for each month issued by the IUGG. As is often pointed out, these are the five quietest days each month, not necessarily five quiet days. One value of using this list is that diverse researchers can analyze data and then compare results for predetermined days, independently chosen. International quiet days are published in *Solar Geophysical Data* (U.S. Department of Commerce, NOAA/NESDIS, 325 Broadway, Boulder, CO 80303, USA) as they are available, which is approximately four months after the actual month of observation. These days are selected according to a scheme that ranks the days of each month based on the Kp index. Kp is presently determined by the Institut für Geophysik der Universität Göttingen, FR Germany, and is composed of data from 13 stations distributed between geomagnetic latitudes of 44.0 degrees and 62.2 degrees (SIEBERT, 1985). Two of the stations are in the Southern Hemisphere. Three criteria are used to select and order quiet days: the sum of the eight values of Kp, the sum of the squares of these values, and the maximum Kp (LINCOLN, 1967; MAYAUD, 1980; MAYAUD also carefully defines all geomagnetic indices used in this paper). There are absolute criteria that distinguish a quiet day from other days in the list of quietest days: the Ap index must be six or less, and only one Kp can be as large as 30 or only two as large as $3-$ (MAYAUD, 1980). Similar criteria distinguish truly disturbed days in a parallel list of five international disturbed days.

If the international quiet days list is not available or is not suitable, any of several geomagnetic indices might be used to order geomagnetic data. In the response to the survey the Kp index itself was widely used to choose quiet intervals. The reported upper bounds for "quietness" varied between $Kp \leq 1$ (the most stringent), to $Kp < 5$ (the most relaxed). Many also imposed limits on "sum Kp" to assure a quiet day, with the least lower bound being 2 and the greatest upper bound being 20. Other indices in active use and the typical upper limit for quietness reported by responders to a questionnaire are listed below.

ap, a "linearized" index value with units of 2 nT, corresponding to each "quasi-logarithmic" Kp value (the ap index for $Kp = 2+$ is 9).

Ap, a composite index calculated from the 8 Kp values of a UT day (it is actually the average of the 8 ap values); quiet conditions are generally less than or equal to 7 although responses varied between 3 and 15.

Ks, Kn, Km, and Am, indices similar to Kp and Ap except that a more global distribution of stations are used by the Institut du Physique du Globe de Paris, France, to compose the indices; the quietness limits for the K indices are the same as for Kp, but the Am unit is nT, so the upper bound is double that for Ap and could extend to 30.

aa, a composite of converted K index data from two antipodal stations; no upper bound was suggested but it should be similar to that for the Am index.

C, a daily integer between 0 (quiet) to 2 (disturbed) determined by each reporting station; the figures reported from all stations are averaged to produce a daily, global value. In China, C is now determined from local K indices (GAO, priv. comm., 1987) and the magnetometer traces for $C = 0$ days are then scanned further to choose the most quiet.

Dst, an hourly measure of ring current; quiet conditions are greater than -10 nT. However, the value of Dst is referenced to the average quiet field at each observatory. CLAUER et al. (1980) found that the uncertainty of the quiet field was a function of season and local time, and ranged from approximately ± 23 nT to ± 10 nT near local midnight, which implies that quiet conditions are poorly determined.

AE, an hourly measure of the auroral electrojet current; quiet conditions are generally less than 100 nT although several use 50 nT and one scientist has used 20 nT as a quietness threshold (see discussion below).

Q, a 15-min integer index ranging between 0 (quiet) and 11. A uniform scale is used for all stations to convert the upper limit of the amplitude variation in the most disturbed horizontal component to the corresponding value (LINCOLN, 1967).

Several additional considerations were mentioned as listed below.

For neutral thermosphere studies by a Fabry-Perot interferometer, the change in Dst should be less than 20 over a period of a few hours (BIONDI, priv. comm., 1987). Other responders also mentioned "smoothness" as a quietness criterion.

Several responders thought that for middle and high latitudes, steps should be taken to assure that no substorms had occurred within the chosen interval of study. It was cautioned that the distribution of AE observatories is not ideal (e.g., BAUMJOHANN, 1986). Steps to supplement low AE periods include checking the components of AE (AU and AL), direct inspection of stackplots of high and middle-latitude magnetograms from a worldwide distribution of observatories, use of all-sky camera photos or satellite auroral images, and a check for low values of energetic particle fluxes measured at geosynchronous orbit (BAKER, priv. comm., 1987).

The hourly range (HR) and daily range index (DRX) are becoming an increasingly popular measure of disturbance at remote digital observatories. These are determined from the difference between the highest and lowest relative values measured at individual stations, either during fiducial hours or a UT day. Quietness depends on latitude: for the X component at subauroral latitudes, HR and DRX < 20 nT; for auroral zone latitudes HR and DRX < 60 nT; and in the polar cap HR and DRX < 40 nT (HRUSKA, priv. comm., 1987).

For geomagnetic main-field modeling, natural annual and semi-annual variability in the solar quiet daily variation, Sq, must be considered. CAMPBELL (1987) notes that the best representation for the main-field level should be taken at the midnight hours (because the ionospheric dynamo currents driven by thermal-tidal motions in the E-region largely dissipate at night), and that at polar cap latitudes a correction for the toward or away direction of the solar wind may be necessary for quiet day data, especially for the Z component during summer months.

The hemispheric power input, by precipitating energetic particles, is measured by polar-orbiting satellites. Quiet conditions correspond to less than 5 gW (FOSTER et al., 1986).

Studies of the cross polar cap electric potential, also measured by polar-orbiting satellites, indicate that contributions to the polar cap potential, in the absence of reconnection (i.e., quiet global geomagnetic conditions) can be limited to less than 20 kV (WYGANT et al., 1983; REIFF and LUHMANN, 1986). Potentials of this magnitude correspond approximately to $Kp \leq 1$ (OLIVER et al., 1983).

The z component of the interplanetary magnetic field near the earth in GSM coordinates (B_z) ought to be northward or at least no more than 0.5 nT southward. However, strongly northward B_z has been recently associated with atypical (and therefore not "quiet") conditions in the polar caps (MAKITA et al., 1988).

There should be no sudden impulses or sudden commencements in a quiet period, nor obvious micropulsation activity (even though other index limits may not be violated). In a study of the characteristics of quiet-time geomagnetic field spectra having variations from 5 min to 4 hrs, CAMPBELL (1977) always found measurable spectral contributions, even on days when $Ap = 0$.

For high-resolution aeromagnetic surveys, magnetometer records near the survey sites should show no deviation greater than 2 nT from a straight line over any 5 min period; for low resolution surveys the upper limit is 5 nT/5 min (HRUSKA, priv. comm., 1987).

$ap(\tau)$, a data set derived by applying an exponential smoothing function to the ap data, permits comparison of geophysical parameters (e.g., the ionospheric critical frequency f_0F_2) with an integration of geomagnetic activity over a number of 3-hr intervals (WRENN, 1987). For selected values of τ, analysis of the frequency distribution of the resulting modified ap values permits a statistical definition of quietness so that 0–10% of all days are "very quiet" and 10–20% are "quiet." For the unweighted (i.e., $\tau = 0$) series over the length of a solar cycle, 1974–1984, WRENN (1987) found that roughly half of those ap values fell between 5 and 18, and that for the quietest 10% of the data, ap was less than 3 (which corresponds to a Kp of $1-$).

Perhaps a new composite "index of indices" should be defined. REDDY (priv. comm., 1987) suggests a Geomagnetic Quietness Index (GQI) based on Ap, AE and Dst for a given period (2- or 3-hourly, or using 24-hr averages), and also emphasizes a need to differentiate "quiet days" from "very quiet days."

Finally, to underscore the introductory statement that quietness depends on application, WINCH (priv. comm., 1987) reports that, for analysis of global Sq and L currents, any day other than the international disturbed days will suffice.

The Duration of Quiet

"Quietness" in the magnetic field or any of the ionospheric or magneto-spheric parameters is a transient condition and is certainly not synchronized to the UT day even though many of the quietness selectors (e.g., the five monthly international quiet days) are defined for a UT day. Also, a day that was quiet on average could contain minimal but important disturbances, depending on the sensitivity of the work. Therefore most of the responders to the survey believed that a further check on time scales finer than 24 hrs was required to assure a quiet period.

Depending on the time constants of the phenomena under study, care must be taken to wait for an appropriate length of time after quiet conditions commence to assure that the factors that contributed to previously disturbed conditions are no longer present. As with the indicators of quiet conditions, opinions varied widely with respect to the methods used to select quiet periods (as opposed to quiet days). The most stringent was offered by C. SUCKSDORFF (priv. comm., 1987) of the Finnish Meteorological Institute. For studies of the secular variation of the earth's internal field, they choose only winter days (Nov.-Feb.) with local $K = 0$, $Ap \leq 20$ for the preceding day *and* $Ap \leq 10$ for the preceding 5 days *and* $Ap \leq 20$ for the preceding 10 days; such days do not exist in every year. For studies of the refilling of magnetospheric flux tubes near $L = 4$ just equatorward of the plasmapause, A. RODGER (priv. comm., 1987) suggests a period of 7–10 days without activity. To assure a truly quiet ionosphere, 3 or 4 consecutive days of $Ap < 7$ are needed (WILKINSON, priv. comm., 1987). RYCROFT (priv. comm., 1987) uses a 48-hr period centered on Greenwich noon to select quiet days. R. MCPHERRON (priv. comm., 1987) notes that 24 hrs is a long time for any given ionospheric or magnetospheric condition to persist. He suggests that for the phenomenon under study, e.g. substorms, that the characteristic response function be constructed to determine the time constant, after which that time interval be removed from both ends of any collected quiet interval before data analysis proceeds. He also cautions that quiet intervals should be grouped with regard to calendar date, perhaps even consecutive 2-week periods, because of seasonal effects in ionospheric and magnetospheric parameters.

Further Discussion

Short-term geomagnetic variability (the order of days or less) at any one location depends on the combined effects of all ionospheric and magnetospheric currents and any telluric currents induced underneath. As was pointed out by FAIRFIELD (priv. comm., 1987), some currents dominate during disturbed times (e.g., the ring current and the auroral electrojets), masking the effects of other currents (e.g., the Sq current) that control variability during less disturbed times. In particular, currents at the magnetopause (OLSON, 1970) respond to a variable solar wind pressure producing effects at middle and low latitude stations that may separate "quiet" from "very quiet" conditions. Ionospheric currents also vary with changes in ionospheric conductivity, a situation influenced by solar radiative output, solar proton events, and even travelling ionospheric disturbances that may have their origin below the troposphere (e.g., the "oceanic effect" (TAKEDA, 1985), or volcanic/seismic events (ROBERTS et al., 1982)).

The Australian IPS Radio and Space Services agency has no customers requesting "warnings" of quiet conditions (WILKINSON, priv. comm., 1987). Besides the difficulty of determining when geophysical disturbances are absent, stringent requirements for extremely quiet conditions based on all possible measures of activity may lead to so few cases (or even none) that the results may be irreproducible or misleading. For most general applications, the absence of a geomagnetic storm is sufficient to assure nominal operations, and after all, extremely quiet conditions are not nominal.

Perhaps the only general conclusion that can be drawn from this exercise is the obvious one: careful thought about influencing factors must go into any geophysical study, especially those investigating the "ground-state" of a physical system. Magnetic disturbances or even evidence of a lack of magnetic disturbance should be used cautiously as a surrogate for the physical factors influencing the magnetosphere and the ionosphere.

Summary

Based on published literature and the responses to a questionnaire sent to geomagnetic main-field, ionospheric, and magnetospheric researchers, several methods of choosing periods of quiet conditions based on geomagnetic records, as well as other observed parameters, have been identified.

Caveats with respect to using geomagnetic indices to select quiet periods include the following:
1. Geomagnetic disturbances are strongly local. Even if the data from all available observatories indicate quiet behavior, there is the distinct possibility that some other location, not sampled, may be disturbed.
2. Geomagnetic indices are convenient but imperfect indicators of geomagnetic activity. Indices based on a quiet-day reference level have uncertainties comparable to the threshold value for quiet conditions. Indices representing average conditions during a 24-hr UT day may not be appropriate.
3. Geomagnetic activity does not fully reflect the range of possible factors that influence the ionosphere or magnetosphere.

Note Added in Proof

A study of electrodynamic patterns in the northern polar region during periods that were exceptionally quiet both magnetically and aurorally revealed that the AE index and the direction of the B_z component are each insufficient for identifying quiet periods [HOFFMAN, R. A., SUGIURA, M., MAYNARD, N. C., CANDEY, R. M., CRAVEN, J. D., and FRANK, L. A. (1988), *Electrodynamic patterns in the polar region during periods of extreme magnetic quiescence*, J. Geophys. Res., *93*, 13515–14541]. In particular, they show that a northward component of the interplanetary magnetic field is not equivalent to auroral quiescence. However, these authors conclude that if a careful evaluation of ground-based magnetograms, including some from stations at higher latitudes than the AE stations, shows that no horizontal component is varying by more than 20 nT for a couple of hours, quiet periods can be identified with a confidence of 80% to 90%.

REFERENCES

BAUMJOHANN, W., *Merits and limitations of the use of geomagnetic indices in solar wind-magnetosphere coupling studies*, In *Solar Wind-Magnetosphere Coupling* (eds. Kamide, Y., and Slavin, J. A.) (Terr. Sci. Publ. Co., Tokyo 1986) pp. 3–15.

CAMPBELL, W. H. (1977), *Spectral Characteristics of Field Variations during Geomagnetically Quiet Conditions*, J. Geomag. Geoelectr. *29*, 29–50.

CAMPBELL, W. H. (1979), *Occurrence of AE and Dst Geomagnetic Index Levels and the Selection of the Quietest Days in a Year*, J. Geophys. Res. *84*, 865–881.

CAMPBELL, W. H. (1987), *Some Effects of Quiet Geomagnetic Field Changes upon Values Used for Main Field Modeling*, Physics of the Earth and Planetary Interiors *48*, 193–199.

CLAUER, C. R., McPHERRON, R. L., and KIVELSON, M. G. (1980), *Uncertainty in Ring Current Parameters due to the Quiet Magnetic Field Variability at Mid-Latitudes*, J. Geophys. Res. *85*, 633–643.

FOSTER, J. C., HOLT, J. M., MUSGROVE, R. G., and EVANS, D. S. (1986), *Ionospheric Convection Associated with Discrete Levels of Particle Precipitation*, Geophys. Res. Lett. *13*, 656–659.

JOHNSTON, H. F. (1943), *Mean K-indices from Twenty-one Magnetic Observatories and Five Quiet and Five Disturbed Days for 1942*, Terr. Magn. Atmos. Elec. *47*, 219–227.

LINCOLN, J. V., *Geomagnetic indices*, In *Physics of Geomagnetic Phenomena, Vol. 1* (eds. Matsushita, S. and Campbell, W. H.) (Academic Press, New York 1967) pp. 67–100.

MAKITA, K., MENG, C.-I., and AKASOFU, S.-I. (1988), *Latitudinal Electron Precipitation Patterns During Large and Small IMF Magnitudes for Northward IMF Conditions*, J. Geophys. Res. *93*, 97–104.

MAYAUD, P. N. (1980), *Derivation, Meaning, and Use of Geomagnetic Indices*, Geophysical Monograph *22*, American Geophysical Union, Washington, D.C., 85–96.

OLIVER, W. L., HOLT, J. M., WAND, R. H., and EVANS, J. V. (1983), *Millstone Hill Incoherent Scatter Observations of Auroral Convection over $60° < \lambda < 75°$. 3. Average Patterns versus Kp*, J. Geophys. Res. *88*, 5505–5516.

OLSON, W. P. (1970), *Variations in the Earth's Surface Magnetic Field from the Magnetopause Current System*, Plan. Space. Sci. *18*, 1471–1484.

REID, G. C., *Solar cosmic rays and the ionosphere*, In *Physics of the Earth's Upper Atmosphere* (eds. Hines, C. O., Paghis, I., Hartz, T. R., and Fejer, J. A.) (Prentice Hall Inc., Englewood Cliffs, New Jersey 1965) pp. 245–270.

REIFF, P. H., and LUHMANN, J. G., *Solar wind control of the polar-cap voltage*, In *Solar Wind-Magnetosphere Coupling* (eds. Kamide, Y., and Slavin, J. A.) (Terr. Sci. Publ. Co., Tokyo 1986) pp. 453–476.

RICHMOND, A. D., and VENKATESWARAN, S. V. (1971), *Geomagnetic Crochets and Associated Ionospheric Current Systems*, Radio Sci. *6*, 139–164.

ROBERTS, D. H., KLOBUCHAR, J. A., FOUGERE, P. F., and HENDRICKSON, D. H. (1982), *A Large-amplitude Traveling Ionospheric Disturbance Produced by the May 18, 1980, Explosion of Mount St. Helens*, J. Geophys. Res. *87*, 6291–6301.

ROSTOKER, G. (1972), *Geomagnetic Indices*, Rev. Geophys. Space Phys. *10*, 935–950.

SIEBERT, M. (1985), *Report on Kp, Ap, Cp*, presented to IAGA Division V (Observatories, Instruments, Indices and Data) Working Group 5 on "Geophysical Indices," Prague, Czechoslovakia.

TAKEDA, M. (1985), *UT Variations of Internal Sq Currents and the Oceanic Effect during 1980 March 1–18*, Geophys. J. R. Astr. Soc. *80*, 649–659.

WRENN, G. L. (1987), *Time-weighted Accumulations ap(τ) and Kp(τ)*, J. Geophys. Res. *92*, 10125–10128.

WYGANT, J. R., TORBERT, R. B., and MOSER, F. S. (1983), *Comparison of S3–3 Polar Cap Potential Drops with the Interplanetary Magnetic Field and Models of Magnetopause Reconnection*, J. Geophys. Res. *88*, 5727–5735.

(Received February 10, 1988, revised/accepted May 18, 1988)

PAGEOPH, Vol. 131, No. 3 (1989)

0033–4553/89/030343–13$1.50 + 0.20/0

Very Quiet Intervals of Geomagnetic Activity and the Solar Wind Structure

Jaroslava Hruska[1] and Antonin Hruska[2]

Abstract—Extended periods of very low geomagnetic activity as described by very quiet intervals (VQI's) occur only at those times when the solar wind velocity V has a generally decreasing trend, i.e., they mainly occur either after the velocity peak of a high speed solar stream has passed the Earth, or at times when the Earth is immersed in a low speed solar plasma provided that the daily mean value of dV/dt is negative. The VQI's most frequently start when $dV/dt < 0$ and $dB_Z/dt > 0$ (B_Z is the geocentric solar magnetrospheric-GSM Z-component of the IMF) and end most likely when $dV/dt > 0$ and $dB_Z/dt < 0$. The temporal trends of the solar wind (SW) velocity affect the variation of the a_p index only when the level of geomagnetic activity is generally low.

It is suggested that a gradual expansion or contraction of the magnetosphere, associated with a slow variation of the SW pressure, plays a role in the modification of the reconnection-driven magnetohydrodynamic (MHD) fluctuations in the magnetosphere.

Key words: Geomagnetic activity, very quiet intervals, solar wind, SW pressure, magnetosphere.

1. Introduction

We have identified those properties of the interplanetary medium which are related to the occurrence of extended periods of very low geomagnetic activity. The very quiet intervals (VQI's), which are somewhat arbitrarily defined by IAGA as time periods lasting eight or more three-hour intervals during which $K_p \leq 1 + (a_p \leq 5)$, are used as a suitable description of these extended periods of low activity.

There have been extensive discussions in the literature of the relations between the geomagnetic indices, AE, a_p and Dst, and the quantities characterizing the interplanetary medium, the solar wind velocity V, the particle density n, the IMF B and the GSM Z-component of B (Wilcox and Colburn, 1969; Arnoldy, 1971; Sawyer and Haurwitz, 1976; Perreault and Akasofu, 1978; Schwenn, 1981;

---

[1] Geophysics Division, Geological Survey of Canada Contribution No. 43687, 1 Observatory Crescent, Ottawa, Canada K1A 0Y3.

[2] Herzberg Institute of Astrophysics, National Research Council of Canada, 100 Sussex Drive, Ottawa, Canada K1A 0R6.

BAKER et al., 1981; SARGENT, 1985; BAGATZE et al., 1985; SAVAUD et al., 1987; etc.). In these studies the emphasis is mostly on evaluation of correlations between the values of interplanetary quantities and the corresponding geomagnetic disturbances. Here we discuss mainly effects of the temporal variations of the solar wind (SW) quantities on the geomagnetic activity and search for conditions under which geomagnetic activity diminishes. We emphasize at the outset, that periods of low activity should not be intuitively linked to the existence of undisturbed conditions in the solar wind.

Statistical analysis of the relations between the SW parameters and VQI's is the topic of Section 2, possible interpretations of statistical results are briefly discussed in Section 3 and the results are summarized in the last Section.

2. Interplanetary Conditions under which the VQI's Occur

2.1 General Description

Inspection of the lists of VQI's (IAGA BULLETINS, 1968–1976) indicates that VQI's have been recorded 16% of the time during solar cycle 20 (1965–1975) and that they tend to occur in the vicinity of heliosheet crossings. Using 98 well-defined heliosheet crossings taken from the listing by WILCOX (1973) or determined from the Composite Interplanetary Medium Data Book (KING, 1977), we found that VQI's were recorded for 22% of the time, during the three-day intervals immediately preceding the crossings, whereas they occurred for 13% of the time during the three-day intervals immediately following the crossings. This distribution of the VQI's in the vicinity of the crossings is consistent with the general trends of the a_p-index during these periods (e.g., WILCOX and COLBURN, 1969). VQI's are recorded also within the well-defined SW sectors and show a 27-day recurrence pattern reminiscent of the recurrence pattern noted by SAWYER (1976) for the disturbed periods associated with the high velocity solar streams.

For years 1964 to 1975 we inspected hourly values of the SW data before, during and after the VQI's (KING, 1975, 1978). The data sufficient for determining the general trends of the SW parameters were available for 180 VQI's and the data coverage was almost complete (not more than one three-hour interval of data was missing) for 43 VQI's. The following qualitative description of the properties of the interplanetary medium at the time of occurrence of the VQI's is based on the full set of data, while the statistics using the superposed epochs analysis are based on the 43 well documented VQI's.

Figure 1 shows three examples of variations of the SW parameters during VQI's. The VQI's always begin at the time when the SW velocity V has a decreasing trend, usually soon after the velocity peak of a high speed solar plasma has passed the Earth. They also begin during the times when the Earth is immersed in a low speed

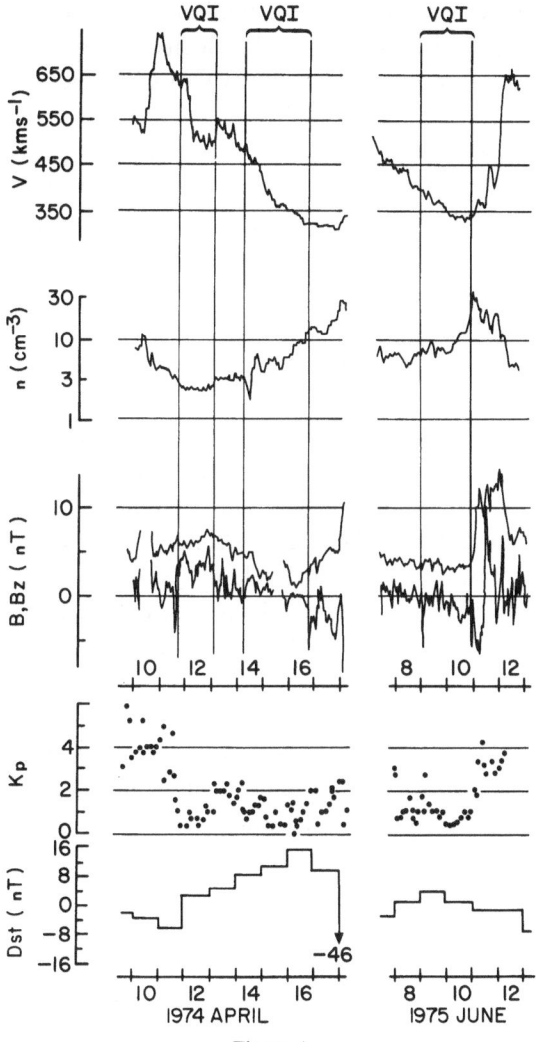

Figure 1

Examples of variations of the SW parameters, K_p indices and the equatorial Dst index during the VQI's in April 1974 in June 1975.

solar wind for prolonged periods of time and when the daily mean value of dV/dt is negative. If V is decreasing smoothly after reaching a peak, then the onset of VQI's coincides with, or follows immediately after, a small amplitude burst of irregular fluctuations of V. (The fluctuations are usually less marked than those shown in Figure 1 for the VQI of April 11, 1974.) The VQI's may be divided into two groups (A and B) according to the behaviour of V at the time of their ending. The VQI's of group A end while V still has a generally decreasing trend and when another small amplitude, time limited fluctuation of V occurs. The VQI's of group

B end close to the time when V reaches a minimum or when the general trend of V becomes positive, i.e., when another high velocity stream approaches the Earth. The distinction between group A and group B VQI's is rather formal and is largely due to our artifical definition of the very quiet intervals. The example of the two VQI's shown on the left side of Figure 1 illustrates this point. If there was no disturbance of the SW parameters on April 13, 1974, both VQI's would coalesce into one long lasting VQI; we may say that the VQI's of group A are actually the VQI's of group B which were "prematurely terminated" by a disturbance of the SW parameters. The onsets of the VQI's are not restricted to a small range of V. They range from $V \sim 390$ km s^{-1} to $V \sim 689$ km s^{-1}. With one exception from 43 cases, the values of V at the beginning are higher than the values of V at the ends of VQI's.

In all analyses of correlations between the geomagnetic indices and the SW parameters, it is usual to view V, n and B as independent parameters although there are causal relationships between these quantities determined by the dynamics of the solar wind. In particular, during the intervals following the passage of the peaks in the high velocity streams, the density and field magnitude show generally decreasing trends; later they show increasing trends which start well before the velocity reaches a minimum preceding the encounter of the Earth with the next high velocity stream. Theoretical description of this phenomenon follows from the SW models (e.g. PIZZO, 1983). The onsets of the VQI's occur usually at the time when n and B either show negative trends, or no strong trends, while the ends of VQI's occur at the times when general trends of n and B are positive. In the individual cases there are, of course, deviations from this idealized description (see Figure 1).

As expected, the occurrence of the VQI's is substantially affected by the B_Z (GSM Z-component of B). For example, the VQI of April 11, 1974 shown on the left of Figure 1, starts when B_Z turns northward and ends when it turns southward. The ends of the other two VQI's shown in Figure 1 also may be attributed to a strong southward turning of B_Z, but the southward trends of B_Z are well pronounced even during the lifetimes of these VQI's.

The quantity B_Z has a somewhat special place in our set of four interplanetary parameters. While V, n and B and their temporal variations are dependent quantities because of the large-scale dynamics of the solar wind, the fluctuations of B_Z reflect mainly the smaller scale turbulence carried by the solar wind and do not depend noticeably on the large-scale structure of the SW plasma. Any systematic large-scale variations of the GSM B_Z, which also may be present, are, for all practical purposes, obscured by the effects of smaller scale turbulence. Thus, for our purposes, B_Z may be considered as an interplanetary parameter independent of the large-scale properties of solar wind.

Figure 1 also shows daily mean values of the equatorial Dst-index. The occurrence of high values of Dst usually coincides with the VQI's and an inspection of the hourly Dst values before, during and after the VQI's indicates that suitably defined periods of high *Dst* (say, $Dst \geq -1$ nT) mostly start within one three-hour interval from the onset of the VQI's and always last longer than the VQI's. The time lag between the end of the VQI's and the ends of the periods of high *Dst* is usually ≤ 1 day, though longer time lags (~ 2 days) also occasionally occur. The decay of the ring current, as measured by the equatorial *Dst*, continues to be noticeable even when the Earth is already immersed in the next high solar stream.

2.2 Statistical Results

Results of the superposed epoch analysis, based on the well documented VQI's, are summarized in Figure 2 which shows the variation of medians of V, n, B and B_Z during a 15-hour interval starting 9 hours before the beginning of the VQI's (T_B) and ending 6 hours after T_B. Also shown are the variations during a 12-hour interval centered on the end of the VQI's (T_E). Dashed lines indicate 50% scatter of data around the medians. The well documented set of the VQI's consists of twenty A-type and twenty-three B-type VQI's. Median values of the SW parameters for the A- and B-type VQI's are shown separately by open and full circles for the time intervals centered at T_E. Figure 2 suggests that it may be useful to characterize the conditions during which the VQI's begin or end by the relative time rates of change of the SW parameters rather than by the values of those quantities. The relative time rate of change Ω_X of quantity X ($X = V, n, B, B_Z$) is given by $|X|^{-1} dX/dt$ and shall be here approximated by the expression

$$\frac{2}{3} \frac{X_i - X_{i-1}}{|X_i + X_{i-1}|}$$

where the subscripts i and $i-1$ indicate three-hour intervals for which X's are computed. The resulting Ω_X is then in (hours)$^{-1}$. The relative time rates, or the time scales $\tau_X = \Omega_X^{-1}$, computed for different quantities are not equivalent. For example, the time scale of the velocity change $\tau_V = \Omega_V^{-1}$, is a measure which includes the large spatial variations of the SW structure and the smaller scale, time dependent, fluctuations of V. The time scale $\tau_{B_Z} = \Omega_{B_Z}^{-1}$, is a measure characterizing mainly the time scale of the turbulence in the solar wind plasma as it appears in the fluctuations of B_Z. Both time scales are computed from the three-hour averages of V and B_Z and should be viewed as parameters approximately describing the temporal trends of the quantities rather than the actual time scales in the precise mathematical sense.

Figure 2 suggests that the VQI's are likely to begin (end) when B_Z turns northward (southward), but about 23% of the VQI's begin at the time when B_Z is turning southward and 35% of the VQI's end when B_Z is turning northward. Clearly, onsets and ends of VQI's are not determined by B_Z alone.

Consider first the onsets of the VQI's. The values of n, B and V are quite randomly scattered and the quantities n and B do not show any strong trends at

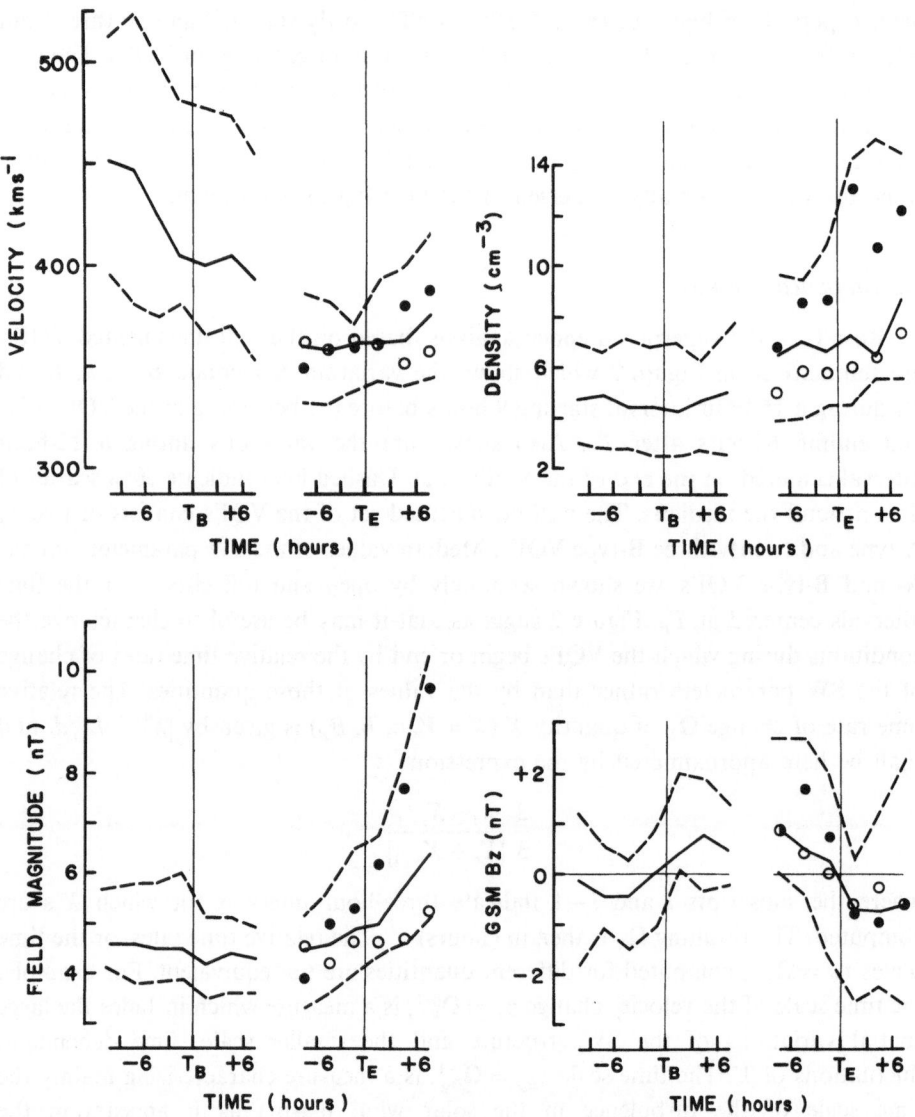

Figure 2

Variations of the SW velocity, density, magnetic field magnitude, and GSM Z-component of the IMF around the beginnings (at the time T_B) and ends (at the time T_E) of the VQI's. Results from the superimposed epochs analysis are shown; medians are denoted by the full lines, 50% scatter of the data by the broken lines; the open (full) circles denote the medians determined from the A-type (B-type) VQI's.

that time. The values of Ω_V, which could range from large negative to large positive values, are in 50% of all cases confined to the relatively small interval from -0.002 to -0.014 (hours)$^{-1}$. Moreover, all VQI's start when the daily mean value of Ω_V is negative. As for the ends of the VQI's, we note that V's show some increasing trends and that the positive Ω_V at the ends of the A-type VQI's occur only if the fluctuations of V at that time are sufficiently strong to overcome the average decreasing trends. Southward turning of B_Z is probably the most usual cause of "switching off" of the A-type VQI's . The particle density and the absolute value of magnetic field clearly increase at the end of the B-type VQI's and also have a tendency to increase for the A-type VQI's. It is not clear whether the ends of the VQI's are causally related to these increases or whether the ends of VQI's simply occur at the times when n and B always increase because of the SW dynamics.

To shed some light on this ambiguity we shall look for those combinations of each of the parameters Ω_V, Ω_n, Ω_B and Ω_{B_Z} for which the VQI's are most likely to begin or end (Ω_n and Ω_B are defined in the same way as Ω_V and Ω_{B_Z}). We shall also look for the corresponding changes Δa_p of the a_p index. We expect that $|\Delta a_p|$ will have the largest values for those combinations of Ω's which are the most favourable for the occurrence of the onsets or ends of the VQI's. Table 1 shows the distribution of T_B and T_E events among all possible combinations of the positive and negative values of Ω_V, Ω_n and Ω_B with positive and negative Ω_{B_Z}. The numbers in parentheses indicate the averages of $|\Delta a_p|$. Table 1 also shows the numbers of VQI's corresponding to positive or negative $\Omega_{(nV^2)}$. Those cases for which one of the Ω's is zero are omitted.

We see from Table 1 that the onsets of the VQI's are most clearly concentrated in the quadrant $\Omega_V < 0, \Omega_{B_Z} > 0$ of the (Ω_V, Ω_{B_Z}) plane and that the largest decreases of a_p are most likely to occur in that quadrant. Similar, although less well pronounced, features of the distribution of the data points are noticeable in the planes (Ω_V, Ω_{B_Z}), (Ω_n, Ω_{B_Z}) and $(\Omega_{(nV^2)}, \Omega_{B_Z})$. For the ends of the VQI's, the data points are distributed between the individual quadrants of the planes (Ω_X, Ω_{B_Z}) $(X = V, n, B, nV^2)$ in a similar manner. The largest densities of the points are in the quadrants $\Omega_X > 0, \Omega_{B_Z} < 0$. The large Δa_p are clearly concentrated in one quadrant for the (Ω_V, Ω_{B_Z}) plane, whereas a somewhat less clear concentration of the large Δa_p is noticeable for the $(\Omega_{nV^2}, \Omega_{B_Z})$ plane. The scatter of the data points and of the corresponding $|\Delta a_p|$ in the plane (Ω_V, Ω_{B_Z}), which shows the "best" (i.e., the most uneven) distributions is illustrated by Figure 3. Our conclusions from Table 1 and Figure 3 are of course restricted to qualitative estimates since the set of the well documented VQI's is small. Visual inspection of the SW parameters and K_p's during the VQI's raises a strong suspicion that the scatter of the points into the quadrant $\Omega_V < 0, \Omega_{B_Z} > 0$ of the (Ω_V, Ω_{B_Z}) plane at the ends of the VQI's may be mostly due to our artificial definition of the VQI's. In particular, according to the adopted definition, the VQI ends when K_p increases above the value $1+$ for one interval. Although K_p may then return to the values $\leq 1+$ for several intervals, its decrease does not have any effect on the length of the VQI. If we ignored such fluctuations of K_p, we would find that Ω_V's at the "new ends" of the VQI's are almost always shifted towards the higher values.

Table 1

Numbers of the VQI's for various combinations of Ω's and the average values of $|\Delta a_p|$ (in parenthesis)

		$\Omega_V < 0$	$\Omega_V > 0$	$\Omega_n < 0$	$\Omega_n > 0$	$\Omega_B < 0$	$\Omega_B > 0$	$\Omega_{(m^2)} < 0$	$\Omega_{(m^2)} > 0$
ONSET	$\Omega_{Bz} < 0$	9 (3.4)	0	2 (3.0)	5 (2.8)	5 (4.0)	3 (2.7)	6 (4.0)	3 (2.3)
	$\Omega_{Bz} > 0$	26 (7.3)	7 (3.7)	17 (9.1)	12 (3.7)	20 (6.5)	11 (6.5)	22 (8.4)	11 (3.0)
END	$\Omega_{Bz} < 0$	11 (4.3)	18 (7.9)	11 (7.2)	16 (5.6)	9 (6.6)	18 (6.7)	8 (5.0)	19 (6.8)
	$\Omega_{Bz} > 0$	7 (2.9)	6 (4.7)	3 (5.0)	10 (3.3)	3 (6.7)	7 (2.9)	3 (3.3)	10 (4.3)

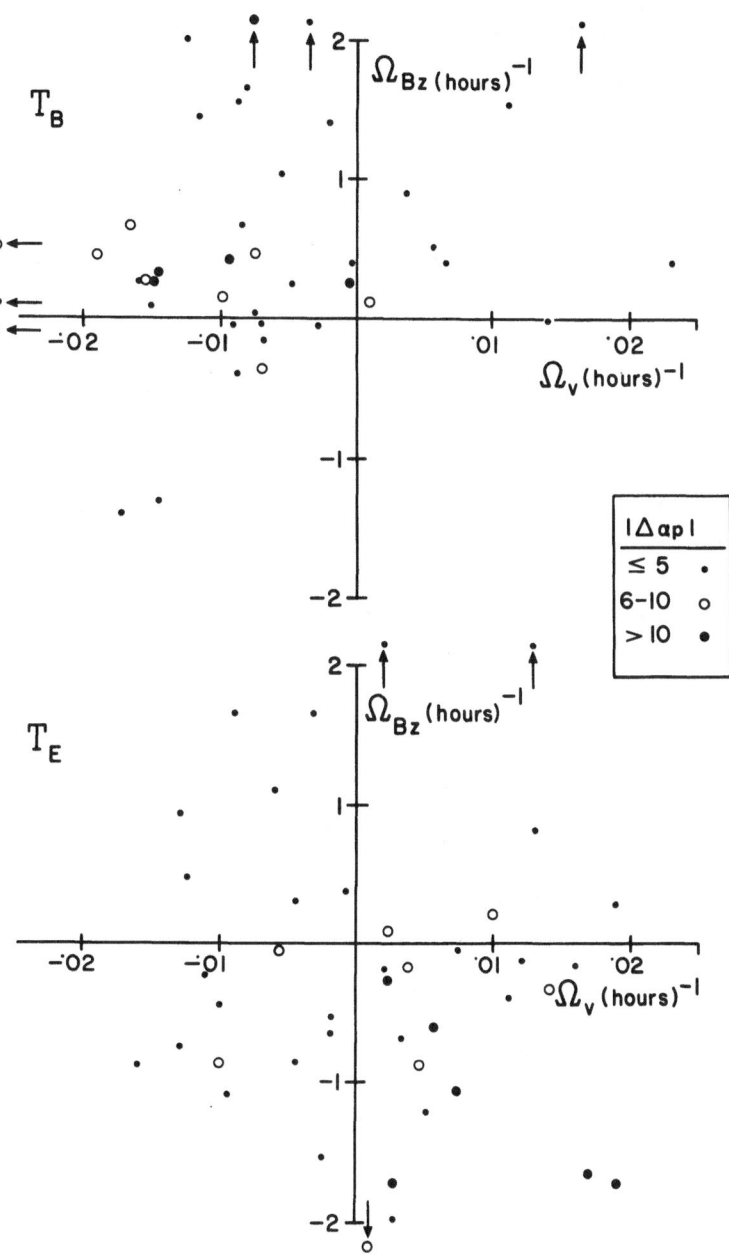

Figure 3
Scatter of the data in the (Ω_V, Ω_{B_Z}) plane and the corresponding values of $|\Delta a_p|$ for the beginning (T_B) and ends (T_E) of the VQI's.

2.3 Relations between Ω_V and a_p

The dependence of the a_p-index on the slow ($|\tau_V| \gg 0$) trends of the SW velocity is in many respects a peculiar phenomenon. With the exception of studies of the relations between the interplanetary shocks ($\tau_V \to 0$) and magnetic storms, the effect of variations of V on geomagnetic activity has never attracted much attention. To check the possibility that some statistical relations between Ω_V and a_p exists, we have chosen at random 300 three-hour intervals ("0-intervals") from the years 1966 to 1975. First, we computed Ω_V's from the hourly values of V's corresponding to the first and second hour of these "0-intervals" and computed the correlation coefficients between Ω_V and a_p's corresponding to this interval, to one interval preceding it ("-1 interval") and to two following intervals ("1 and 2 interval"). Also computed were the correlation coefficients between Ω_V and the changes $(\Delta a_p)_i = (a_p)_i - (a_p)_{i-1}$ ($i = 0, 1, 2$). No correlation coefficient exceeded 0.1, as one might expect. The inspection of the plots of a_p vs Ω_V and Δa_p vs Ω_V, however, suggested that for small values of a_p there is a difference between the changes of a_p during the times when $\Omega_V < 0$ and the changes of a_p during the times when $\Omega_V > 0$. We selected from our set those intervals for which $a_p \leq 5$ and computed the change Δa_p between the selected interval and the interval preceding it. For i in order 0, 1, 2, we have found

$$\langle \Delta a_p \rangle_0 = -1.54 \pm 0.42$$

$$\langle \Delta a_p \rangle_1 = -1.61 \pm 0.54 \quad \text{when } \Omega_V < 0$$

$$\langle \Delta a_p \rangle_2 = -0.93 \pm 0.40$$

and

$$\langle \Delta a_p \rangle_0 = -0.33 \pm 0.38$$

$$\langle \Delta a_p \rangle_1 = -0.80 \pm 0.32 \quad \text{when } \Omega_V > 0.$$

$$\langle \Delta a_p \rangle_2 = -0.85 \pm 0.66$$

Clearly, the values of $(\Delta a_p)_i$ corresponding to randomly chosen intervals with a small a_p must have a negative average value which is approximately -0.9 for our data set. We see that there is a noticeable difference between $\langle \Delta a_p \rangle$'s computed separately for $\Omega_V < 0$ and for $\Omega_V > 0$. The temporal decrease (increase) of V enhances (diminishes) the absolute value of Δa_p, i.e. the positive (negative) Ω_V tends to increase (decrease) the geomagnetic activity. The effect disappears in the case of the "2 intervals" (for the time lags > 6 hours). For the high levels of activity any effect due to Ω_V, if it exists, is completely obscured by the scatter of Δa_p.

No effects were found when the described procedure was repeated for Ω_n and Ω_B. The implications following from the analysis of the VQI's, which was limited by the smallness of the set of well documented VQI's, are compatible with the above results and indicate that at the times when the geomagnetic activity is generally low, activity increases when $\Omega_V > 0$ and decreases when $\Omega_V < 0$.

3. Possible Interpretations of the Statistical Results

The a_p index is known to follow the trends of the AE-index and is often perceived as an acceptable, though not flawless, substitute for the AE-index. Being derived from the observed ranges of geomagnetic field in the sub-auroral region, it may be also viewed as an imperfect measure (see e.g., MAYAUD, 1980) of the level of the low frequency turbulence in the inner magnetosphere. These two perceptions of a_p are complementary rather than contradictory. The reconnection-related processes in the plasma sheet determining the value of AE may also trigger the turbulence in the inner magnetosphere and increase a_p. Thus, following the current trends of thought, we may say, that a_p is primarily determined by the reconnection rate, i.e. by the electromagnetic process rather than by mechanical processes such as Kelvin-Helmholtz instability, energy transfers by "viscous" interactions on the magnetopause or penetration of plasma "lumps". We may speculate whether the variation of V modifies the reconnection process or whether we have to find a secondary process, occurring only when $\Omega_V \neq 0$, which modified the existing, reconnection generated, irregular fluctuations in the inner magnetosphere. We probably may disregard modifications by $\Omega_V \neq 0$ of the mechanical processes since they themselves play at most a secondary role in the excitation of the fluctuations. This also applies to the ring current related drift instabilites.

Returning to the reconnection process, we note that the reconnection rate can only be affected by variations of V through the changes of the dawn to dusk electric by field (if B_Z is southward). Clearly, with $\Omega_V \ll \Omega_{B_Z}$, the variation of the product VB_Z is determined almost exclusively by the fluctuations of B_Z. The variations of V are therefore not expected to modify the reconnection rate and the generation of fluctuations in the magnetosphere. Consequently, we are lead to search for a plausible mechanism, which is associated with the variations of V, and which may modify the amplitude of the reconnection generated fluctuations already existing in the magnetosphere.

We note that the only direct effect of the slow variations of V on the magnetosphere is the gradual changing of its size. The nose distance of the magnetopause is proportional to $(nV^2)^{1/6}$ (e.g., BEARD, 1973) and the slow changes of the solar wind pressure ($\sim nV^2$) cause gradual compression or expansion of the linear dimensions Λ of the magnetosphere. An obvious consequence of the change of Λ is, of course, the change of the resonance frequency of the fluctuations in the magnetospheric cavity. We are, however, concerned about the change of the amplitude of the fluctuations and not about the change of the frequency. To address this problem in the simplest way, we assume that the magnetospheric fluctuations are characterized by some elementary, typical MHD wave of frequency $\omega \gg \Omega_V$ and wave length $\lambda = 2\pi/\kappa \ll \Lambda$; we ask how the amplitude of such a wave is changed by compression or expansion. It is not difficult to find that the application of the methods similar to the analyses of waves in a static medium with parameters weakly varying in space (e.g., STIX, 1962; STIX and SWANSON, 1983) leads, even for the waves with the simplest possible dispersion relations, to the solutions with complex ω's and k's in some regions of the space-time. (Amplification or damping of the waves is measured by $\text{Im}\{\omega\}$ which may acquire a wide range of values depending on the exact setup of boundary and initial conditions.) The problem is mathematically tractable, but attempts to solve it would be meaningful only if the exact properties of the magnetospheric waves in the static limit and the properties of the expansion or contraction of the magnetosphere as described by the Euler deformation rate tensor, were reasonably well known. These conditions are not satisfied at the present and there is no firm ground which could serve as a basis for a precise mathematical analysis.

4. Summary

We have identified several special features of the solar wind which are favourable for the occurrence of the VQI's:

(a) The VQI's occur only when the SW velocity has a generally decreasing trend; they occur after a peak velocity of high speed solar stream has passed the Earth and they also occur at times when the Earth is immersed in a low velocity solar plasma and when the daily average of the SW velocity decreases with time.

(b) VQI's frequently start at the time when a short-lived (a few hours long) irregular fluctuation of V, superimposed on the general decreasing trend of V, occurs. They end either at the time of another short-lived fluctuation of V (when V is still generally decreasing) or near the time when the SW velocity reaches its minimum, or starts to show a generally increasing trend. The occurrence of the VQI's is not correlated with the value of V; the SW velocity at the onset of the VQI is higher than the velocity at its end.

(c) The VQI's begin when at least one of the conditions $\Omega_V < 0$ and $\Omega_{B_Z} > 0$ is satisfied and when they end usually at least one of the conditions $\Omega_V > 0$ and $\Omega_{B_Z} < 0$ is met. Clearly the beginnings (ends) of the VQI's are frequently associated with northward (southward) turning of the B_Z-component of the IMF, but the turnings of B_Z are not sufficient or necessary conditions for the beginnings or ends to occur.

(d) The VQI's are not characterized by steady, undisturbed conditions in the interplanetary medium. The geomagnetic activity, characterized by the a_p-index, is low for extended periods of time, when the conditions in (a) above occur.

Slow expansions or compressions of the magnetosphere indicated by the parameter Ω_V (or perhaps better by the parameter $\Omega_{(nV^2)}$) play a role in modification of the reconnection-driven geomagnetic activity measured by the a_p-index. The effect of Ω_V is noticeable and even substantial only when the general level of the geomagnetic activity is low.

Acknowledgements

The authors would like to thank J. R. Burrows, R. L. Coles, H-L. Lam and J. K. Walker for their comments.

REFERENCES

ARNOLDY, R. L. (1971), *Signature of the Interplanetary Medium for Substorms*, J. Geophys. Res. *76*, 5189–5201.

BAKER, D. N., HONES, E. W., PAYNE, J.B., and FELDMAN, W. C. (1981), *A High Time Resolution of Interplanetary Parameter Correlations with AE*, J. Geophys. Res. Lett. *8*, 179–182.

BAGATZE, L. F., BAKER, D. N., McPHERON, R. L., and HONES, E. W. (1985), *Magnetospheric Impulse Response for Many Levels of Geomagnetic Activity*, J. Geophys. Res. *90*, 6387–6394.

BEARD, D. B. (1973), *The Interactions of the Solar Wind with Planetary Magnetic Fields: Basic Principles and Observations*, Planet. Space Sci. *21*, 1475–1496.

IAGA BULLETINS, edited by D. van Sabben (1968–1976), *Geomagnetic Data 1965–1975*, Nos: 12th–12x, 32a–32f.

KING, J. H. (1975), *Interplanetary Magnetic Field Data 1963–1974*, Report UAG–46 (World Data Center).

KING, J. H. (1977), *Interplanetary Medium Data Book—Appendix*, NSS DC/WDC–A–RS–77–04a.

MAYAUD, P. N. (1980), *Derivation, Meaning, and Use of Geomagnetic Indices*, Geophysical Monograph *22*, AGU, Washington, D. C.

PIZZO, V. J. (1983), *Quasi-steady solar wind dynamics*, In *Solar Wind 5* (ed. Neugebauer, M.) (NASA Conf. Publ.) 2280.

PERREAULT, P., and AKASOFU, S.-I. (1978), *A Study of Geomagnetic Storms*, Geophys. J., R. A. S. *54*, 597–572.

SARGENT III, H. H. (1985), *Recurrent Geomagnetic Activity: Evidence for Long-lived Stability in Solar Wind Structure*, J. Geophys. Res. *90*, 1425–1428.

SAVAUD, J. A., TREILHAN, J. P., SAINT-MARC, A., DAUDOURAS, J., RÈME, H., KORTH, A., KREMSER, G., PARKS, G. K., ZAITZEV, A. N., PETROV, V., LAZUTINE, L., and PELLINEN, R. (1987), *Large-scale Response of the Magnetosphere to a Southward Turning of the Interplanetary Magnetic Field*, J. Geophys. Res. *92*, 2365–2376.

SAWYER, C. (1976), *High Speed Streams and Sector Boundaries*, J. Geophs. Res. *81*, 2437–2441.

SAWYER, C., and HAURWITZ, M. (1976), *Geomagnetic Activity at the Passage of High Speed Streams in the Solar Winds*, J. Geophys. Res. *81*, 2435–2436.

SCHWENN, R. (1981), *Solar Wind and its Interactions with the Magnetosphere: Measured Parameters*, Adv. Space Res. *1*, 3–17.

STIX, T. H., *The Theory of Plasma Waves* (McGraw-Hill, New York 1962).

STIX, T. H., and SWANSON, D. G., *Propagation and mode conversions for waves in nonuniform plasmas*, In *Handbook of Plasma Physics, Vol. 1* (eds. Rosenbluth, M. N., and Sagdeev, R. Z.) (North-Holland Publ. Co., Amsterdam 1983).

WILCOX, J. M. (1973), *Solar Activity and the Weather*, NASA Rep. AD–772069.

WILCOX, J. M., and COLBURN, D. S. (1969), *Interplanetary Sector Structure in the Rising Portion of the Sunspot Cycle*, J. Geophys. Res. *74*, 2388–2392.

(Received February 10, 1988, revised/accepted April 25, 1988)

PAGEOPH, Vol. 131, No. 3 (1989)

0033–4553/89/030357–14$1.50 + 0.20/0

Least Squares and Integral Methods for the Spherical Harmonic Analysis of the *Sq*-Field

Koji Kawasaki[1], S. Matsushita[2], and J. C. Cain[3]

Abstract—Spherical harmonic coefficients (SHCs) for the daily magnetic variation fields (solar and lunar) and the main field of the earth are usually estimated by the method of least squares applied to a truncated spherical harmonic series. In this paper, an integral method for computing the SHCs for the solar quiet daily magnetic variation field *Sq* is described and applied to *Sq* data for May and June 1965. The *Sq* SHCs thus derived are then compared with the results obtained using both unweighted and weighted versions of the least squares method. The weighting used tends to orthogonalize the least squares terms. The integral and weighted least squares results agree closely for terms up to order 4 and degree 30, but both disagree considerably for the higher degree terms with the results of the unweighted least squares. Errors introduced by the numerical integration can be shown to be small, hence the disagreement between integral and unweighted least squares coefficient sets arises from improper weighting. Also, it is concluded that discrepancies between the geomagnetic northward and eastward component-derived coefficient sets arise from either time-dependent external sources that produce non-local-time, based fields or nonpotential sources and not from truncation of the spherical harmonic series as has previously been suggested.

Key words: Spherical harmonic analysis, *Sq*.

Introduction

Under certain conditions some geophysical phenomena may be practicably represented by an expansion in spherical harmonic functions. Gauss (1839) was the first to apply such functions to the measurements of the geomagnetic field and showed that, to within the limits of accuracy of the data taken up to that time, this field was mostly of internal origin. This method has been applied to the geomagnetic field up to the present time, though to greatly expanded data bases and by increasingly sophisticated means of computation.

Some fifty years after Gauss' initial application of spherical harmonics, Schuster (1889) applied the method to the daily variation part of the geomagnetic field.

---

[1] Geophysical Institute, University of Alaska, Fairbanks, AK 99775-0800, U.S.A.

[2] High Altitude Observatory, National Center for Atmospheric Research, Boulder, CO 80307, U.S.A. Deceased.

[3] Department of Geology, Florida State University, Tallahassee, FL 32306, U.S.A.

Contrary to the part of the field that does not vary with local time (the main field), he found the daily variation field to be primarily of external origin. Schuster's analysis also indicated that just the first few terms were required to give what was then thought to be an adequate representation of the daily variation field at middle and low latitudes.

The spherical harmonic equations applied to the measured geomagnetic elements of the solar daily variation field (*Sq*) have generally been treated by the method of least squares. Various analyses of *Sq* including modern day ones, have shown there is a discrepancy between the spherical harmonic coefficients (SHCs) obtained separately from the two horizontal components, although potential theory shows that for a source-free region the coefficient sets derived from either component should be identical. CHAPMAN and BARTELS (1940) believed this arises principally from truncation of the spherical harmonic series in the least squares method which does not take into account higher order terms of slowly diminishing magnitude and alternating signs. We have investigated this discrepancy by examining *Sq* during two summer months in 1965 by the least squares method and an integral method.

Magnetic Potential of Sq in Spherical Harmonics

Although the treatment of *Sq* in terms of spherical harmonics has been dealt with adequately elsewhere (CHAPMAN and BARTELS, 1940; MATSUSHITA, 1967), for completeness the method is outlined here.

It is assumed that *Sq* is completely derivable from a potential *V*. This assumption implies there are no electric currents where the potential is defined. To make the problem of representing *Sq* in spherical harmonics more tractable, it has generally been assumed that stations on the earth at the *same latitude* will observe identical local time-based variations. This latter assumption is correct only to a first approximation, as has been shown by WALKER (1913) and others, and more fully by MATSUSHITA and MAEDA (1965). However, it is possible to obtain a reasonably consistent spherical harmonic representation of *Sq* in a particular longitude range. Alternatively, the SHCs may be obtained for an instant of universal time from a set of worldwide *Sq* data. The instantaneous SHCs derived in this way will vary as a function of UT, and will include the longitudinal inequality or dependence of the *Sq* system (SUGIURA and HAGAN, 1967).

For the purposes of this report we shall adopt the above simplifying assumption that *Sq* is only local-time dependent which allows us to write the total potential of *Sq* due to both internal and external sources as:

$$V = C + \sum_{n=1}^{\infty} \sum_{m=0}^{n} \left\{ \left(e_{n,a}^m \frac{r^n}{R^{n-1}} + i_{n,a}^m \frac{R^{n+2}}{r^{n+1}} \right) \cos m\phi \right.$$
$$\left. + \left(e_{n,b}^m \frac{r^n}{R^{n-1}} + i_{n,b}^m \frac{R^{n+2}}{r^{n+1}} \right) \sin m\phi \right\} P_n^m(\theta) \tag{1}$$

where $P_n^m(\theta)$ is a numerical multiple of $P_{n,m}(\theta)$ (see CHAPMAN and BARTELS, 1940)—the associated Legendre polynomial function, e_n^m and i_n^m are constant coefficients referring to the external and internal parts of the total potential, respectively, R is the radius of the earth, (r, θ, Φ), is the point of observation in geomagnetic spherical polar coordinates and C is a constant. Here Φ (degrees) $= 15t$, where t (hours) is the local geomagnetic time.

The potential is related to the Sq-field, Sq, in the usual way:

$$Sq = -\nabla V. \tag{2}$$

Thus the potential can be related to the components of Sq observed at the earth's surface $(r = R)$.

The daily variation of the Sq components (the geomagnetic X', Y', or Z' component; hereinafter, we shall drop the primes on these components writing them simply as X, Y, and Z), on the other hand, can be expanded in a Fourier series and equated to each respective component of Equation (2). X, Y, and Z are taken positive northward, eastward and downward, respectively. These correspond to decreasing values of θ, increasing values of Φ and decreasing values of r, respectively.

Since $P_n^m = 0$ for $m > n$, and since all integral values of m are possible from $m = 0$ to $m = \infty$, equating the Fourier series representation of the real field to the spherical harmonic representation of the field components leads to the following relations:

$$x_{ma}(\theta) = \sum_{n=m}^{\infty} a_n^m \frac{\partial P_n^m}{n \, \partial \theta}; \qquad x_{mb}(\theta) = \sum_{n=m}^{\infty} b_n^m \frac{\partial P_n^m}{n \, \partial \theta} \tag{3}$$

$$y_{ma}(\theta) = -\sum_{n=m}^{\infty} b_n^m \frac{m}{n \sin \theta} P_n^m; \quad y_{mb}(\theta) = \sum_{n=m}^{\infty} a_n^m \frac{m}{n \sin \theta} P_n^m \tag{4}$$

$$z_{ma}(\theta) = \sum_{n=m}^{\infty} \alpha_n^m P_n^m; \qquad z_{mb}(\theta) = \sum_{n=m}^{\infty} \beta_n^m P_n^m \tag{5}$$

where x_{ma}, x_{mb}, y_{ma}, y_{mb}, z_{ma}, and z_{mb} are the Fourier amplitudes and

$$a_n^m = n(e_{n,a}^m + i_{n,a}^n); \quad b_n^m = n(e_{n,b}^m + i_{n,b}^m)$$
$$\alpha_n^m = ne_{n,a}^m - (n+1)i_{n,a}^m; \quad \beta_n^m = ne_{n,b}^m - (n+1)i_{n,b}^m. \tag{6}$$

In this derivation, the $m = 0$ Fourier terms and hence the $m = 0$ spherical harmonic terms are set to zero, because the baseline for Sq is not precisely known.

Calculation of the Spherical Harmonic Coefficients

The spherical harmonic representation of the external and internal parts of Sq may be calculated from either the set of Equations (3) and (5) or the set (4) and (5), suitably truncated. Since (3) involves derivatives of the modified associated Legendre polynomials, it is mathematically simpler to use the set (4) and (5) rather than

(3) and (5). From a statistical viewpoint, however, it is better to solve all three equations, simultaneously, using an overdetermined system of equations. On the other hand, nonpotential fields in the data may not be recognized, as such, if the data is treated in this manner. From the standpoint of understanding the physical causes of the Sq variation, it is thus preferable to solve the sets (3) and (5) and (4) and (5) separately, first, to determine whether large systematic discrepancies exist in the coefficients calculated from the two sets of equations.

It is usual to limit the calculations of the first few coefficients. For example, the highest order may be taken as p and the highest degree as q ($p \leq q$), with terms beyond these assumed to be small. Thus (3), (4) and (5) are reduced to sets of finite equations, one set for each Fourier coefficient from each latitude at which there are daily variation data. The method of least squares can then be applied to this set of equations provided the number of latitudes r at which the daily variation is known is larger than the maximum degree q.

The choice of the highest order and degree to which to carry the calculations of the spherical harmonic coefficients in the least squares method is highly subjective. This choice has not always been based on physical considerations (which are not adequately known in the first place) nor on a good statistical foundation, but rather on the constraints imposed by the number of stations available and the tractability of the equations to some sort of computation scheme. Unfortunately, serious errors in estimating the magnitudes of the coefficients may result from truncating the set of equations. A somewhat more objective approach to determining the coefficients would be to carry the truncation out to variable order and degree and examine the variation of the coefficients. When the variation becomes small, one can assume one is approaching a limiting value; however, this method depends on the number of available stations and also on proper weighting of the station with latitude. SCHMITZ and CAIN (1983) have addressed the latter point.

Integral equations for calculating the coefficients, which avoid subjectivity in the choice of order and degree and thus avoid difficulties in truncation and choice of weighting, can be derived using the property of orthogonality of the associated Legendre polynomials. For the Z-component the relevant integral equation may be found in CHAPMAN and BARTELS (1940), page 613. Thus, the Z data give the SHCs α_n^m and β_n^m in the following forms:

$$\alpha_n^m = \frac{2n + 1}{4} \int_0^\pi z_{ma}(\theta) \sin \theta \, d\theta \tag{7}$$

$$\beta_n^m = \frac{2n + 1}{4} \int_0^\pi z_{mb}(\theta) P_n^m(\theta) \sin \theta \, d\theta. \tag{8}$$

To obtain the integral equations for a_n^m and b_n^m from the Y-component data, we multiply both sides of (4) by $P_n^m(\theta) \sin^2 \theta d\theta$ and integrate along a meridian, again using the orthogonality of the polynomials. The results are (KAWASAKI, 1967):

$$a_n^m = \frac{2(n+1)}{4m} \int_0^\pi y_{mb}(\theta) P_n^m(\theta) \sin^2\theta \, d\theta \qquad (9)$$

$$b_n^m = \frac{2(n+1)}{4m} \int_0^\pi y_{ma}(\theta) P_n^m(\theta) \sin^2\theta \, d\theta. \qquad (10)$$

A spherical harmonic representation of the X-component may also be obtained by means of an integral equation. Inspection of Equation (3) leads to the following equation for the a_n^m

$$a_n^m \frac{\sqrt{(n+1)^2 - m^2}}{2n+1} - a_{n+2}^m \frac{(n+3)\sqrt{(n+2)^2 - m^2}}{(n+2)(2n+5)} = \frac{2n+3}{4} \int_0^\pi X_{ma} P_{n+1}^m \sin^2\theta \, d\theta$$

$$(11)$$

and a similar equation for the b_n^m. Note the linear combination of two coefficients on the left-hand side of (11); this arises because the derivative $dP_n^m/d\theta$ can be written in terms of the sum of two Legendre polynomials having the same order, but one of degree $(n+1)$ and the other of degree $(n-1)$. The forms (9), (10) and (11) are not explicitly shown by CHAPMAN and BARTELS (1940).

Unfortunately, although (11) can be used to provide a spherical harmonic expansion for the X-component, the coefficients themselves will always appear in a linear combination with each alternate coefficient, except for the coefficient a_{m+1}^m which results from the case $n = m - 1$. This means that one cannot explicitly solve for coefficients of the form $a_{m+2\gamma}^m$ where $\gamma = 0, 1, 2, \ldots$. However, since a_{m+1}^m can be obtained explicitly, coefficients of the form $a_{m+2\gamma+1}^m$ where $\gamma = 0, 1, 2 \ldots$, may be generated recursively.

Equations (7)–(11) can be integrated exactly if the Fourier coefficients y_{ma}, y_{mb}, z_{ma}, z_{mb}, x_{ma} and x_{mb} are known integrable functions of the colatitude θ. Polynomial or other curves may be fitted to obtain an estimate of the magnitudes of the Fourier coefficients for latitudes between observation stations if the data are sparse.

Comparison of Least Squares and Integral Methods

For the purpose of illustration, we have applied the integral Equations (7)–(11) to the quiet day data of May and June 1965. Essentially the same set has previously been analyzed by MATSUSHITA et al. (1973), but to improve the data coverage we have added the data from four southern polar observatories, giving a total of 44 observing stations distributed from pole to pole. During May and June 1965, there were 37 moderately quiet days with three-hourly $K_p \leqslant 2+$ which were taken as basic set for our analysis. The daily curves were averaged, transformed to geomagnetic components and Fourier analyzed in local geomagnetic time. The data were further averaged globally, i.e., the Fourier coefficients derived were plotted as a function of latitude and a smooth curve drawn through the points. Smoothing was needed to eliminate the more erratic changes with latitude, especially those near the auroral zone that may be attributable to selecting quiet days containing some small but significant substorm related effects. Although subjective, the smoothing process itself is not relevant to the major results and conclusions of this report. Figure 1 is an example of a smoothed curve for the Fourier coefficient z_{1b} plotted as a function of latitude, which shows this coefficient tends to be larger near the auroral zone and equator. This curve shown here for illustrative purposes is derived from the 40 original observing stations used by MATSUSHITA *et al.* (1973) and did not include the additional 4 stations mentioned above.

The smoothed curves of the Fourier amplitudes as a function of the colatitude were then divided in two degree increments starting at $1°$ colatitude, giving a total of 90 points which formed the basis for a trapezoidal-rule, numerical integration of Equations (7)–(10). The resulting SHC to degree 30 and order up to 40 are shown in Tables 1–4 where they appear in the columns labelled "$I(\ )$". Because the Sq curves are adequately represented by low order Fourier terms, we chose cutoff at $m = 4$. We computed the SHC by the integral technique to degree $n = 30$, because we wished to compare them to the SHC derived by least squares on series representations at various degree of truncation.

Least squares analyses, treating each of the 90 points equally, that is, un-weighted, were done on the Y-component to obtain the a_n^1; these may be found in Table 1 in the columns labelled "$U(Yn)$", where n is the degree at which the analysis was truncated. Examination of the three sets of a_n^1 derived for truncations at $n = 6$, 18 and 30 clearly shows that the coefficients of lowest degree tend to agree with each other and with the results of the integration method, but that those of highest degree disagree with each other.

As more and more terms are included in the finite series, the value of a particular coefficient tends to converge to a fixed value (see also ALLDREDGE and KAWASAKI, 1981). This is illustrated for β_4^1, β_5^1 and β_{11}^1 in Figure 2 from data plotted in Figure 1 for truncation degrees $n = 6$, 12, 18, 24, 30 and 42. Note that these βs should not be compared with those of Table 1, because the latter values

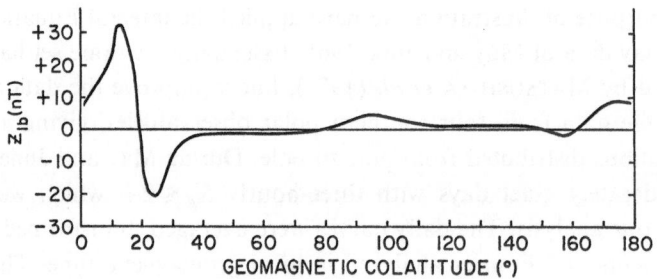

Figure 1

Fourier coefficient z_{1b} in nT as a function of colatitude in $(°)$ base on 40 station data of MATSUSHITA *et al.* (1973).

Table 1

Integral and least squares-derived SHC's for m = 1 Carried to Various n

Harmonic $m=1$	$I(Y)$ a_n^m	$W(Y)$ a_n^m	$U(Y30)$ a_n^m	$U(Y18)$ a_n^m	$U(Y6)$ a_n^m	$I(Y)$ b_n^m	$W(Y)$ b_n^m	$U(X30)$ b_n^m	$(UX30)$ a_n^m	$I(Z)$ α_n^m	$I(Z)$ β_n^m
$n=1$	6.11	6.11	6.11	6.11	6.10	-2.94	-2.94	-3.66	5.43	2.42	0.93
2	15.35	15.35	15.35	15.35	15.27	-2.86	-2.86	-7.33	13.27	5.26	-1.05
3	2.16	2.16	2.16	2.16	1.70	1.88	1.88	1.02	1.18	2.13	-0.49
4	-4.40	-4.40	-4.40	-4.40	-6.09	-3.65	-3.65	1.05	-6.65	-0.13	1.49
5	3.97	3.97	3.97	3.97	8.22	4.61	4.61	3.87	1.21	4.67	2.97
6	1.21	1.21	1.21	1.20	12.77	7.71	7.71	6.46	8.66	3.83	2.19
7	2.91	2.91	2.91	2.91		1.15	1.15	6.25	4.10	3.87	3.55
8	5.75	5.75	5.75	5.73		3.30	3.30	5.48	5.82	2.66	4.96
9	2.83	2.83	2.83	2.81		6.66	6.66	7.27	3.89	4.72	7.31
10	5.16	5.16	5.16	5.12		8.82	8.82	7.33	6.32	4.14	4.86
11	-0.21	-0.21	-0.21	-0.26		8.53	8.53	8.07	5.92	2.67	7.03
12	1.11	1.11	1.11	1.00		2.00	2.00	6.05	6.02	1.99	5.02
13	1.01	1.01	1.01	0.86		5.79	5.79	3.71	4.42	2.17	6.08
14	-0.44	-0.44	-0.44	-0.79		6.72	6.72	2.45	2.12	1.03	2.16
15	0.01	0.01	0.01	-0.96		-0.67	-0.67	1.53	-1.97	-0.56	2.48
16	-0.18	-0.18	-0.18	-2.40		-1.19	-1.19	3.25	-2.78	-0.62	-0.01
17	-3.47	-3.47	-3.47	-0.32		-0.89	-0.89	1.94	-3.85	-1.68	-0.71
18	-0.82	-0.82	-0.82	6.52		-3.90	-3.90	0.98	-1.72	-1.31	-3.22
19	0.06	0.06	0.06			-5.88	-5.88	-0.04	-3.38	-3.12	-3.09
20	0.71	0.71	0.71			-6.05	-6.05	-0.97	-4.19	-1.51	-4.09
21	0.21	0.21	0.22			-5.41	-5.41	-0.16	-4.19	-2.01	-4.05
22	1.36	1.36	1.34			-3.00	-3.00	0.18	-3.65	-1.29	-4.25
23	0.78	0.78	0.80			-1.66	-1.66	-1.13	-1.60	-1.06	-3.44
24	-0.82	-0.82	-0.85			-3.57	-3.56	-1.78	-0.87	-0.36	-2.60
25	0.34	0.34	0.42			-0.77	-0.77	-3.41	-1.36	0.32	-2.00
26	3.34	3.35	3.28			4.43	4.44	-3.21	-1.25	0.49	-0.99
27	0.47	0.47	0.99			3.75	3.75	-3.23	-2.03	0.88	0.19
28	0.47	0.47	0.24			1.76	1.76	-2.65	-0.85	1.39	0.72
29	1.35	1.35	-0.01			3.82	3.82	-3.26	0.40	1.27	1.60
30	0.59	0.60	0.95			7.12	7.13	-3.47	0.93	1.20	1.55

NOTE: $I(Y, Z)$ = Integral Method on (Y, Z)-Component; $W(Y)$ = Weighted Least Squares Method on Y-Component; $U(Xn, Yn)$ = Unweighted Least Squares Method on (X, Y)-Component Carried to Degree n.

Table 2

Integral and least squares derived SHCs for m = 2 carried to n = 30

Harmonic $m = 2$	$I(Y)$ a_n^m	$U(X30)$ a_n^m	$I(Y)$ b_n^m	$U(X30)$ b_n^m	$I(Z)$ α_n^m	$I(Z)$ β_n^m
$n = 2$	-1.48	-1.68	5.32	8.25	-0.80	1.14
3	-12.33	-13.64	4.48	4.74	-5.65	0.92
4	-3.90	-4.30	3.19	4.56	-2.22	0.81
5	4.92	-1.02	2.53	0.88	-2.44	1.98
6	0.97	-3.00	1.13	0.73	-2.54	2.33
7	-0.68	-2.03	3.81	2.21	-3.00	2.27
8	-0.81	1.76	1.70	4.86	-1.98	2.21
9	7.31	-0.42	4.78	1.34	-1.10	2.23
10	2.38	-0.35	5.07	7.55	-0.45	2.69
11	1.20	1.48	0.93	3.51	-1.51	0.99
12	2.79	1.40	1.89	2.05	1.86	0.97
13	2.67	1.69	-0.55	4.04	0.73	-0.01
14	-0.20	3.53	0.60	2.48	1.90	0.60
15	-1.09	0.58	-1.25	2.07	0.02	-0.64
16	-4.97	-0.93	-3.34	2.50	0.36	0.08
17	-3.81	1.74	-1.55	0.65	0.05	0.01
18	-3.89	-0.54	-0.54	-1.97	-1.24	1.27
19	-4.39	-0.72	0.06	-1.90	-1.74	0.65
20	-4.11	1.25	0.07	-1.43	-1.55	1.71
21	-2.56	-0.28	0.09	-2.98	-1.48	1.54
22	-1.52	-1.95	0.92	-1.45	-1.39	2.20
23	0.28	-0.60	2.29	-0.70	-1.46	1.45
24	2.11	-1.36	0.80	-2.88	-0.67	1.37
25	3.56	-2.05	0.21	-1.17	0.04	1.08
26	2.17	-0.51	-0.94	-0.39	-0.01	0.77
27	3.52	-0.60	0.08	-0.81	0.10	0.04
28	3.32	-2.40	0.33	1.10	0.50	-0.21
29	2.18	-1.62	-1.31	2.01	0.51	-0.42
30	0.24	-0.83	-1.75	1.06	0.43	-0.23

are obtained for the full set of 44 stations. Also shown in Figure 2 are the corresponding values obtained by the integral method.

Relative errors introduced by numerically integrating Equations (7)–(10) over a finite set of equally-spaced points can be estimated. For the α_n^m coefficient the computed value denoted by $\alpha_n^{m'}$ will be

$$\alpha_n^{m'} = \frac{(2n + 1)\pi\Delta}{720} \left\{ \alpha_n^m \sum_{i=1}^{180/\Delta} [P_n^m(\theta_i)]^2 \sin(\theta_i) \right.$$

$$\left. + \sum_{\beta=1}^{\infty} \sum_{i=1}^{180/\Delta} \alpha_{n \pm 2\beta}^m P_n^m(\theta_i) P_{n \pm 2\beta}^m(\theta_i) \sin(\theta_i) \right\} \tag{12}$$

where $n - 2\beta > m$, α_n^m is the actual coefficient value, Δ is the interval between

Table 3

Integral and least squares derived SHCs for m = 3 carried to n = 30

Harmonic $m = 3$	$I(Y)$ a_n^m	$U(X30)$ a_n^m	$I(Y)$ b_n^m	$U(X30)$ b_n^m	$I(Z)$ α_n^m	$I(Z)$ β_n^m
$n = 3$	−1.08	−1.08	−5.24	−15.66	0.41	−1.11
4	4.78	4.47	−1.31	−4.79	1.98	−0.04
5	1.41	−0.13	−0.83	−6.86	0.55	−0.33
6	0.97	−1.47	−0.71	−2.95	0.10	1.57
7	0.47	1.12	0.43	−3.68	0.65	1.88
8	0.23	−0.90	−1.14	−2.23	−0.66	0.29
9	0.05	0.78	0.53	−5.96	−0.17	1.13
10	0.30	−2.20	−2.00	−2.40	−1.76	1.89
11	0.33	−0.64	−0.50	−4.16	0.50	1.25
12	0.06	−2.37	−0.97	−0.87	−1.72	1.65
13	0.55	1.47	−0.70	−2.17	−1.23	0.56
14	−0.60	0.32	−0.57	−0.66	−1.69	2.19
15	−0.05	−0.44	−1.58	−1.58	−1.38	0.57
16	0.01	−0.71	−0.39	−0.58	−1.69	0.07
17	−0.80	−1.19	−0.92	0.59	−2.33	0.43
18	0.06	0.55	−0.95	0.15	−1.91	−0.85
19	−0.66	0.71	−0.55	−0.54	−1.26	−0.46
20	−0.32	1.51	−1.15	−0.30	−2.08	−2.25
21	−0.82	−0.35	−0.11	−1.17	−1.00	−1.39
22	−0.53	0.26	−0.78	0.74	−1.25	−1.74
23	−0.54	0.54	−0.23	−0.22	0.00	−2.30
24	−0.81	1.46	0.14	0.54	−0.26	−1.82
25	−0.34	0.69	−0.49	−1.20	0.10	−1.75
26	−0.29	0.29	0.22	−0.57	1.01	−1.12
27	−0.53	−0.30	−0.00	−0.35	0.87	−0.90
28	−0.34	−0.11	0.21	0.05	1.13	−0.86
29	−0.70	1.26	0.50	−0.19	1.17	0.36
30	−0.18	0.45	0.03	−0.32	1.22	0.14

latitudinal points and i is an index running over the $180/\Delta$ data points. Alternate coefficients of degrees higher and lower than n appear in the computed value because the products $P_n^m P_{n \pm 2\beta}^m$ are symmetric with respect to the equator, but do not have symmetry in the intervals $0 < \theta < \pi/2$ and $\pi/2 < \theta < \pi$ relative to the angles $\theta = \pi/4$ and $\theta = 3\pi/4$, respectively. In the limit as $180/\Delta \to \infty$ the second term containing such products approaches zero, while the first term approaches α_n^m. For $\Delta = 2°$ the factor multiplying α_n^m in the first term differs from unity by less than one part in 100 for $m = 1$, $n = 30$. The cross-product in the second term does not exceed 10^{-4} for $m = 1$, $n = 30$, so that even if the first 30 coefficients have the same sign and have a magnitude of order 10 nT, the computed $\alpha_n^{m'}$ will differ α_n^m by, at most, a few parts in 100. This assumes that coefficients beyond $n = 30$ have relatively small magnitudes compared with those up to degree 30.

Table 4

Integral and least squares derived SHCs for m = 4 carried to n = 30

Harmonic $m = 4$	$I(Y)$ a_n^m	$U(X30)$ a_n^m	$I(Y)$ b_n^m	$U(X30)$ b_n^m	$I(Z)$ α_n^m	$I(Z)$ β_n^m
$n = 4$	1.65	9.84	1.27	0.98	0.20	0.85
5	−0.42	0.54	−0.10	−1.67	0.03	−0.38
6	0.81	6.05	1.20	0.35	0.41	0.54
7	−0.05	1.29	−0.48	−0.02	0.16	−0.59
8	1.04	4.76	0.06	−0.38	0.76	0.11
9	0.15	0.61	−0.33	0.93	0.22	−0.61
10	0.88	4.52	−0.42	−0.08	1.04	−0.20
11	0.03	0.40	0.08	0.29	0.22	−0.40
12	0.12	3.02	−0.25	0.78	1.18	−0.41
13	0.52	0.77	0.34	0.03	0.29	−0.06
14	−0.32	1.94	0.55	0.74	1.20	−0.39
15	1.21	0.23	−0.76	0.17	0.43	0.22
16	−3.89	1.79	0.93	0.76	1.08	−0.21
17	0.87	−0.26	−0.74	−0.07	0.48	0.30
18	0.34	1.47	0.29	0.78	0.86	0.12
19	0.23	−0.07	0.30	0.00	0.39	0.31
20	1.00	1.12	−0.67	0.45	0.65	0.45
21	−0.33	0.10	0.99	0.12	0.22	0.29
22	1.25	0.87	−0.83	0.29	0.50	0.67
23	−0.26	0.23	0.40	0.07	0.04	0.36
24	0.68	0.58	0.15	0.38	0.40	0.76
25	0.26	0.22	−0.44	−0.10	−0.15	0.40
26	0.14	0.46	1.00	0.40	0.28	0.68
27	0.48	0.05	−0.93	−0.05	−0.28	0.43
28	0.11	0.57	0.89	0.32	0.14	0.60
29	0.23	−0.11	−0.34	0.06	−0.32	0.45
30	0.17	0.63	−0.21	0.24	0.00	0.47

The difficulty with the usual least squares analysis as has been applied to spherical harmonic expansions on the Sq-field in the recent past is the data near the poles ($\theta = 0, \pi$) are given the same weight as those closer to the equator. Thus, it would appear that the disagreement between the integral and unweighted least square-derived coefficient sets arises from improperly weighting, or more correctly, here, not weighting the truncated equations on which the least squares analyses were conducted.

SCHMITZ and CAIN (1983) have shown that the weighting factor $w_i = \sin \theta_i$ (indicating a particular latitude) will tend to remedy this situation by allowing the data to be treated in an equal area scheme. ALLDREDGE and KAWASAKI (1981) earlier showed that this was the case for zonal harmonics ($m = 0$). This is actually a factor which tends to orthogonalize the terms with respect to the data.

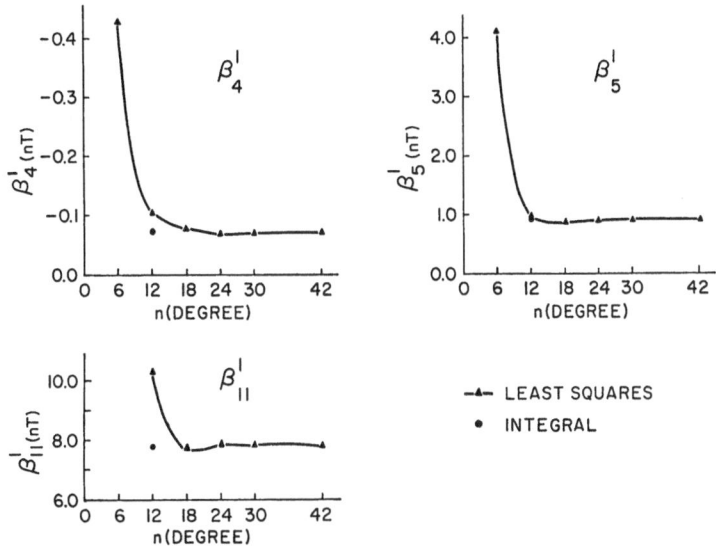

Figure 2

Examples (triangles) of unweighted least squares SHC values for Figure 1 in nT as a function of truncation degree and corresponding values by integral method (solid circles).

The weighting factor $w_i = \sin \theta_i$ in the notation of SCHMITZ and CAIN (1983) and $w_i^2 = \sin \theta_i$ in the notation of ALLDREDGE and KAWASAKI (1981) was applied in a least squares analysis of the Z-data such that the θ part of the right-hand side of (5) becomes $P_n^m(\theta)\sqrt{\sin \theta}$, which in analogy to (7) and (8) are orthogonal over the $180/\Delta$ points of data in the interval $0 < \theta < \pi$. For the Y-data, the weighting factor $w_i^2 = \sin^3 \theta_i$ (in the notation of ALLDREDGE and KAWASAKI (1981)) was used in the least squares analysis.

The least squares analyses were then run on the modified sets of data for coefficients from $m = 1$ to 4; $n = 1$ to 30. Only the results for the Y-data labelled "$W(Y)$" are tabulated in Table 1. In most cases the agreement between the results of the weighted least squares and the integral method is better than two decimal places and in many cases as much as five decimal places. Thus only the weighted least squares for a_n^1 and b_n^1 labelled "$W(Y)$" in Table 1 are shown separately. The entire set of a_n^1 and b_n^1 is tabulated even though there are only 5 cases in which they differ from the integral-derived coefficient. The other columns labelled "$I(\ )$" in Table 1 as well as Tables 2–4 should be understood to also represent the results of the weighted least squares analyses for each corresponding coefficient, good to two decimal places.

Finally, we have also analyzed the X-component with the unweighted version of the least squares method. The results are listed in the Tables in the columns labelled "$U(X30)$". A comparison with Y-component derived coefficients shows that there are substantial differences between the two sets even for many of the terms of lowest degree, e.g., the coefficients a_4^4, b_2^1, and b_3^3. That these differences are real can be confirmed by using Eqation (11) to derive b_2^1 and, recursively, b_4^1, etc. from the X-data. These integral-derived coefficient values agreed well with the values listed under column "$U(X30)$" in Tables 1–4 for the low-degree of coefficients. Also we are reasonably confident for the present Sq data, that the first few terms of an unweighted least squares analysis for the X-component data will behave similarly to the Y-component data, which have been shown to agree in the values of the coefficients of lowest degree for the unweighted and weighted least squares analyses as well as the integration analysis.

The above results indicate that the differences between the X and Y derived SHCs are real and not due to truncation of a series having coefficients with alternating signs and amplitudes which slowly diminish as suggested by CHAPMAN and BARTELS (1940), p. 687.

Some other possible sources of this discrepancy are (1) errors in data reduction, (2) externally caused, temporal variations which introduce non-local-time effects in the data and (3) sources which produce fields which cannot be represented by a scalar potential. Although data reduction, namely smoothing, many have introduced some error in the analysis, it is not likely to have produced the large differences observed in the *low degree* coefficients such as a_4^4, b_2^1, b_3^3 and others. Indeed, the effect of smoothing the Fourier coefficients, x_{ma} and y_{ma}, for example, should be to produce greater discrepancy in *higher degree* coefficients and lesser in those of *lower degree* for the same order m. Thus it is concluded that either or both the sources (2) and (3) listed above produce the major part of the discrepancy between the coefficient sets derived from X and Y separately. More on this will be discussed in the next section.

Summary and Discussion

The accuracy of a numerical integration of the integral equations can be improved by increasing the number of observation latitudes. However, the latitudinal intervals between the data points need not be equal in the integration and one can numerically integrate Equations (7)–(10) without much difficulty over unequally-spaced data. On the other hand, the question of proper weighting makes application of least squares on such a set of data somewhat more difficult. For unequally-spaced data, it is clear one cannot directly use the simple weighting factors (as has been done on the equi-latitudinal data synthesized in the report) alone, but must also include another factor to account for the latitudinal separation.

In principle, the accuracy of the results of the integration method can be further improved by inverting equations of the type (12) which are applied to the Z-data. Using the approximate values $\alpha_{n \pm 2\beta}^m$, $\beta = 0, 1, 2, \ldots$, calculated according to (7) and the sum of products of the form $P^m_{n \pm 2\beta} P^m_n \sin \theta_i$ at each latitudinal data point, a corrected α^m_n may be calculated from (12).

Since the primary purpose of this report was to outline an integral method of spherical harmonic analysis and to compare it with the least squares method we have not attempted to separate the internal from external sources and to place physical significance on our results. Nevertheless, it is interesting to speculate on the physical reasons for the large discrepancies between X-component and Y-component-generated spherical harmonic coefficient that occur in some anlayses. We do not believe from the present analysis that they arise wholly from error in data reduction nor from truncation of the spherical harmonic series as suggested by CHAPMAN and BARTELS (1940).

A source of nonpotential field, vertical currents through the spherical surface of observation, may produce differing X-derived and Y-derived coefficients. There are three known sources of earth-to-air currents—the fair-weather electric field, fields induced by magnetospheric processes, and thunderstorm activity. Since the current densities produced by the fair-weather electric field are small and relatively uniform over the earth (MÜHLEISEN, 1971), it does not appear that such a field could account for the discrepancy. Further, magnetospherically-induced fields, of which, those associated with the auroral regions should be largest, are about one order of magnitude smaller than the steady fair-weather field (MOZER, 1971). Thunderstorms, on the other hand are known to be associated with intense, nonuniform electric fields near the ground and moreover their occurrence tends to be local time dependent, e.g., in some regions, they are principally an afternoon-evening phenomenon. Thus, except for the possibility of some intermittent local time effects due to thunderstorms, Sq is not likely to be disturbed significantly enough by any of the above-mentioned earth-air currents for them to be the chief causes of the discrepancy.

A more likely possibility is that this discrepancy arises from fields of external sources which, when viewed from an earth-rotating frame, contain both local-time dependent and independent parts that are also universal-time dependent, e.g., main phase storms. As an example, consider an axially-symmetric external field, which is the major component of the ring current field. This field will have only northward and vertical components and since it is a universal-time-dependent phenomenon, it can introduce a non-local-time variation in the northward component of the daily variation, which is contrary to the initial assumption on which the analysis is based, namely, that Sq is only a local-time dependent phenomenon. Hence if such temporal disturbances exist in the data, they can produce incompatible X- and Y-data, such that the spherical harmonic coefficients derived therefrom will differ.

Acknowledgments

This paper appears in part in the M.S. thesis of the first author who was supported by NSF Grant (SP-5544) and NASA Grant (MSG201–62) during the period of preparation of the thesis. He is also grateful for the support of the U.S. Geological Survey, Denver, Colorado, where the present work was started and the Geophysical Institute, University of Alaska Fairbanks where it was completed.

We are grateful to W. H. Campbell, S.-I. Akasofu, the late Sydney Chapman, L. R. Alldredge, the late F. C. Frischknecht, N. W. Peddie and D. R. Schmitz for fruitful discussions and comments on various stages of the manuscript. We would like to thank J. H. Allen and H. R. Kroehl of NOAA for providing the supplementary data and helping with data reduction. We also thank two referees whose comments were found to be helpful in revising this manuscript.

The National Center for Atmospheric Research is sponsored by the National Science Foundation.

REFERENCES

ALLDREDGE, L. R., and KAWASAKI, K. (1981), *Spherical Harmonic Analysis in the Presence of High Harmonics*, J. Geomag. Geoelectr. *33*, 503–515.

CHAPMAN, S., and BARTELS, J., *Geomagnetism*, Vol. 2 (Oxford Univ. Press, London 1940).

GAUSS, C. F., *Allgemeine Theorie des Erdmagnetismus*, In *Resultate aus den Beobachten des Magnetischen Vereins im Jahre 1838* (C. F. Gauss and W. Weber, Weidmann Leipzig 1839) pp. 1–57.

KAWASAKI, K., *Solar Daily Geomagnetic Variations on Quiet Days at High Latitudes* (M.Sc. Thesis, Univ. of Alaska, 1967).

MATSUSHITA, S., *Solar quiet and lunar daily variation fields*, In *Physics of Geomagnetic Phenomena* (eds. Matsushita, S., and Campbell, W. H.) (Academic Press, New York 1967) Vol. 11, pp. 301–424.

MATSUSHITA, S., and MAEDA, H. (1965), *On the Geomagnetic Solar Quiet Daily Variations Field During the IGY*, J. Geophys. Res. *70*, 2535–2558.

MATSUSHITA, S., TARPLEY, J. D., and CAMPBEL, W. H. (1973), *IMF Sector Structure Effects on the Quite Geomagnetic Field*, Radio Science *8*, 963–972.

MOZER, F. S. (1971), *Balloon Measurement of Vertical and Horizontal Atmospheric Electric Fields*, Pure Appl. Geophys. *84*, 32–45.

MÜHLEISEN, R. P. (1971), *New Determination of Air-earth Current over the Ocean and Measurements of Ionosphere Potential*, Pure Appl. Geophys. *84*, 112–115.

SCHMITZ, D. R., and CAIN, J. C. (1983), *Geomagnetic Spherical Harmonic Analyses I, Techniques*, J. Geophys. Res. *88*, 1222–1228.

SCHUSTER, A. (1889), *The Diurnal Variation of Terrestrial Magnetism*, Phil. Trans. London (A) *180*, 467–518.

SUGIURA, M., and HAGAN, M. P. (1967), *Universal-time Changes in the Geomagnetic Solar Daily Variation Sq*, Sci. Rep. NSF GA 478, Dept. of Atmospheric Sciences, Univ. of Washington.

WALKER, G .W. (1913), *The Diurnal Variation of Terrestrial Magnetism*, Proc. Roy. Soc. London (A) *89*, 379–392.

(Received February 15, 1988, revised/accepted March 15, 1988)

PAGEOPH, Vol. 131, No. 3 (1989)

0033–4553/89/030371–23$1.50 + 0.20/0

Polar Region Sq

WEN-YAO XU[1]

Abstract—Geomagnetically quiet day variations in the polar region are reviewed with respect to geomagnetic field variation, ionospheric plasma convection, electric field and current. Persistently existing field-aligned currents are the main source of the polar region Sq. Consequently, the morphology and variability of the polar region Sq largely depend upon both field-aligned currents and ionospheric conductivity. Since field-aligned currents are the major linkage between the ionosphere and the magnetosphere, the latter is controlled by solar wind state, in particular, the interplanetary magnetic field, the polar region Sq exhibits remarkable IMF dependence.

Key words: S_q^p, polar ionospheric current, field-aligned current, IMF sector effect, ionosphere-magnetosphere coupling.

1. Introduction

Geomagnetic variations in the polar region have been studied extensively by many scientists (CHAPMAN and BARTELS, 1940; MATSUSHITA and CAMPBELL, 1967; AKASOFU and CHAPMAN, 1972; NISHIDA, 1978; KAMIDE and WOLF, 1987, and references therein). In the polar region, there are various geomagnetic variations and corresponding ionospheric electric currents, which are described by a series of evolving terms, such as DS, S_q^p, $DP1$, $DP2$, L^p, DPY and DPZ. All of these terms are used to designate different contributions to the total geomagnetic field and current system. S_q^p, as the polar part of the global quiet day variation Sq, is a basic and most simple type.

Although a close relationship exists between S_q^p and mid-low latitude Sq, there are differences in many respects.

First of all, the concept of "quiet condition" is different. At mid and low latitudes it is easy to determine the quiet days according to the K_p index. In the polar region, however, the quiet-time magnetosphere-ionosphere occurs only when the energy transfer from solar wind to the magnetosphere is at a minimum. In the open magnetosphere model, energy transfer is minimal when both the antiparallel

[1] Institute of Geophysics, Academia Sinica, P.O. Box 928, Beijing, China.

merging region and the merging rate are small, the conditions for which are associated with near zero or small positive IMF B_z, and low solar wind velocity. In fact, even during such quiet conditions, the geomagnetic field at high latitudes is continuously disturbed. The solar wind flow past the magnetosphere creates never-resting boundary layers; the polar region is practically never divorced of aurora. A minimal, but not zero, cross-cap electric field potential (5–10 kV), and therefore, low earthward convection velocity in the plasmasheet persist. Consequently, the choice of quiet days in the polar region is more or less different from that at mid-low latitudes.

The second difference between S_q^p and mid-low latitude Sq is at their origin. The electric field and current system at mid-low latitudes on magnetically quiet days are believed to be produced basically by the dynamo action of the thermosphere winds. In the polar region, however, the electric field and current system are thought to be driven mainly by field-aligned currents, derived either directly from the solar wind or internally from energetic particles in the outer magnetosphere. It is suggested by recent observations of large-scale electric fields at ionosphere altitudes that over the polar cap the electric field results for a large-fraction from a nearly direct mapping of the interplanetary electric field along the complicated geometry of reconnected interplanetary and geomagnetic field lines and possibly for a small-fraction from viscous-like interaction on the flanks of a magnetosphere.

The third difference between S_q^p and mid-low latitude Sq, related with the second one, is their correlation with the magnetosphere. The mid-low latitude Sq is basically an internal process in the ionosphere, although the solar wind state, interplanetary magnetic field and magnetospheric processes would influence Sq configuration (for instance, IMF sector effects on Sq intensity and focus). On the contrary, S_q^p in the polar region largely depends upon solar wind parameters and magnetospheric state. The polar ionosphere has been described as a "television screen", upon which we see projected the manifestation of the complex workings of the distant magnetosphere, such as aurora, electric field, current, plasma convection, and dynamic processes. Only by considering all of the detailed *in situ* magnetospheric data, in addition to the two-dimensional television image, can we hope to understand the complexities of the magnetosphere-ionosphere system, and their coupling processes of mass, energy and momentum.

In recent years great attention has been attracted to studying plasma convection, large-scale electric fields in the magnetosphere and ionosphere-magnetosphere coupling (KAMIDE and WOLF, 1987). As an important linkage between the ionosphere and the magnetosphere, the field-aligned current ($j_\parallel$) along with consequent polar plasma convection, electric potential, field and current have been studied extensively. As a basic state in the polar ionosphere, S_q^p lays the ground for studying the magnetosphere-ionosphere system under different activity levels.

Several theoretical and empirical models describing the magnetosphere-ionosphere coupling and the formation of S_q^p have been proposed on the basis of observations by satellites, incoherent scatter radar, and ground-based magnetometers. A few powerful computer codes have been designed for organizing and analyzing these data and computing plasma convection, electric field and current systems.

2. *Morphology of* S_q^p

On analyzing the IGY geomagnetic data in both Arctic and Antarctic regions, NAGATA and KOKUBUN (1962) found that the geomagnetic daily variation field in the polar cap on geomagnetically quiet days consists not only of the well established

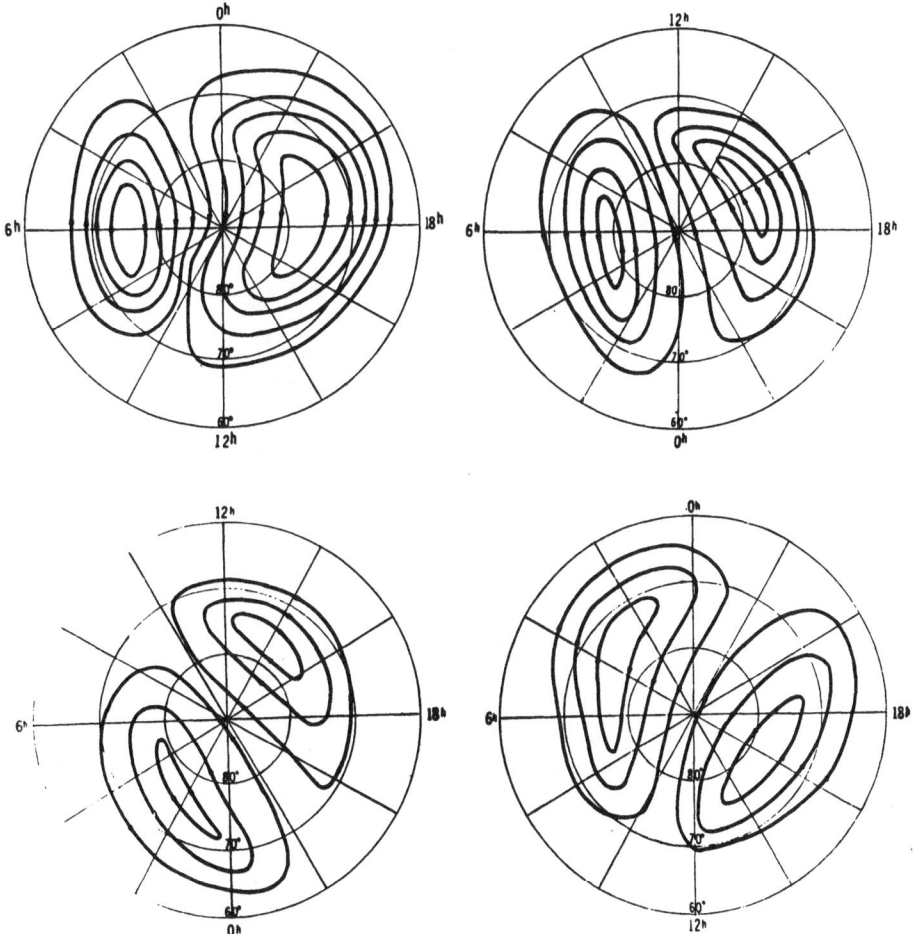

Figure 1
Equivalent current systems of the additional geomagnetic variation S_q^p in sunlight (top) and darkness (bottom) for Northern (left) and Southern (right) Hemispheres (after NAGATA and KOKUBUN, 1962).

S_q^o field which is a smooth extrapolation to the polar region of Sq defined at mid and low latitudes, but also of an additional field, S_q^p field, which is very similar to that of the D_s^c-field. The equivalent current pattern of the S_q^p field (as shown in Figure 1) seems to have some definite regular characteristics, namely (a) the S_q^p current system consists of two vortices centered roughly in morning and evening meridians; the direction of the currents over the central part of the polar cap is from 23 h toward 11 h in local geomagnetic time; (b) the current intensity shows a strong seasonal variation. The total current amounts to about 15×10^4 A in the sunlit polar cap, but it is reduced to less than one third of its value when the polar cap becomes dark, while its mode of distribution form is kept almost invariant as compared with the sunlit polar region S_q^p.

The existence of S_q^p was confirmed by succeeded studies (e.g., FAIRFIELD, 1963; KAWASAKI and AKASOFU, 1967; MATSUSHITA *et al.*, 1973; AKASOFU *et al.*, 1980;

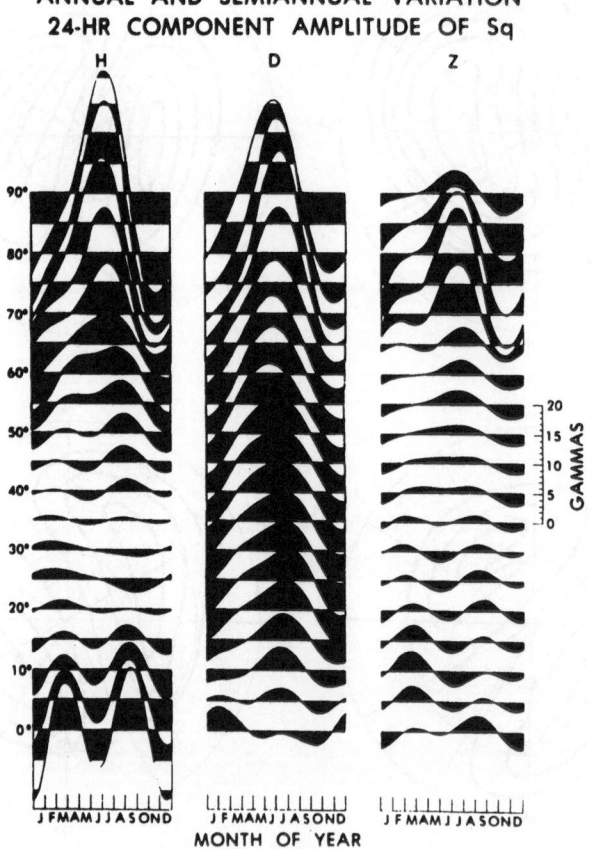

Figure 2
Annual and semi-annual variations in 24-hr component amplitude of Sq for 5 degrees geomagnetic latitudes (after CAMPBELL, 1982).

AKASOFU and AHN, 1981; CAMPBELL, 1982; MATSUSHITA and XU, 1982a,b,c). CAMPBELL (1982) studied the variation of *Sq* in the North American region by using 2.5-mn digitized magnetic data on quiet days and demonstrated (Figure 2) that: (1) in the polar region, the 24-hour component of the daily field is dominant; (2) the annual variation of the 24-hour component shows a summertime maximum. These characteristics, in general, agree with the expectation for ionospheric current sources by NAGATA and KOKUBUN (1962). Besides, the electric field and convection measurements by incoherent scatter radar and satellites also revealed the basic feature of S_q^p. Figure 3 presents a schematic representation of convection equipotential contours (plasma convection streamlines) in both the ionosphere and magnetosphere from the empirical Chatanika radar electric field model of FOSTER (1983) and mapped into the equatorial plane of the magnetosphere by FOSTER (1984). In this figure a predominantly two-celled structure is clearly displayed.

Many authors constructed equivalent ionospheric S_q^p current systems for different seasons and different geomagnetic activity levels that revealed more detailed characteristics of S_q^p (KAWASAKI and AKASOFU, 1967; MATSUSHITA *et al.*, 1973; AKASOFU *et al.*, 1980; AKASOFU and AHN, 1981; MATSUSHITA and XU, 1982c). In Figure 4, one of their results is depicted for comparison with Nagata and Kokubun's result, in which a seasonal variation of intensity and pattern can be clearly seen. It should be pointed out that in addition to its seasonal variation, S_q^p depends on magnetic activity and IMF sector polarity. When magnetic activity

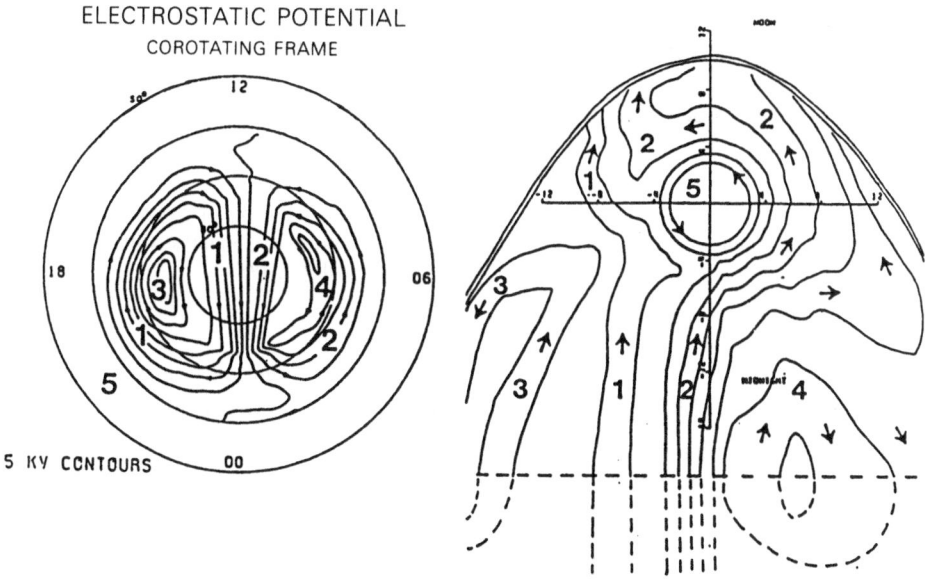

Figure 3
Equipotential contours of large-scale plasma convection in the ionosphere and in the equatorial plane of the magnetosphere (after FOSTER, 1984).

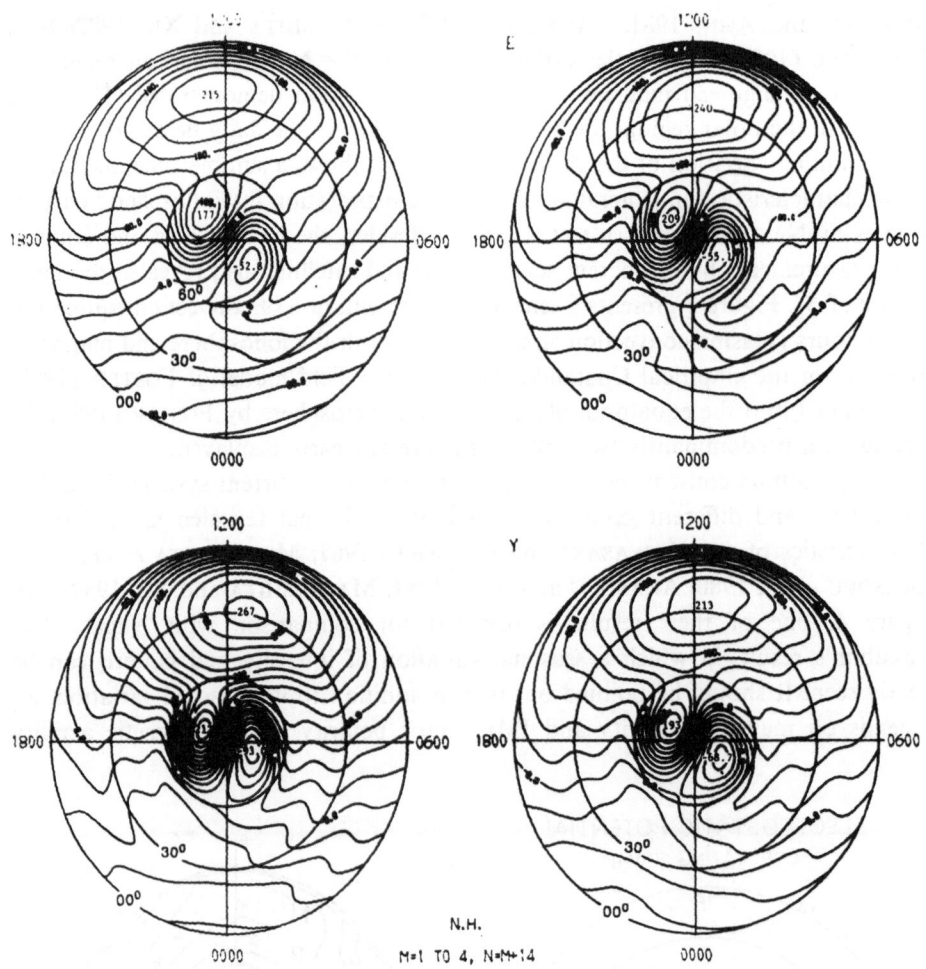

Figure 4
External S current systems in the Northern Hemisphere with a contour interval of 20 kA in December solstitial (top left), equinoctial (top right), and June solstitial (bottom left) months, with the yearly average (bottom right) (after MATSUSHITA and XU, 1982c).

increases, the S_q^p current pattern is distorted, a current enhancement along the auroral belt appears. IMF sector effects cause the S_q^p current vortices to shift toward morning side or evening side according to sector polarities. More detailed discussion will be presented in Section 4.

Although NAGATA and MIZUNO (1955) claimed that the Sq-field in the polar region on absolutely quiet days is considered as an extension of the Sq-field defined at the mid and low latitudes, the derivation of the S_q^p field has been repeated by several authors by utilizing the data on more quiet days than those studied by Nagata and Kokubun. KAWASAKI and AKASOFU (1967) showed an entirely

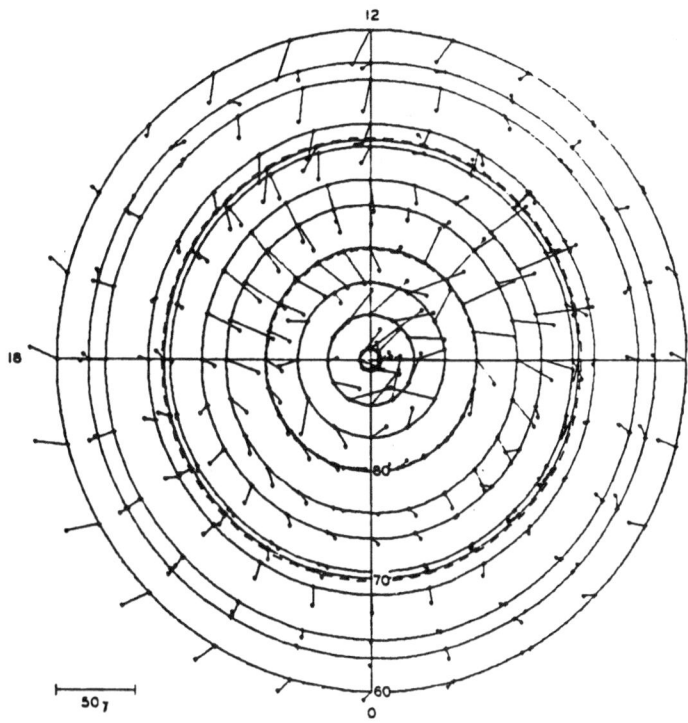

Figure 5(a)
The polar plot of the residual for May 8, 1964, an extremely quiet day (after KAWASAKI and AKASOFU, 1967).

different current direction inside the latitude circle of 85 degrees on extremely quiet days (Figure 5a). MAEZAWA (1976) concluded that an antisunward current develops over the entire polar cap when IMF is northward, the current being most intense in the noon sector around 82 degrees geomagnetic latitude (Figure 5b). Assuming that ionospheric Hall currents were responsible for the magnetic variation, he suggested that for a northward IMF the magnetospheric convection is directed sunward across the polar cap. It was found by BURKE *et al.* (1979) on the basis of S3–2 satellite data and BYTHROW *et al.* (1985) on the basis of MAGSAT data that at times of northward interplanetary magnetic field B_z the polar electric field may sometimes assume a four-celled pattern, with sunward convection near the pole, flanked by lobes of tailward convection on the duskside and dawnside (Figure 6). This pattern was seen in about 15% of the passes with $B_z > 0$, as confirmed by AE-C. A similar result was claimed by HORWITZ and AKASOFU (1979), using ground magnetic data from the northern polar cap, taken mostly in July and August, when the cap is sunlit and the electric field there is expected to drive appreciable currents (Figure 7). This special kind of polar current is named NBZ (IIJIMA *et al.*, 1984).

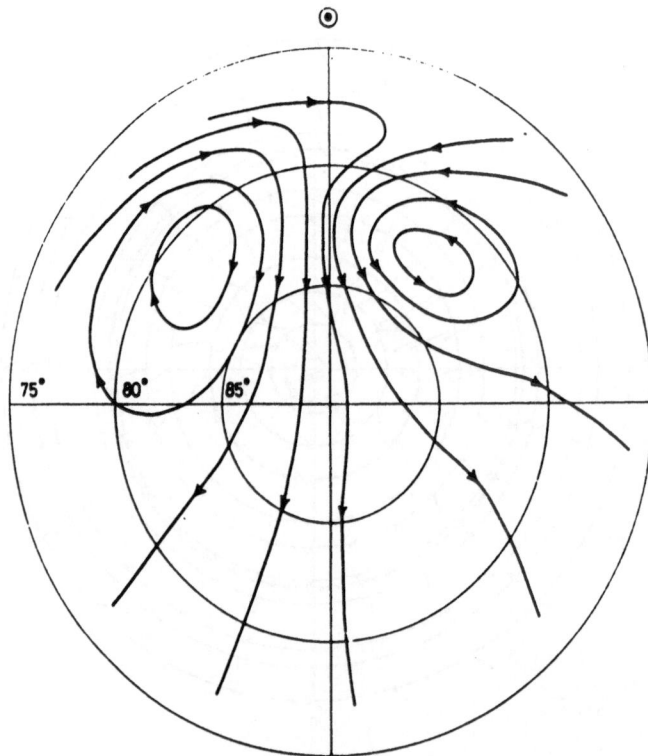

Figure 5(b)

Equivalent current pattern in the polar cap when IMF $B_z > 0$. The total current induced between two consecutive streamlines by a 1 nT increase in B_z is about 4×10^3 A (after MAEZAWA, 1976).

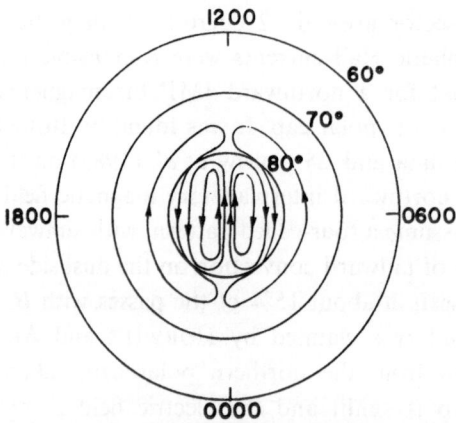

Figure 6(a)

Four-celled pattern of polar electric equipotentials associated with northward interplanetary B_z (after BURKE *et al.*, 1979).

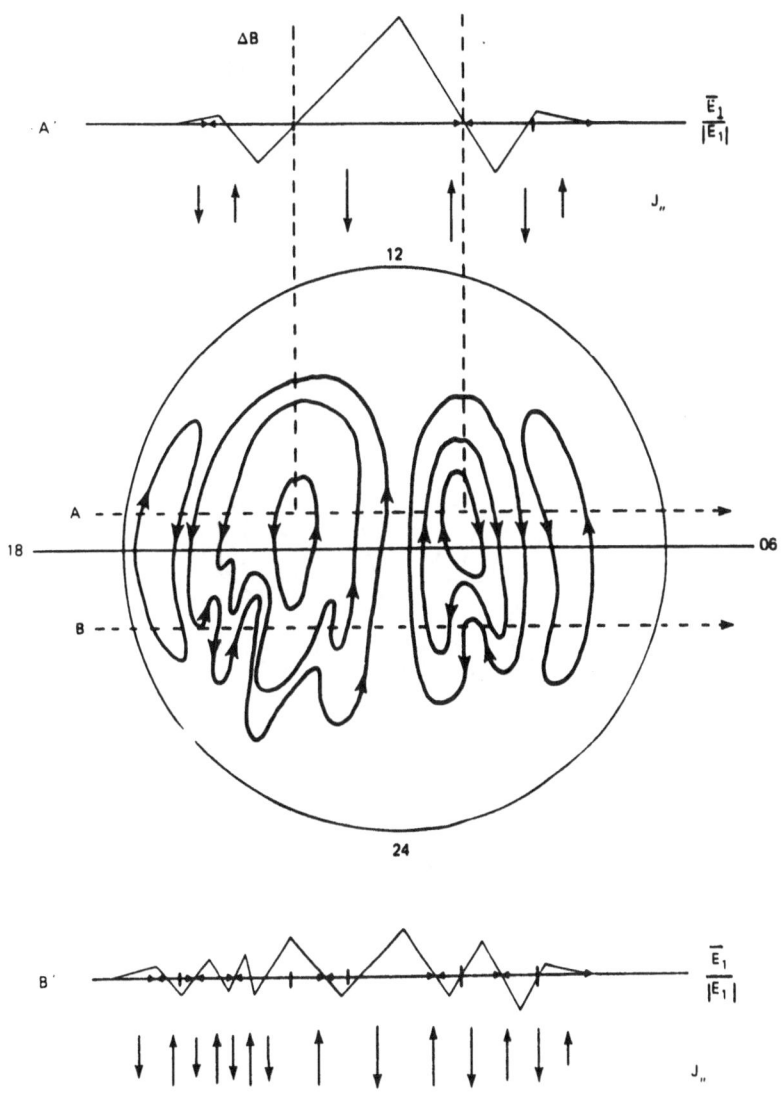

Figure 6(b)

A conceptual schematic of the polar cap convection pattern and its magnetic and electric field signatures during periods of northward IMF (after P. F. BYTHROW *et al.*, 1985).

3. Origin of S_q^p

The electric field and current in the polar region are mainly caused by field-aligned currents and reflect electrodynamics and hydrodynamics processes in the magnetosphere.

Field-aligned currents flowing along magnetic lines which lead to the polar and auroral regions constitute the major link between the ionosphere and the magnetosphere. They were first proposed by Birkeland in 1908 (CHAPMAN and BARTELS,

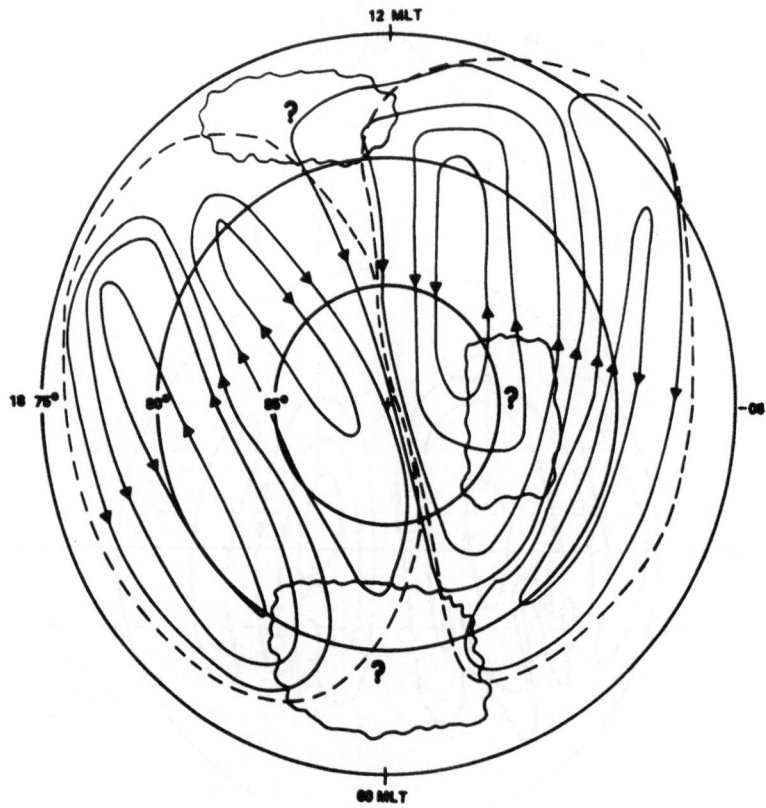

Figure 7
A qualitative polar ionospheric current system in the summer hemisphere during northward IMF (after
HORWITZ and AKASOFU, 1979).

1940; BOSTROM, 1968), but were only observed directly from space by ZMUDA and
ARMSTRONG (1974a,b) and independently by SUGIURA (1975). Several important
features of the field-aligned currents have been determined from a wealth of satellite
observations (POTEMRA *et al.*, 1979; BURKE, 1982; SAFLEKOS *et al.*, 1982). Five
distinct types of quiet-time $j_{\parallel}$ are distinguished, as shown in Figure 8 (IIJIMA and
POTEMRA, 1976a,b; STERN, 1983; IIJIMA *et al.*, 1984):

1. 'Region 1' current consists of current sheets which approximately follow the
   auroral oval. The current flows into the ionosphere at dawn and out of it at
   dusk. On the average, the total region 1 current in each flank of each polar
   cap is about 1.5×10^6 A at quiet times. The current density is on the order
   of $1\,\mu\text{A/m}^2$ with the maxima on the dayside at MLT 14–16 and 06–09.
   Current sheets are typically 2.5 degrees wide of latitude. The 'Region 1'
   currents show relatively stable behaviour and persist during geomagnetically
   quiet periods.

Southward IMF

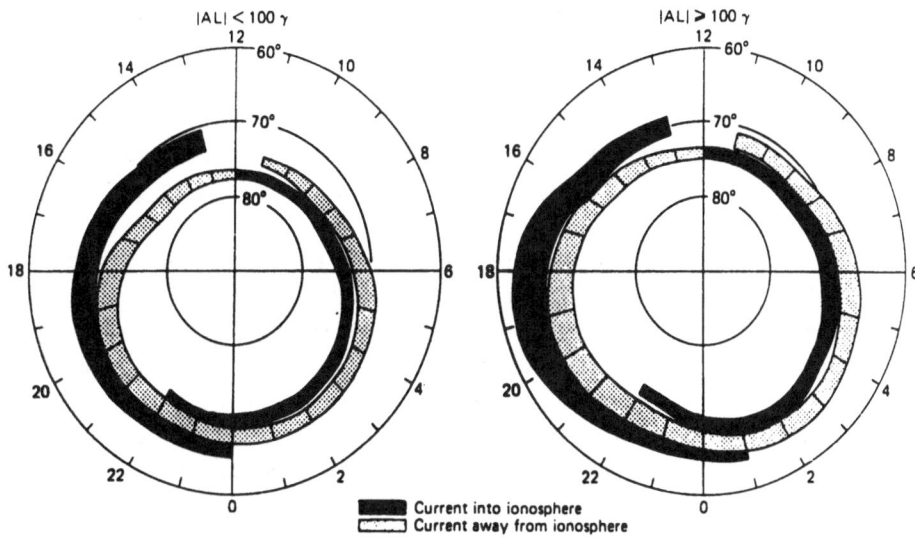

(a)

Northward IMF

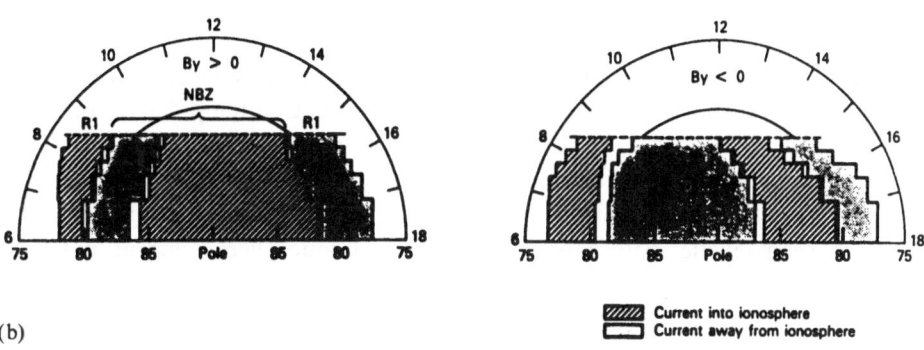

(b)

Figure 8
Field-aligned currents of regions 1 and 2 (a), and NBZ (b) (after IIJIMA and POTEMRA, 1976a, IIJIMA
et al., 1984).

2. 'Region 2' currents adjoin 'Region 1' currents on the equatorward side with opposite polarities. They have typically 75% of the intensity of 'Region 1' currents with maxima near midnight. The 'Region 2' currents show great variability and are closely associated with auroral electrojet activity.
3. Cusp currents are located poleward of the 'Region 1' in the cusp region and are also of opposed direction.

4. A three-layer overlap of current sheets exists in the Harang discontinuity region, near midnight, with an outward current flanked by two earthward current sheets.
5. NBZ currents are an additional set of field-aligned currents poleward of the 'Region 1' and 'Region 2' currents commonly observed in the auroral zone and generally have constant densities and sharp boundaries.

There exist two main classes of sources driving $j_\parallel$: dynamo action (in the boundary layer and plasmasheet) and charge separation (due to gradient and curvature drifts).

When field-aligned currents flow into and out of the ionosphere, they must be balanced by the divergence of ionospheric conduction currents. The appropriate distribution of the conduction currents requires a specific distribution of space charge and ionospheric electric field, which may be computed from given field-aligned currents. This field must in turn be consistent with the driving magnetospheric electric field. Several scientists tried to explain the complete loop process of the ionosphere-magnetosphere coupling with a self-consistent manner (e.g., VASYLI-UNAS, 1972; JAGGI and WOLF, 1973; SOUTHWOOD and WOLF, 1978). In this case the ionospheric convection, electric potential, field and current depend not only on boundary conductions in the magnetosphere but also on the ionosphere conductivity model. Other workers broke the chain into individual links and attempted to understand them in more detail (e.g., FEJER, 1964; YASUHARA et al., 1975; YASUHARA and AKASOFU, 1977; VOLLAND, 1978; NISBET et al., 1978; KAMIDE and MATSUSHITA, 1979a,b; BEULER et al., 1982; ZANETTI et al., 1984; RASMUSSEN and SCHUNK, 1987). Two types of boundary conditions have been used in the later approach: electric potential distribution along a given latitude circle (e.g., YA-SUHARA et al., 1975), or field-aligned current distribution in the polar region (e.g., KAMIDE and MATSUSHITA, 1979a,b; RASMUSSEN and SCHUNK, 1987).

When comparing computed results with geomagnetic observation data, the relationship between the equivalent ionospheric current system, responsible for observed geomagnetic disturbances, and the real ionospheric current must be noticed. As pointed out by FUKUSHIMA (1976), contributions to ground magnetic disturbances from $j_\parallel$ tend to cancel those arising from the related ionospheric Pedersen currents. The ground magnetic signature of the current system involving $j_\parallel$ is believed to be mostly due to Hall currents, with only the nightside part of the polar cap showing a large difference between the Hall current and the equivalent current intensity (FRIIS-CHRISTENSEN, 1984).

Figure 9 shows some of the computer simulation of the polar region currents caused by $j_\parallel$ (KAMIDE and MATSUSHITA, 1979a). It is noted in the figure that the current patterns depend on the distribution and intensity of the field-aligned currents, the ionospheric conductivities and the boundary conditions. Under quiet conditions, the 'Region 1' and 'Region 2' currents produce a well-defined two-celled ionospheric current system. Along with increasing magnetic activity, an auroral enhancement of ionospheric conductivity appears and distorts the ionospheric

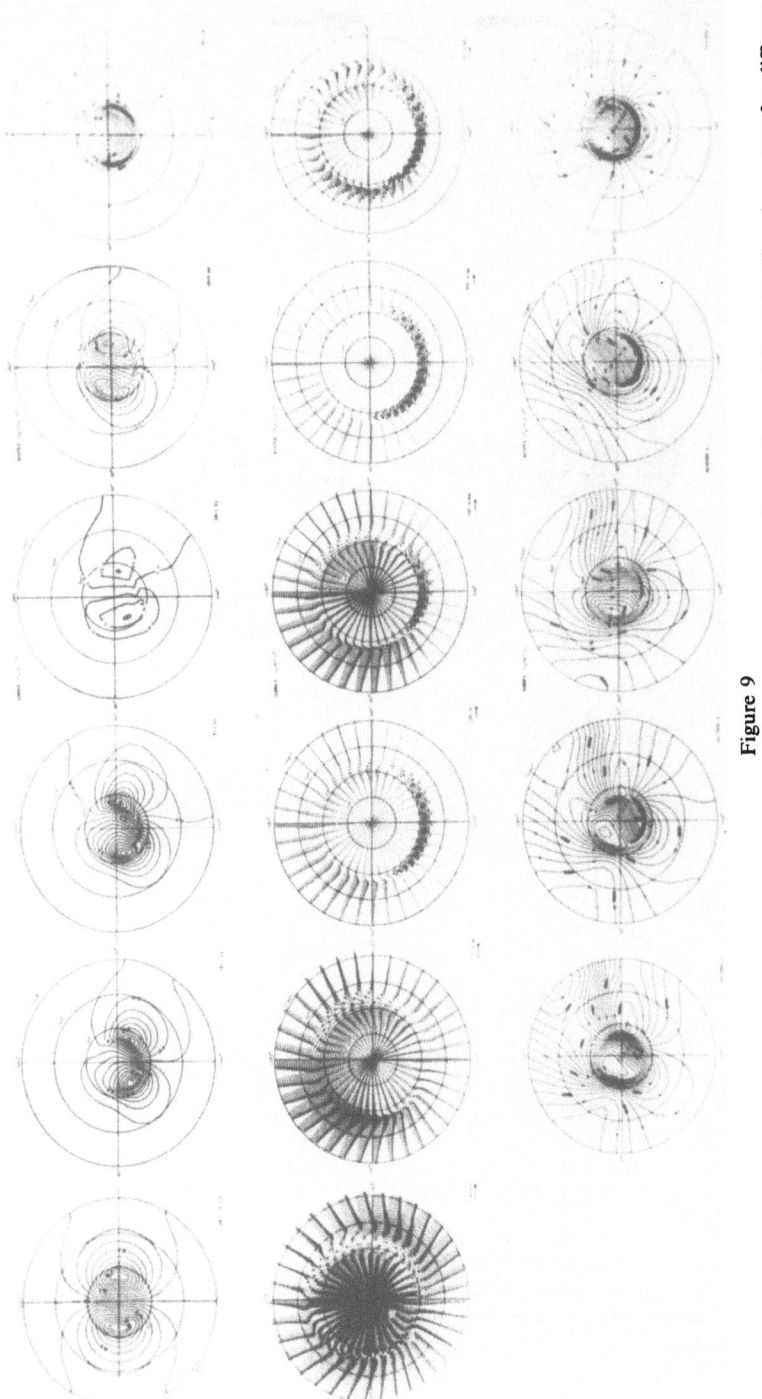

Figure 9

Ionospheric electrical potential (top), current vectors (middle), and equivalent current systems (bottom) caused by field-aligned currents for different simulation models: (a) simplest model with no conductivity gradient; (b) very quiet case without auroral conductivity enhancement; (c) quiet period model with light auroral enhancement; (d) and (e) seasonal variation; (f) double field-aligned currents.

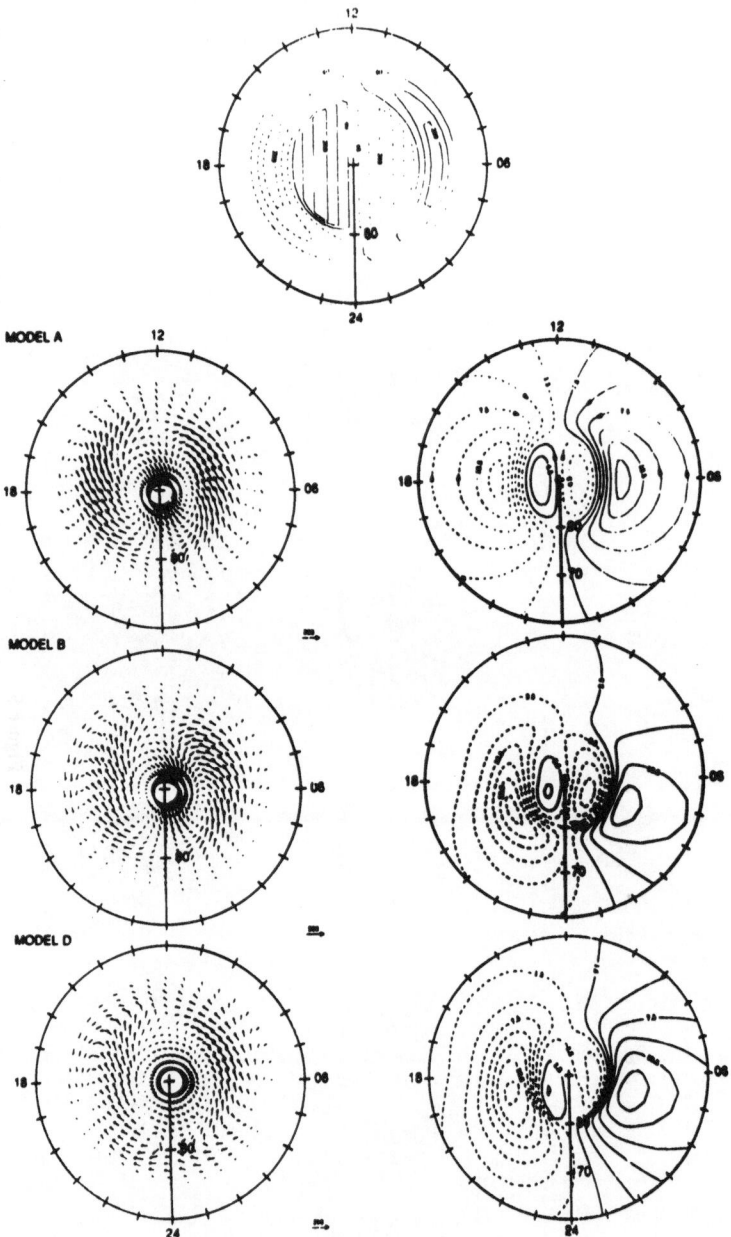

Figure 10
Simple model of field-aligned currents of Region 1 and NBZ (top); distributions of the ionospheric currents (left) and electric potentials (right) produced by the field-aligned currents (after RASMUSSEN and SCHUNCK, 1987).

currents, forming a remarkable auroral electrojet. When considering the NBZ currents, a characteristic four-celled ionospheric current pattern is obtained, as shown in Figure 10 (RASMUSSEN and SCHUNK, 1987). In this figure, an assumed field-aligned current distribution is given in the top panel. Electric currents and potentials producted by this $j_\parallel$ system are demonstrated in the lower panel for three ionospheric conductivity models (see Figure 1 in their paper).

In addition to above-mentioned direct computer simulation, reverse problems have been studied for estimating ionospheric field and current, as well as $j_\parallel$ from ground-based and/or space based magnetic observations. Two main schemes have been proposed for calculating three-dimensional current systems by KAMIDE *et al.* (1981) and KISABETH (1979). In Figure 11 the comparison of field-aligned currents computed by these two methods are shown (AKASOFU *et al.*, 1981). It is noted that the two methods produce similar $j_\parallel$ patterns.

4. *IMF Effects on the Polar Region Sq*

The configuration of the magnetosphere depends on the orientation of the interplanetary magnetic field. In particular, the directions of B_z and B_y components of the IMF are especially critical. Furthermore, since the ionosphere is coupled to the magnetosphere by the high conducting nature of the earth magnetic field lines, the ionosphere shares in this dependence.

Ground-based measurements gave the first hints that the interplanetary magnetic field is a controlling factor of $j_\parallel$, the polar region electric field and currents (DUNGEY, 1961; SVALGAARD, 1969; MANSUROV, 1969; MATSUSHITA and XU, 1982b). MATSUSHITA *et al.* (1973) examined the magnetic data at 40 stations located from the northern polar cap through the southern subauroral zone during quiet days in 1965, and investigated the S_q^p equivalent ionospheric current systems corresponding to toward- and away-sector structures. A clear shift of the average S_q^p pattern toward the afternoon (or early morning) side of the polar cap for the toward- (or away-) sector is observed as shown in Figure 12. This IMF sector effect was confirmed by many scientists (MATSUSHITA and XU, 1982b; FRIIS-CHRISTENSEN, 1984; FOSTER, 1987). FRIIS-CHRISTENSEN (1984) analyzed the polar geomagnetic fields with respect to various combinations of B_z and B_y components and constructed electric potential contour maps for these cases, as shown in Figure 13. In this figure the pattern shift can be clearly seen when $B_z < 0$ (southward IMF). Another evidence is the convection pattern constructed by FOSTER (1983, 1987), as using Chatanika incoherent scatter radar data (Figure 14). By using the KRM method, MATSUSHITA and XU (1982b) estimated the distribution of $j_\parallel$ for different sector polarities, depicted in Figure 15 only for the away sector. Similar results were obtained by many other scientists.

As mentioned above, when northward component B_z of IMF increases, a

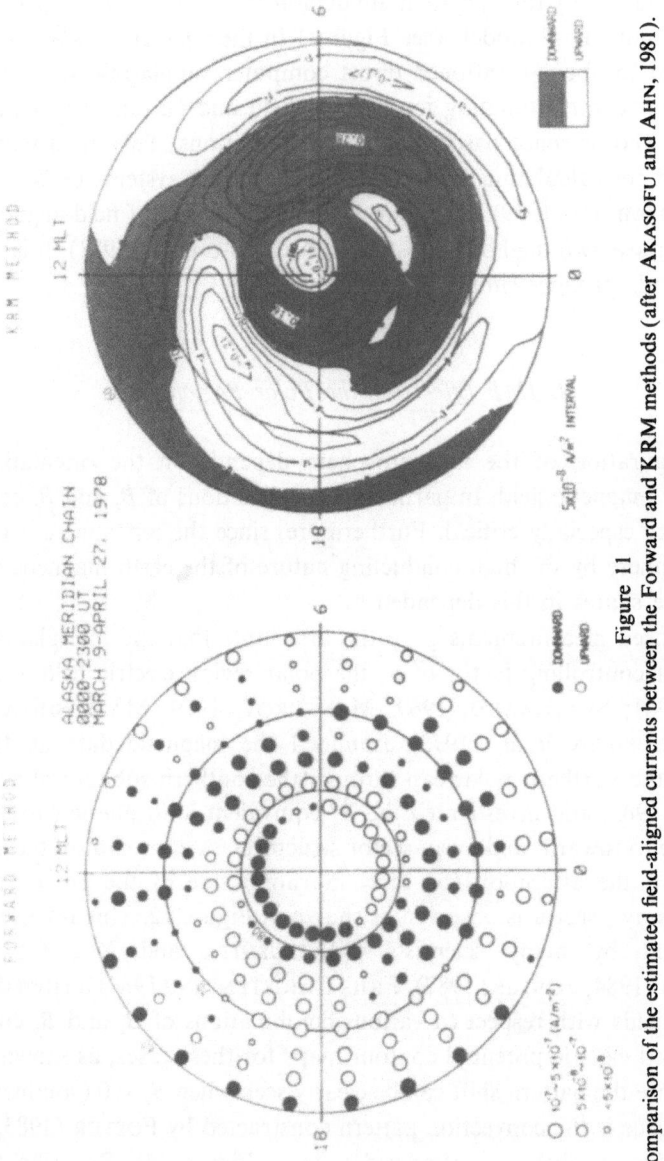

Figure 11
Comparison of the estimated field-aligned currents between the Forward and KRM methods (after AKASOFU and AHN, 1981).

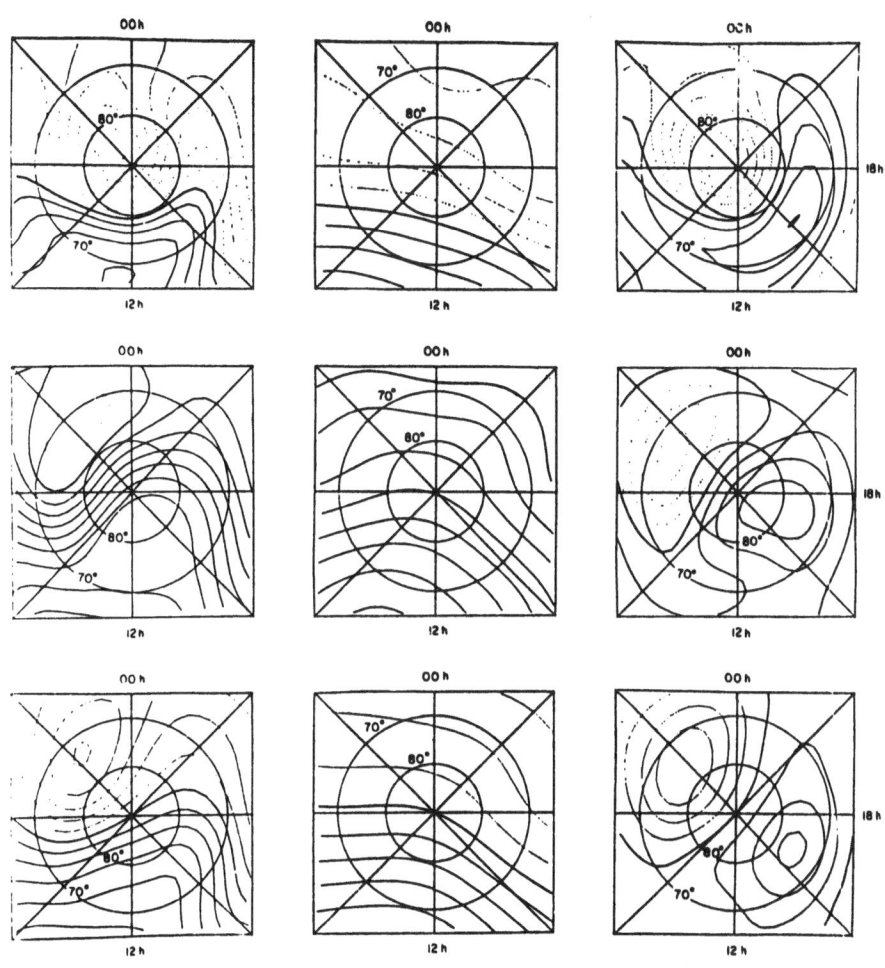

Figure 12

Equivalent overhead current systems in the northern polar region on quiet days during May and June of 1965 corresponding to toward (top) and away (middle) sectors and all average (bottom). The diagrams from left to right are Sq $(= S_q^0 + S_q^p)$, S_q^0 and S_q^p, which show counterclockwise (solid lines) and clockwise (dotted lines) current systems with a contour interval of 10^4 A (after MATSUSHITA et al., 1973).

sunward convection of ionospheric plasma, corresponding to antisunward currents in the central portion of the polar cap has been observed, as analyzing the data from ground-based magnetometers and satellite (FRIIS-CHRISTENSEN and WIL-HJELM, 1975; MAEZAWA, 1976; HORWITZ and AKASOFU, 1979; HEELIS et al., 1986). (See Figures 5, 6, and 7.) The characteristic convection pattern and current system during the period of $B_z > 0$ have also been suggested by special distribution of NBZ $j_{\parallel}$ (MCDIARMID et al., 1977; IIJIMA et al. 1984; FRANK et al., 1986). This direction of convection during northward IMF, opposite to the direction that occurs during southward IMF orientation, is supported by electric field measure

ELECTRIC POTENTIAL

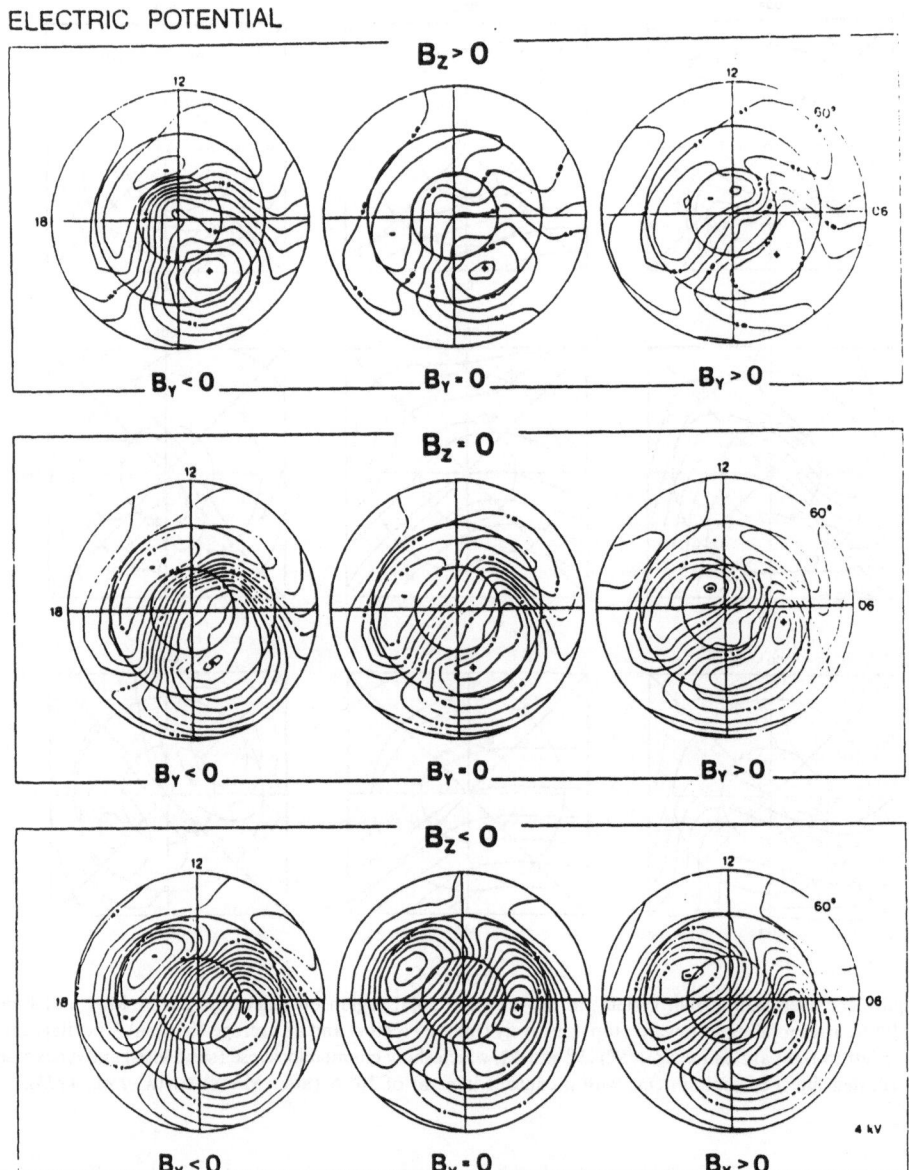

Figure 13
Electric potential contours corresponding to various combinations of IMF B_z and B_y (after FRIIS-CHRISTENSEN *et al.*, 1984).

ments from balloons (MOZER and GONZALEZ, 1973) and from satellites (BURKE *et al.*, 1979; REIFF, 1982), theta aurora (FRANK *et al.*, 1986) and neutral motion observation (KILLEEN *et al.*, 1985). Although statistical studies indicate that the sunward convection occurs predominantly in the summer hemisphere, in particular,

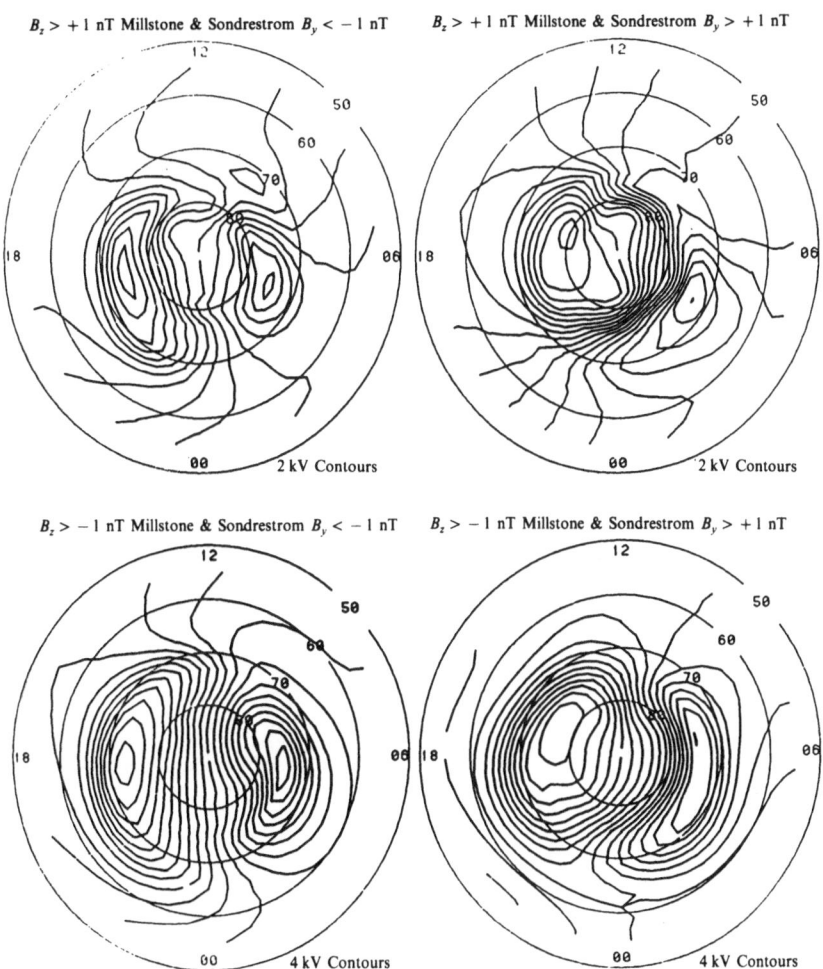

Figure 14
Equipotential contours of the average convection electric field patterns for IMF away sectors (left) and toward sectors (right) both for IMF $B_z > 0$ (top) and $B_z < 0$ (bottom) conditions (after FOSTER, 1987).

in the region of the polar cap sunward of the dawn-dusk meridian (BURKE *et al.*, 1979; MAEZAWA, 1976; IIJIMA *et al.*, 1984), some evidence has been found for large-scale four-celled convection in both hemispheres and in both sides of the dawn-dusk meridian (BYTHROW *et al.*, 1985; REIFF, 1982; KILLEEN *et al.*, 1985).

5. Summary

As a basic state in the polar ionosphere, S_q^p variation there lays the ground for studying various processes under different activity conditions. Persistently existing $j_\parallel$ is the main source of the polar region *Sq*.

FIELD-ALIGNED CURRENTS

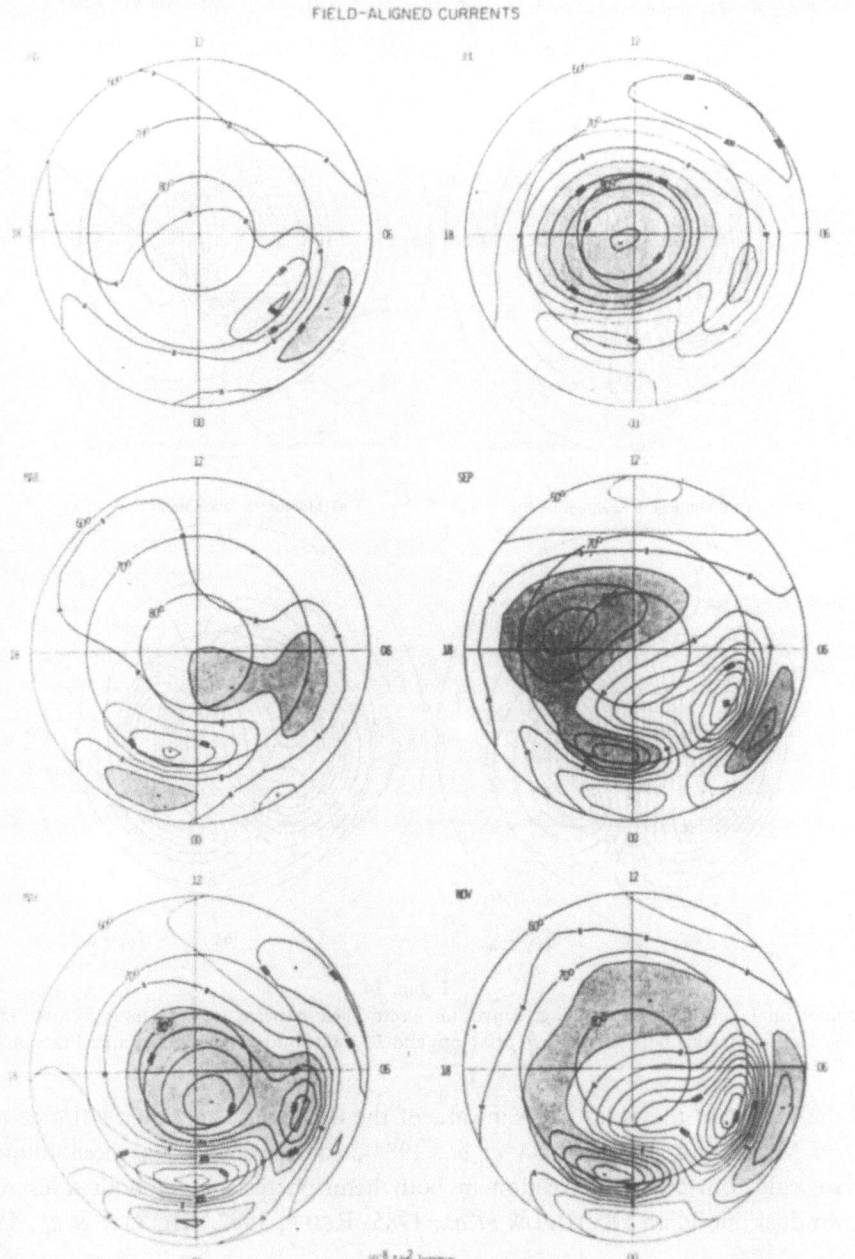

Figure 15
Estimated field-aligned current density distributions with a contour interval of 10^{-8} A/m^2 for away
sector (after MATSUSHITA and XU, 1982b).

Consequently, the morphology and variability of S_q^p largely depend upon both $j_{\parallel}$ and ionosphere conductivity, the latter, in turn, is closely related to particle precipitation associated with $j_{\parallel}$. Since $j_{\parallel}$ is the major link between the ionosphere and the magnetosphere, the latter is controlled by a solar wind state, in particular, IMF, the polar region *Sq* exhibits remarkable IMF dependence. When $B_z < 0$, two-celled patterns are predominant in plasma convection, electric potential, field and current; when $B_z > 0$, sometimes, four-celled patterns appear.

Acknowledgements

This work was supported by the National Natural Science Foundation of China (NSFC).

REFERENCES

AKASOFU, S.-I., and AHN, B. H. (1981), *Distribution of the Field-aligned Currents, Ionospheric Currents, and Electric Field in the Polar Region on a Very Quiet Day and a Moderately Disturbed Day*, J. Geophys. Res. *86*, 753–760.

AKASOFU, S.-I., and CHAPMAN, S., *Solar-terrestrial Physics* (Oxford University Press, Oxford 1972).

AKASOFU, S.-I., KAMIDE, Y., and KISABETH, J. (1981), *Comparison of Two Modeling Methods for Three-dimensional Current System*, J. Geophys. Res. *86*, 3389–3396.

AKASOFU, S.-I., KISABETH, J., AHN, B.-H., and TOMICK, G. J. (1980), *The S_q^p Magnetic Variation, Equivalent Current, and Field-aligned Current Distribution Obtained from the IMS Alaska Meridian Chain of Magnetometers*, J. Geophys. Res. *85*, 2085–2091.

BEULER, E., LI, C. H., and NISBET, J. S. (1982), *Relationships between Birkeland Currents, Ionospheric Currents, and Electric Fields*, J. Geophys. Res. *87*, 757–776.

BOSTROM, R. (1968), *Current in the Ionosphere and Magnetosphere*, Ann. Geophys. *24*, 681–694.

BURKE, W. J. (1982), *Magnetosphere-ionosphere Coupling: Contributions from IMS Satellite Observations*, Rev. Geophys. Space Phys. *20*, 685–708.

BURKE, W. J., KELLEY, M. C., SAGALYN, R. C., SMIDDY, M., and LAI, S. T. (1979), *Polar Cap Electric Field Structure with Northward Interplanetary Magnetic Field*, Geophys. Res. Lett. *6*, 21–24.

BYTHROW, P. F., BURKE, W. J., POTEMRA, T. A., ZANETTI, L. J., and LUI, A. T. Y. (1985), *Ionospheric Evidence for Irregular Reconnection and Turbulent Plasma Flows in the Magnetotail during Periods of Northward Interplanetary Magnetic Field*, J. Geophys. Res. *90*, 5319–5325.

CAMPBELL, W. H. (1982), *Annual and Semiannual Changes of the Quiet Day Variations (Sq) in the Geomagnetic Field at North American Locations*, J. Geophys. Res. *87*, 785–796.

CHAPMAN, S., and BARTELS, J., *Geomagnetism* (Clarendon Press, Oxford 1940).

DUNGY, J. W. (1961), *Interplanetary Magnetic Field and Auroral Zones*, Phys. Rev. Lett. *6*, 47–48.

FAIRFIELD, D. H. (1963), *Ionospheric Current Pattern in High Latitudes*, J. Geophys. Res. *68*, 3589–3602.

FEJER, J. A. (1964), *Theory of Geomagnetic Daily Disturbance Variations*, J. Geophys. Res. *69*, 123–137.

FOSTER, J. C. (1983), *An Empirical Electric Field Model Derived from Chatanika Radar Data*, J. Geophys. Res. *88*, 981–987.

FOSTER, J. C. (1984), *Ionospheric Signatures of Magnetospheric Convection*, J. Geophys. Res. *89*, 855–865.

FOSTER, J. C., *Radar-deduced Models of the Convection Electric Field*, in *Quantitative Modeling of Magnetosphere-ionosphere Coupling Processes* (eds. Kamide, Y., and Wolf, R. A.) (Koyto Sangyo University, Kyoto, Japan 1987) pp. 71–76.

FRANK, L. A., CRAVEN, J. D., GURNETTI, D. A., SHAWHAN, S. D., WEIMER, D. R., BURCH, J. L., WINNINGHAM, J. D., CHAPPELL, C. R., WAITE, J. H., HEELIS, R. A., MAYNARD, N. C., SUGIURA, M., PETERSON, W. K., and SHELLEY, E. G. (1986), *The Theta Aurora*, J. Geophys. Res. *91*, 3177–3224.

FRIIS-CHRISTENSEN, E., and WILHJELM, J. (1975), *Polar Cap Currents for Different Directions of the Interplanetary Magnetic Field in the Y-Z Plane*, J. Geophys. Res. *80*, 1248–1260.

FRIIS-CHRISTENSEN, E., *Polar cap current systems*, In *Magnetospheric Currents* (ed. Potemra, T. A.) (AGU, Washington D.C. 1984) pp. 86–95.

FUKUSHIMA, N. (1976), *Generalized Theorem for No Ground Magnetic Effect of Vertical Currents Connected with Pedersen Currents in the Uniform-conductivity Ionosphere*, Rep. Ionos. Space Res. Jap. *30*, 35–40.

HEELIS, R. A., REIFF, P. H., WINNINGHAM, J. D., and HANSON, W. B. (1986), *Ionospheric Convection Signatures Observed by DE-2 during Northward Interplanetary Magnetic Field*, J. Geophys. Res. *84*, 2567–2572.

HORWITZ, J. L., and AKASOFU, S.-I. (1979), *On the Relationship of the Polar Cap Current System to the North-south Component of the Interplanetary Magnetic Field*, J. Geophys. Res. *84*, 2567–2572.

IIJIMA, T., and POTEMRA, T. A. (1976a), *The Amplitude Distribution of Field-aligned Currents at Northern High Latitudes Observed by Triad*, J. Geophys. Res. *81*, 2165–2174.

IIJIMA, T., and POTEMRA, T. A. (1976b), *Field-aligned Currents in the Dayside Cusp Observed by Triad*, J. Geophys. Res. *81*, 5971–5975.

IIJIMA, T., POTEMRA, T. A., ZANETTI, L. J., and BYTHROW, P. F. (1984), *Large-scale Birkeland Currents in the Dayside Polar Region during Strongly Northward IMF: A New Birkeland Current System*, J. Geophys. Res. *89*, 7441–7452.

JAGGI, R. K., and WOLF, R. A. (1973), *Self-consistent Calculation of the Motion of a Sheet of Ions in the Magnetosphere*, J. Geophys. Res. *78*, 2852–2866.

KAMIDE, Y., and MATSUSHITA, S. (1979a), *Simulation Studies of Ionospheric Electric Fields and Currents in Relation to Field-aligned Currents, 1. Quiet Period*, J. Geophys. Res. *84*, 4083–4098.

KAMIDE, Y., and MATSUSHITA, S. (1979b), *Simulation Studies of Ionospheric Fields and Currents in Relation to Field-aligned Currents, 2. Substorms*, J. Geophys. Res. *84*, 4099–4115.

KAMIDE, Y., and WOLF, R. A., *Quantitative Modeling of Magnetosphere-Ionosphere Coupling Processes* (Kyoto Sangyo University 1987).

KAMIDE, Y., RICHMOND, A. D., and MATSUSHITA, S. (1981), *Estimation of Ionospheric Electric Fields, Ionospheric Currents, and Field-aligned Currents from Ground Magnetic Records*, J. Geophys. Res. *86*, 801–813.

KAWASAKI, K., and AKASOFU, S.-I. (1967), *Polar Solar Daily Geomagnetic Variations on Exceptionally Quiet Days*, J. Geophys. Res. *72*, 5363–5371.

KILEEN T. L., HEELIS, R. A., HAYS, P. B., SPENSER, N. W., and Hanson, W. B. (1985), *Neutral Motions in the Polar Thermosphere for Northward Interplanetary Magnetic Field*, Geophys. Res. Lett. *12*, 159–162

KISABETH, J. L. (1979), *On calculating magnetic and vector potential fields due to large-scale currents in an infinitely conducting earth*, In *Quantitative Modeling of Magnetospheric Processes* (ed. Olson, W. P.) (AGU Washington, D. C. 1979) pp. 473–498.

MAEZAWA, K. (1976), *Magnetospheric Convection Induced by the Positive and Negative Z Components of the Interplanetary Magnetic Field: Quantitative Analysis Using Polar Cap Magnetic Records*, J. Geophys. Res. *81*, 2289–2302.

MANSUROV, S. M. (1969), *New Evidence of a Relationship between Magnetic Fields in Space and on Earth*, Geomag. Aeron. *9*, 622–623.

MATSUSHITA, S., and CAMPBELL, W., *Physics of Geomagnetic Phenomena* (Academic, New York 1967).

MATSUSHITA, S., and XU, WEN-YAO (1982a), *Sq and L Currents in the Ionosphere*, Ann. Geophys. *38*, 295–305.

MATSUSHITA, S., and XU, WEN-YAO (1982b), *Equivalent Ionospheric Current Systems Representing IMF Sector Effects on the Polar Geomagnetic Field*, Planet. Space Sci. *30*, 641–656.

MATSUSHITA, S., and XU, WEN-YAO (1982c), *Equivalent Ionospheric Current Systems Representing Solar Daily Variations of the Polar Geomagnetic Field*, J. Geophys. Res. *87*, 8241–8254.

MATSUSHITA, S., TARPLEY, J. D., and CAMPBELL, W. (1973), *IMF Sector Structure Effects on the Quiet Geomagnetic Field*, Radio Sci. *8*, 963–972.

McDIARMID, I. B., BUDZINSKI, E. E., WILSON, M. D., and BURROWS, J. R. (1977), *Reverse Polarity Field-aligned Currents at High Latitudes*, J. Geophys. Res. *82*, 1513–1518.

MOZER, F. S., and GONZALEZ, W. D. (1973), *Response of Polar Cap Convection to the Interplanetary Magnetic Field*, J. Geophys. Res. *78*, 6784–6786.

NAGATA, T., and KOKUBUN, S. (1962), *An Additional Geomagnetic Daily Variation Field (S_q^p Field) in the Polar Region on Geomagnetically Quiet Day*, Rep. Ionos. Space Res. Jan. *16*, 256–274.

NAGATA, T. AND MIZUNO, H. (1955), *Sq-Field in the Polar Region on Absolutely Quiet Days*, J. Geomag. Geoelectr. *7*, 69–74.

NISBET, J. S., MILLER, M. J., and CARPENTER, L. A. (1978), *Currents and Electric Fields in the Ionospheric Due to Field-aligned Auroral Currents*, J. Geophys. Res. *83*, 2647–2657.

NISHIDA, A., *Geomagnetic Diagnosis of the Magnetosphere* (Springer-Verlag, New York, Heidelberg, Berlin 1978).

POTEMRA, T. A., IIJIMA, T., and SAFLEKOS, N. A., *Large-scale characteristics of Birkeland currents*, In *Dynamics of the Magnetosphere* (ed. Akasofu, S.-I.) (D. Reidel, New York 1979) pp. 165–199.

RASMUSSEN, C. E., and SCHUNK, R. W. (1987), *Ionospheric Convection Driven by NBZ Currents*, J. Geophys. Res. *92*, 4491–4504.

REIFF, P. H. (1982), *Sunward Convection in Both Polar Caps*, J. Geophys. Res. *87*, 5976–5980.

SAFLEKOS, N. A., SHEEHAN, R. E., and CAROVILLANO, R. L. (1982), *Global Nature of Field-aligned Currents and their Relation to Auroral Phenomena*, Rev. Geophys. Space Phys. *20*, 709–734.

SOUTHWOOD, D. J., and WOLF, R. A. (1978), *An Assessment of the Role of Precipitation in Magnetospheric Convection*, J. Geophys. Res. *83*, 5227–5232.

STERN, D. P. (1983), *The Origin of Birkeland Currents*, Rev. Geophys. Space Phys. *21*, 125–138.

SUGIURA, M. (1975), *Identification of the Polar Cap Boundary and the Auroral Belt in the High Latitude Magnetosphere: A Model for Field-aligned Currents*, J. Geophys. Res. *80*, 2057–2068.

SVALGAARD, L. (1969), *Sector Structure on the Interplanetary Magnetic Field and Daily Variation of the Geomagnetic Field at High Latitudes*, Dan. Met. Inst. Geophys. Pap., R–16.

VASYLIUNAS, V. M., *The interrelationship of magnetospheric processes*, In *Earth's Magnetospheric Processes* (ed. McCormac, B. M.) (D. Reidel, Hingham, Mass. 1972) pp. 29–38.

VOLLAND, H. A. (1978), *A Model of the Magnetospheric Electric Convection Field*, J. Geophys. Res. *83*, 2695–2699.

YASUHARA, F., and AKASOFU. S.-I. (1977), *Field-aligned Currents and Ionospheric Electric Fields*, J. Geophys. Res. *82*, 1279–1368.

YASUHARA, F., KAMIDE, Y., and AKASOFU, S.-I. (1975), *Field-aligned and Ionospheric Currents*, Planet. Space Sci. *23*, 1355–1368.

ZANETTI, L. J., POTEMRA, T. A., IIJIMA, T., BAUMJOHAN, W., and BYTHROW, P. F. (1984), *Ionospheric Birkeland Current Distributions for Northward Interplanetary Magnetic Field: Inferred Polar Convection*, J. Geophys. Res. *89*, 7453–7458.

ZMUDA, A., and ARMSTRONG, J. C. (1974a), *The Diurnal Variation of the Region with Vector Magnetic Field Changes Associated with Field-aligned Currents*, J. Geophys. Res. *79*, 2501–2502.

ZMUDA, A., and ARMSTRONG, J. C. (1974b), *The Diurnal Flow Pattern of Field-aligned Currents*, J. Geophys. Res. *79*, 4611–4619.

(Received/accepted January 12, 1988)

PAGEOPH, Vol. 131, No. 3 (1989)

0033-4553/89/030395-18$1.50 + 0.20/0

Field-aligned Currents in the Undisturbed Polar Ionosphere

H. W. KROEHL[1]

Abstract—Field-aligned currents, FAC's, which couple ionospheric currents at high latitudes with magnetospheric currents have become an essential cornerstone to our understanding of plasma dynamics in the polar region and in the earth's magnetosphere. Initial investigators of polar electrodynamics including the aurora were unable to distinguish between the ground magnetic signatures of a purely two-dimensional current and those from a three-dimensional current system, ergo many scientists ignored the possible existence of these vertical currents. However, data from magnetometers and electrostatic analyzers flown on low-altitude, polar-orbiting satellites proved beyond any reasonable doubt that field-aligned currents existed, and that different ionospheric regions were coupled to different magnetospheric regions which were dominated by different electrodynamic processes, e.g., magnetospheric convection electric fields, magnetospheric substorms and parallel electric fields. Therefore, to define the "undisturbed" polar ionosphere and its structure and dynamics, one needs to consider these electrodynamic processes, to select times for analysis when they are not strongly active and to remember that the polar ionosphere may be disturbed when the equatorial, mid-latitude and sub-auroral ionospheres are not.

In this paper we will define the principle high-latitude current systems, describe the effects of FAC's associated with these systems, review techniques which would minimize these effects and present our description of the "undisturbed" polar ionosphere.

Key words: Polar ionosphere, field-aligned currents, ionospheric electric fields.

1. A Historical Introduction

The most captivating feature of the polar ionosphere is the aurora borealis. Known even to the Greeks and Romans, it was and is, a rare phenomenon at such low magnetic latitudes. Authors, poets, and philosophers described the aurora as "smoke of straw which is burnt in the country" or as "gulfs," "abysses," "torches" or "beams" of various hues and colors. An association between these optical emissions and variations in the magnetic field was made as early as 1716 by Halley, with supporting reports by Celsius, Hiorter and Wilcke in 1741. As late as 1896, it was an accepted proposition that "there is no necessary connection between the aurora borealis and magnetic perturbations in the Arctic" (ANGOT, 1896).

[1] Solar-Terrestrial Physics Division, National Geophysical Data Center/NOAA, Boulder, CO 80303, USA.

In 1882, Balfour Stewart proposed that the diurnal variations in the earth's magnetic field resulted from electric currents which were external to the earth's surface. Later BIRKELAND (1908) proposed a three-dimensional system of external currents at high latitude with (magnetic) field-aligned currents driving the horizontal currents in the polar ionosphere. The relation between magnetic variations, or currents, and aurora, or enhanced ionization, lead one to consider Ohm's law, Maxwell's equations and the conditions for current continuity, which may be briefly summarized by $\bar{J} = (\Sigma)\bar{E}$, where $\bar{J}$ is current density in amperes/cm^2, Σ is the height-integrated conductivity in mhos and $\bar{E}$ is the electric field in volts, and $J_\parallel = \text{div } J$, where $J_\parallel$ is the field-aligned current. Thus the important plasma parameters are horizontal and vertical current, conductivity and electric field. The first complete ionosphere picture of such current systems described in three-dimensions for the high latitude was presented by BOSTROM (1964).

Needless to say, Birkeland's and Bostrom's concept was not universally accepted because a three-dimensional current system, unique from a two-dimensional horizontal system, could not be unambiguously defined using only ground magnetic recordings in the presence of uniform conductance. However, ZMUDA (1966) flew a magnetometer on the polar orbiting satellite 1963–38C, and the recorded data could only be adequately explained by a three-dimensional current structure. Initial efforts to explain the magnetic disturbances as hydromagnetic waves were proven to be groundless because the latitudinal extent of the disturbances was too small. Thus, the auroral displays were associated with electric currents which caused the magnetic perturbations and the horizontal currents were directly related to vertical currents.

The magnetic field and ionosphere of the earth, and their interaction with the magnetic field and plasma of interplanetary space, determine the structure and intensity of the earth's magnetosphere. Figure 1 presents the principal large-scale structure of plasma, magnetic fields and currents of the magnetosphere. Through this schematic, the importance of field-aligned currents that couple magnetospheric and ionospheric current systems is shown in the polar cap, polar cusp, and auroral regions. In this paper, we will discuss large-scale currents of these regions as though they were separate, though they are not, and our definition of a state of quiescence will depend on the interaction of the appropriate magnetospheric region with the solar wind and with the ionosphere.

Large-scale field-aligned currents are embedded in regions of precipitating plasma which also cause the auroral emissions. The resulting enhanced ionospheric conductivity becomes very important to the solution of Ohm's law. Since we are modeling large-scale current systems, we need the large-scale conductance which is difficult to determine, even though the global extent of the aurora, at least for large auroral displays, was well documented as early as the 1700's. FELDSTEIN and STARKOV (1967) defined an auroral oval which was continuous in local time but moved in latitude as the level of magnetic activity and local time changed. This oval

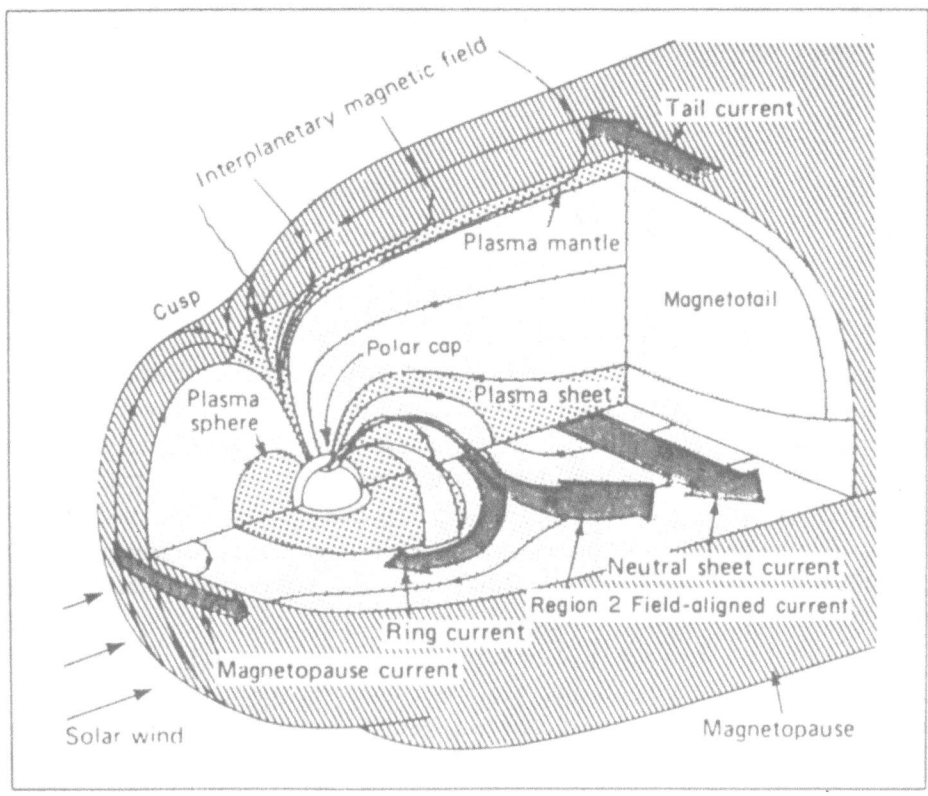

Figure 1
A schematic representation of the earth's magnetosphere. Regions of current and plasma density are
highlighted.

contributed significantly to large-scale models and is still used. (Of course, we
cannot exclude solar radiation as the principal source of conductance on the
dayside even at very high altitudes.)

The large-scale structure of auroral currents was first systematically described
by VESTINE (1944) and HARANG (1946) who reported on the average location
and intensity of magnetic variations which they attributed to electrojet currents.
Subsequent reports described the average location, intensity, structure and dynamics of these auroral electrojet currents as a function of local time, Universal Time
and season as well as parameters of the solar wind, cf. BERTHELIER et al. (1974),
ALLEN and KROEHL (1975), ROSTOKER (1977), KAMIDE (1979), AKASOFU et al.
(1980) and FRIIS-CHRISTENSEN (1984). Recent studies focus on the instantaneous
distribution of high latitude electric fields and currents on global scales as described in KAMIDE et al. (1981), LEVITIN et al. (1982) and RICHMOND et al.
(1988).

Here, we will review empirical models which are derived from statistical studies of measured parameters and relate those to large-scale numerical models which describe the instantaneous distribution of conductance, electric fields and currents in the polar ionosphere

2. Satellite and Ground-based Instrumental Data Analyses

Direct measurements of plasma densities and motions at ionospheric and magnetospheric altitudes are available only from rocket-borne and satellite-borne instruments which are very limited in space and time. Satellite recorded data can be averaged to define a global distribution but rocket-borne data are too limited to yield much information about large-scale structures. The satellite instruments which have proven to be the most useful are electrostatic analyzers (ESA) that provide number fluxes of electrons and ions at specified energies and directions; auroral images, that record auroral intensities at specified wavelengths or energies; drift meters, which record large-scale plasma motions; and magnetometers, that monitor the magnetic field at satellite altitudes.

ESA data have been used successfully to model the large-scale distribution of Hall (perpendicular to the electric field) and Pedersen (the parallel component) conductances of the ionosphere. However, existing large-scale conductance models derived form satellite-borne ESA data contain several severely-limiting assumptions: (1) the activity level occurring during different orbits must be defined from geophysical indices that were designed for different purposes, (2) a three-dimensional model of atmospheric constituent densities into which the precipitating particles will collide and (3) a three-dimensional model of the earth's magnetic field that governs plasma motion. The first model by WALLIS and BUDZINSKI (1981) used a gross measure of the Kp index, then SPIRO et al. (1982) used a gross measure of the AE index, then HARDY et al. (1987) used the complete Kp index with a large data base of ESA data as did FULLER-ROWELL and EVANS (1987) with their index of auroral precipitating particle energy. Detailed ESA data have also been used to compute field-aligned currents to compare with those computed from magnetometer data, but no large-scale models have resulted.

Another very useful technique is being developed which uses images of auroral emissions at specific wavelengths over a specified area. This technique allows one to determine the instantaneous distribution of Hall and Pedersen conductances within the instrumental field-of-view and during the time required to scan the polar ionosphere. Even though this method also suffers from some of the ESA model limitations, it has been successfully demonstrated by AHN et al. (1988).

Models of the large-scale electric field structure in the polar ionosphere have been generated from plasma drift measurements onboard several satellites. The ever-present dawn-to-dusk electric potential drop in the polar cap was modified to reflect the two-celled pattern associated with magnetospheric convection (HEPPNER, 1977). The two-celled pattern was found to be strongly skewed by the east-west component of the IMF and this pattern has been modified by HEPPNER and MAYNARD (1987) to distinguish between different levels of geophysical activity.

Magnetometer data from the Triad satellite was analyzed by IIJIMA and POTEMRA (1978) to produce the large-scale FAC model shown in Figure 2. The concept of Region 1 (poleward) and Region 2 field-aligned currents (FAC's) originated from this study and forms our basic understanding of three-dimensional auroral current systems. This analysis assumed that the currents were vertical and that the current sheets were infinite in length and they intentionally neglected small-scale features.

Several ground-based instruments have been used to remotely sense ionospheric currents, conductances and electric fields. Perhaps the most useful technique has been the incoherent scatter radar, data which are used to complement other data and/or to validate other techniques. Since only a few radars have been installed at high latitudes, radar data have been used primarily to construct statistical models. Using the principle of Doppler-shifting of electromagnetic waves to determine plasma drift velocities, FOSTER (1983) reported on the seasonal and activity level variations in the large-scale patterns of ionospheric electric fields.

A network of ground-based instruments, on the other hand, allows one to construct an instantaneous model, which is not effected by the smoothing required

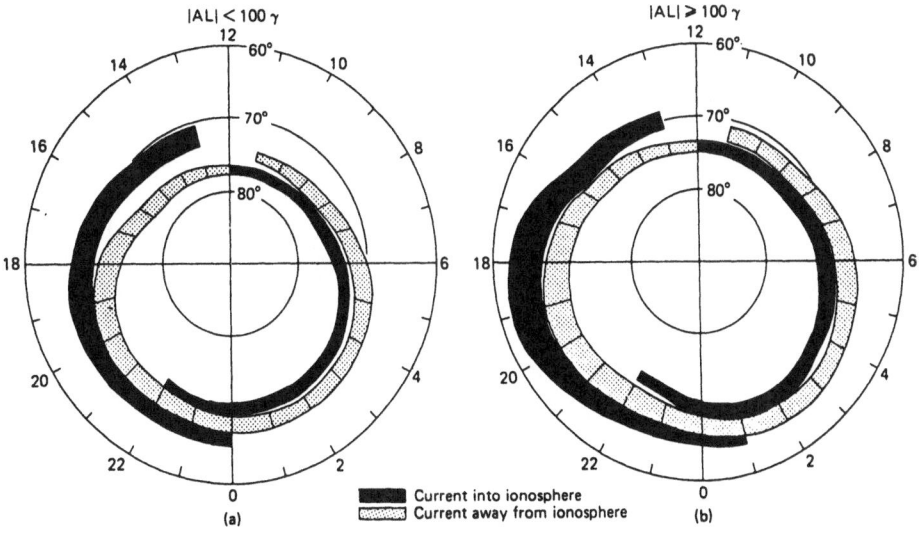

Figure 2

The "frequently-copied" Figure 13 of field-aligned currents for disturbed (b) and less-disturbed (a) auroral conditions from IIJIMA and POTEMRA (1978). Note the continuous structure of the currents in local time.

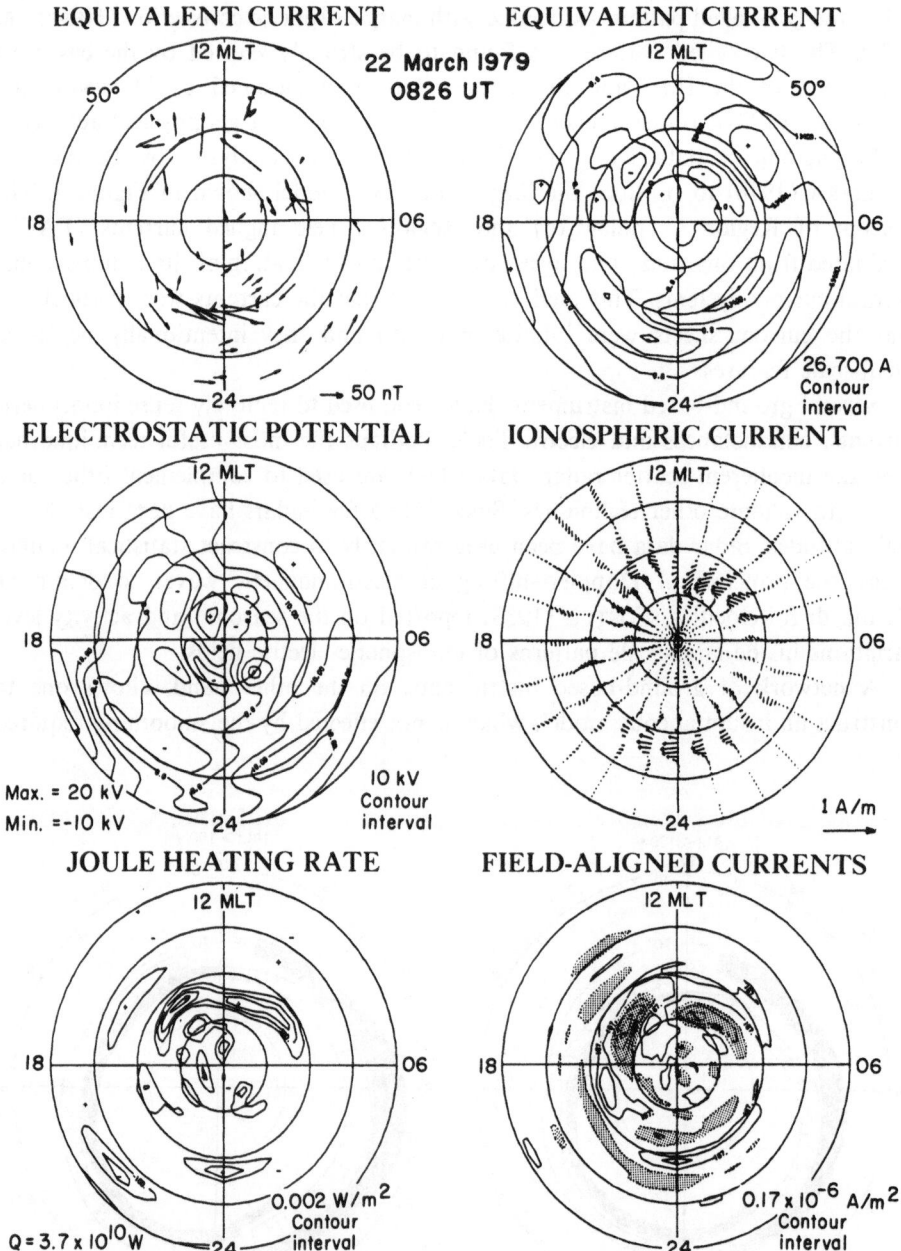

Figure 3

The important electrodynamic parameters computed by the KRM method from ground-based magnetic variations data and a conductance model for the SSC at 0826 UT on March 22, 1979. Note the cusp currents near noon, the convection electrojets near dawn and dusk and the substorm currents near midnight. Also note the difference between this instantaneous picture of FAC's and Figure 2.

by most statistical analysis techniques. Using the technique reported in KAMIDE *et al.* (1981) and RICHMOND *et al.* (1988), the authors describe the instantaneous distribution of vertical and horizontal currents and electric fields in the polar ionosphere, assuming (1) a conductance model, (2) no neutral winds, and (3) vertical magnetic field lines. This technique resulted in the definition of four ionospheric current systems (KROEHL and KAMIDE, 1985) which form the foundation of this paper. They are (1) the substorm-driven, westward auroral electrojet centered near 00 MLT, (2) the convection-driven, westward auroral electrojet centered near 06 MLT, (3) the convection-driven, eastward auroral electrojet centered near 18 MLT and (4) the cusp current centered near 12 MLT. Figure 3 shows the instantaneous distribution of ionospheric currents and fields in corrected geomagnetic coordinates (Gustafsson, 1970) of these four systems during a "storm sudden commencement" on March 22, 1979. (Though these systems appear here to be disconnected in local time, for other UT's and for other conditions they are connected.) Figure 3 also demonstrates that a westward horizontal current is always bounded by a downward FAC on the poleward side and an upward FAC on the equatorward side. This relation is reversed for an eastward current. Note that the instantaneous FAC system is closely associated with the horizontal current system which may or may not yield the simple, well-defined Region 1 and Region 2 structure reported by IIJIMA and POTEMRA (1978).

In summary, two different approaches to large-scale or global modeling of empirical FAC's has evolved; first, a statistical analysis of more direct measures of $J_{\|}$, J, E or Σ and second, an instantaneous simulation of an indirect measure of the important electrodynamic parameters. The statistical approach requires use of a geophysical index and the resulting model is limited by that index, the data, the technique and the statistical methods. The simulation is limited by the technique and the availability of data to drive the model for any instant in time.

We have chosen to pursue the simulation approach with the input of statistical values of conductance when and where direct observations are unavailable. We will consider the polar ionosphere to be that region where plasma densities and dynamics are controlled by magnetosphere-ionosphere (or solar wind-magnetosphere-ionosphere) coupling, at least during disturbed conditions. Finally, we define an "undisturbed polar ionosphere" as one that is not experiencing disturbed conditions.

3. Auroral Zone Currents

When attempting to define the three-dimensional, large-scale current systems of the undisturbed auroral zone, it is important to accept three conditions: (1) The classical auroral zone, or oval, always exists; in other words, there are always precipitating particles. This is seen in low-altitude, polar-orbiting satellite data from ESA's. (2) Auroral FAC's always exist (T. Potemra, private communication).

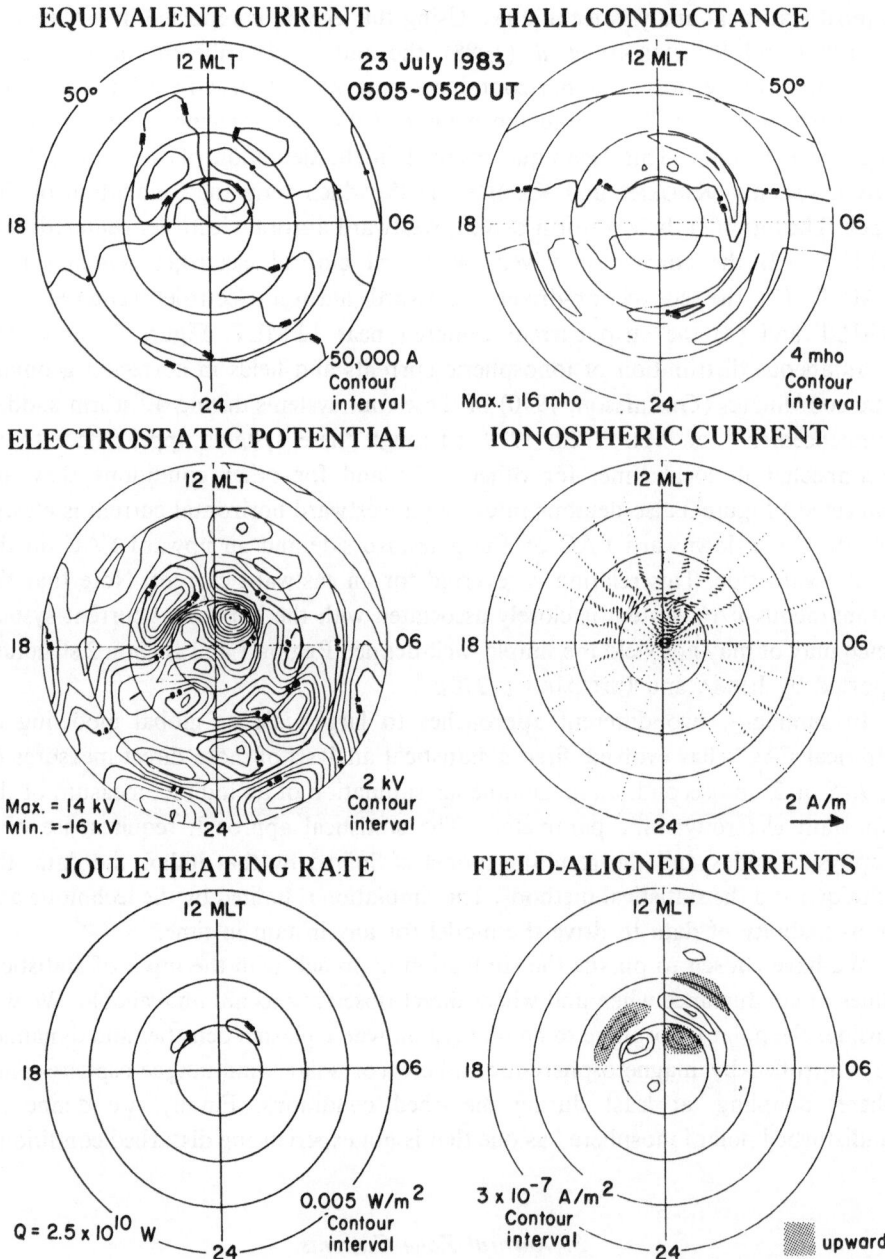

Figure 4
Same method as Figure 3 shown for a time when the polar currents were undisturbed. Note that no significant ionospheric current is evident near midnight.

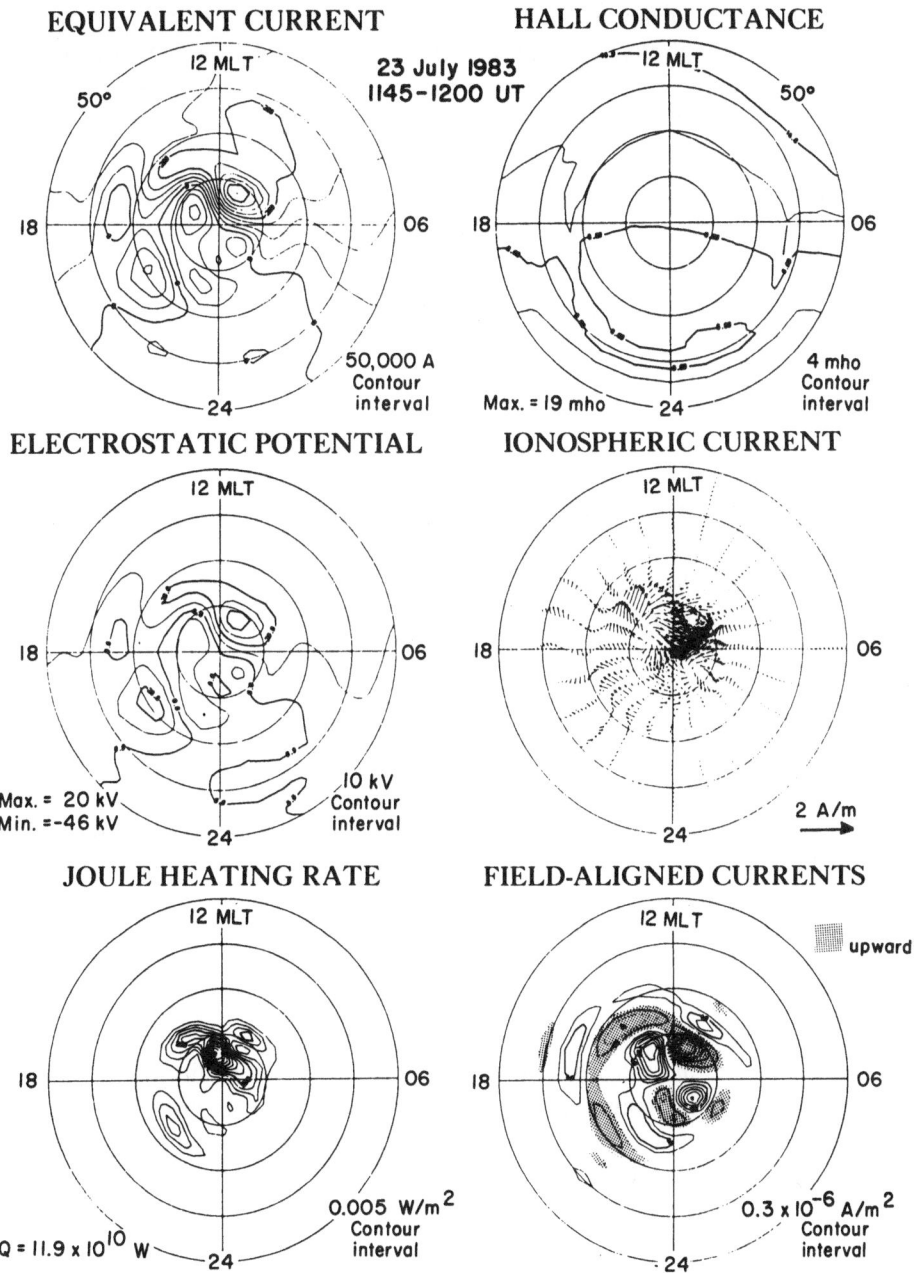

Figure 5
Same method as Figure 3 shown for a time when the IMF was strongly northward. Note the contracted auroral region, the reversed convection currents and the lack of a four-celled structure in the electric potential distribution.

(3) Large-scale electric fields always exist over the polar region, (N. Maynard, private communication) though they may appear to reverse. Therefore, the contribution that these currents may provide to a Sq (or undisturbed or solar-driven) analysis is very important at both high and low latitudes.

In the previous section we introduced three auroral current systems. Substorm expansion currents near 00 MLT exist only during disturbed conditions. However, the two convection current systems always exist because they result from interaction between the solar wind and magnetospheric plasma at the magnetopause. The resulting currents depend upon the solar wind velocity and density and the interplanetary magnetic field (IMF) with the appropriate time delay. Thus, to minimize their effects, the time used to model Sq currents must be carefully selected. It should be noted that LIU *et al.* (1987) showed Region 2 FAC signatures in mid-latitude magnetic variations with the H-component most affected at lower latitudes and the D-component most affected at high latitudes.

In Figure 4, we show a typical instantaneous model of high-latitude current systems during undisturbed conditions, i.e., $v = 420$ km/s, $n = 20$ cm^{-3}, IMF $= 5$ nT and $B_z = 2$ nT. Note the equivalent current plot in which high-latitude current effects are seen at lower latitudes.

When the IMF turns northward, an unusual condition appears to exist at high latitudes, namely, that magnetospheric convection is reversed, the effect of which is shown in Figure 5 and has been reported by MAEZAWA (1976), IIJIMA (1984) and others. This system, which they call the NBZ current system, may be related to a four-celled convection pattern or it may be the high-latitude component of the cusp current system during strongly northward IMF conditions. Thus, the auroral region is least disturbed but never quiet, when the solar wind velocity and density and IMF values are relatively small.

4. *Polar Cusp Currents*

Polar cusps, one in each hemisphere, result from the interaction of solar wind plasma with the dayside magnetosphere and in essence create an open region in the earth's distant magnetic field, thus the cusp currents, and particle produced ionization, are controlled directly by the solar wind and respond quickly to changes in the IMF. For example, the latitude of the cusp moves rapidly equatorward as the southward component of the IMF is increased. In addition, the northern and southern cusps are frequently asymmetric due to seasonal differences, and to differences in orientation of the earth's magnetic field.

During local summer, polar cusp ionization is controlled by the solar radiation, however, during local winter the particle precipitation is the controlling source. Dayside particles from the solar wind are much less energetic than nightside precipitating particles and are absorbed by the atmosphere at much higher altitudes, and thus they enhance the Pedersen conductance more than the Hall conductance. And since the electric potential gradient is usually largest in the eastward direction, the dark cusp current has larger equatorward and Pedersen components than the auroral currents which are dominated by the Hall component.

The east-west component of the cusp current system is controlled by, and will respond very quickly to, changes in the east-west component of the IMF, especially during times when the IMF is northward. Due to the merging of antiparallel components of the IMF with those of the magnetosphere, in the Northern Hemisphere a positive B_y-component produces a northward electric field and an eastward ionospheric current while a negative B_y-component produces a westward ionospheric current. In the Southern Hemisphere, the ionospheric current direction is reversed because the azimuthal projection of the magnetospheric magnetic field is the same for the dawnside (duskside) in the north as it is for the duskside (dawnside) in the south. AKASOFU et al. (1980) report that the ionospheric current responds very quickly to a reversal of the B_y-component of the IMF; in their case, the response time was less than the sample time which was 5 minutes.

Considering that the polar cusps are at high magnetic latitudes, i.e., between 77° and 83° under quiet conditions, a strong seasonal effect exists in the solar-produced, Hall and Pedersen conductances which causes a strong seasonal effect in the resulting field-aligned currents. In Figures 6 and 7, we used the same input magnetic variation data but changed the effective season from summer (6) to winter (7) as reflected in the plot of Hall conductance. The IMF was eastward and northward so no substorm was in progress and magnetospheric convection was minimal.

A comparison of Figures 6 and 7 shows that the computed ionospheric current remains essentially unchanged but that the electrostatic potential is increased significantly because, in Ohm's law, the voltage must be increased to compensate for the increased resistance or decreased conductance. In fact, the "maximum potential difference," or "cross polar cap potential," which is directly related to magnetospheric convection, has increased from 49 kV to 116 kV. This change is also reflected in the Joule heat production, which is simply $E^2(\Sigma_P)$. We also notice a small change in the overall distribution and intensity of FAC's except in the downward FAC maximum near 13 MLT and 77°N. Now on the other hand, if the electric potential is kept constant, or for the same intensity of magnetospheric convection, then decreased conductances (winter condition) yield decreased FAC's. Thus the winter FAC structure is less disturbed in general than is the summer structure.

Exactly how the cusp FAC's relate to the daytime auroral FAC's is not well understood at this time. It does appear, however, that the ionospheric currents from the cusp system respond to changes in the solar wind more quickly than do the convection systems. How these FAC's affect the lower latitudes is also unclear. For this purpose, it appears that the least disturbed cusp system occurs when the auroral zone is also the least disturbed.

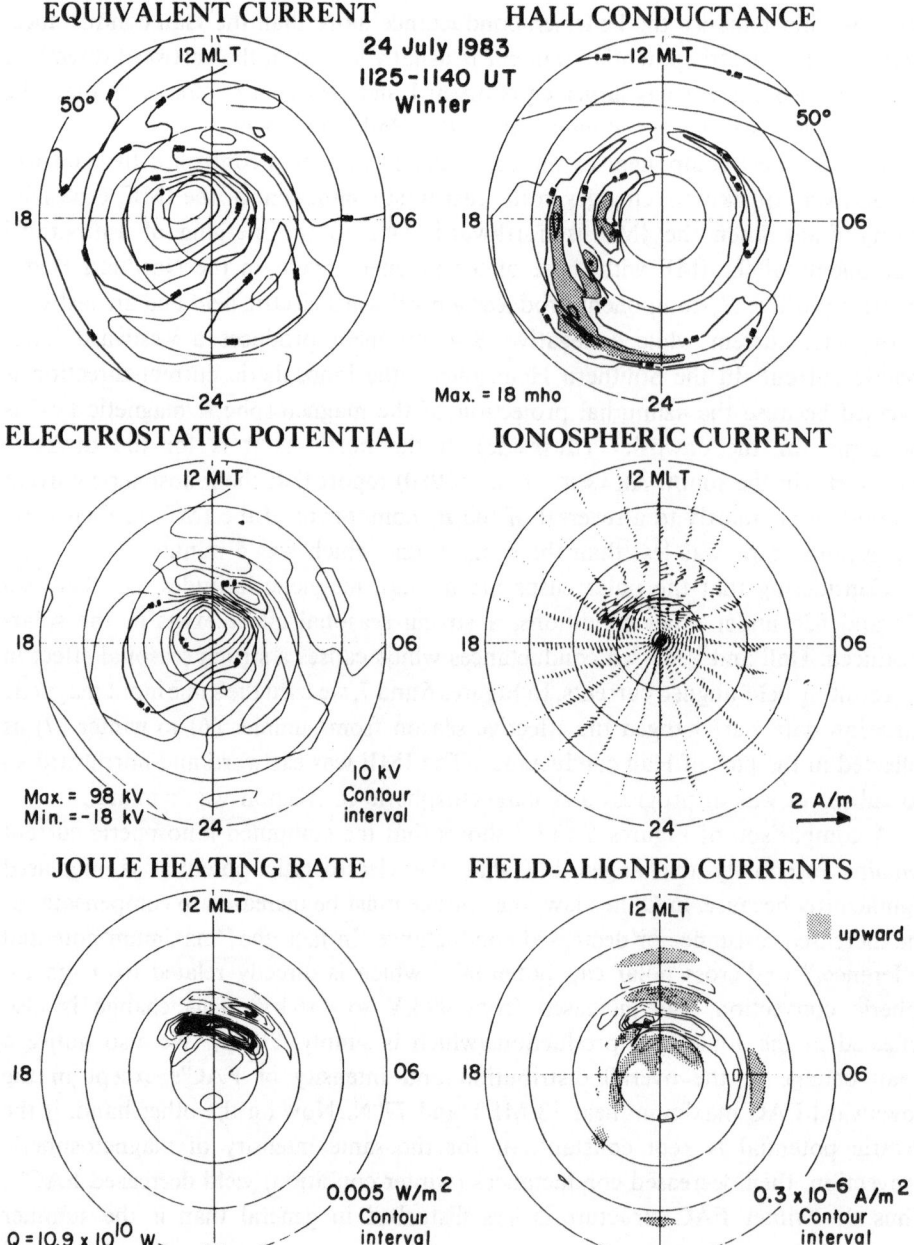

Figure 6
Same method as Figure 3 shown for a time with a relatively strong eastward ($B_y > 0$) cusp current during
local summer.

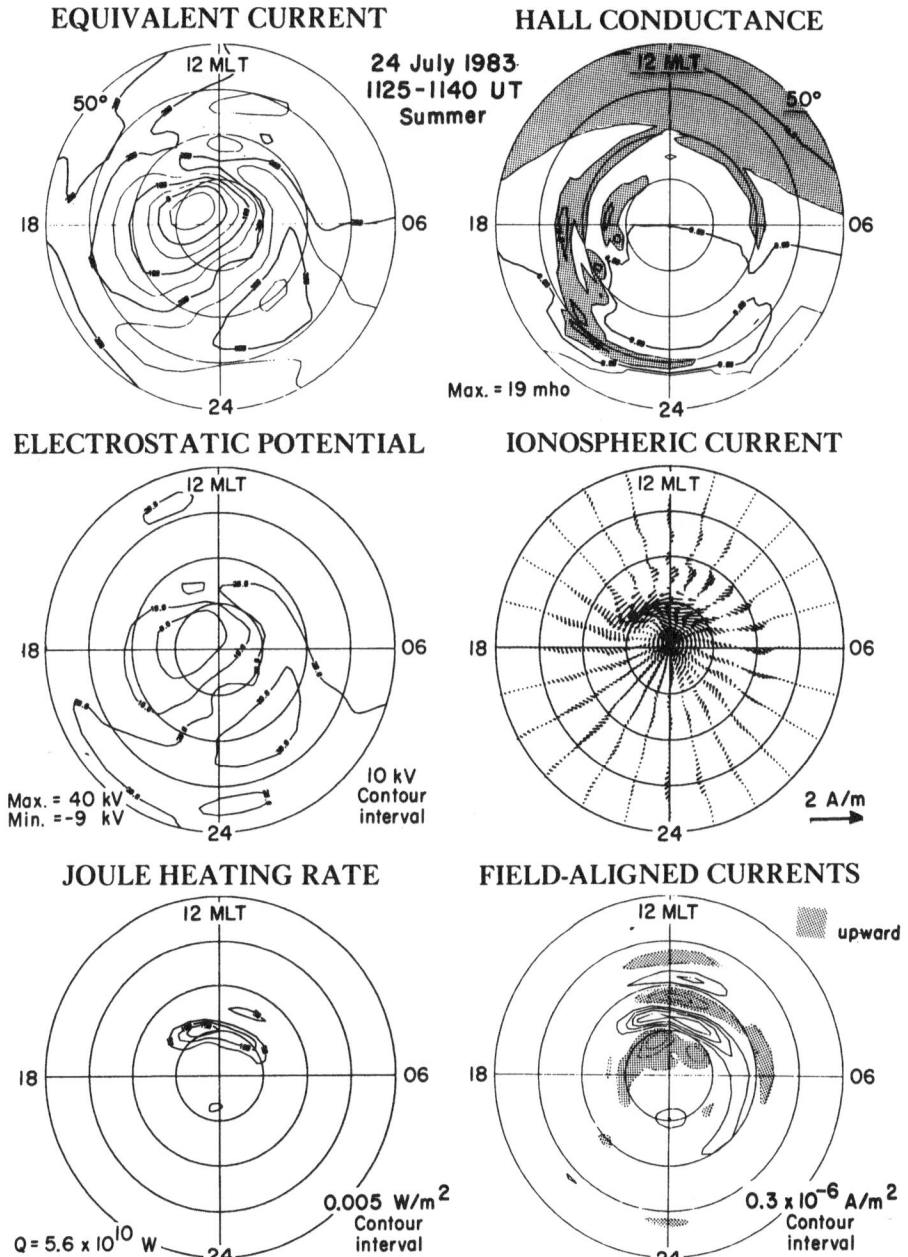

Figure 7

Same as Figure 6 except that local winter is simulated by moving the subsolar point to 20°S as seen in the Hall conductance panel. Note the increase in the maximum FAC from 1.4×10^{-6} A/m to 2.3×10^{-6} A/m, Joule heating and electrostatic potential.

5. Polar Cap Currents

We define the polar cap as that region where the apex of the magnetic field line is open and coincides with the earth's magnetopause or the lobes of the magnetotail as shown in Figure 1. The polar cap extends from the magnetic pole down in latitude to the polar cusp on the dayside and the auroral zone on the nightside. Because many of these field lines merge with the IMF, the resulting ionospheric currents respond quickly to changes in the solar wind and the IMF. Precipitating electron data from polar orbiting satellites show that the polar cap is very active due to the formation of arcs and patches during times of a northward IMF (GUSSENHOVEN, 1982). Since the cross-polar cap electric potential is relatively small with this enhanced conductance, we notice that the resulting ionospheric current and FAC's are also relatively weak.

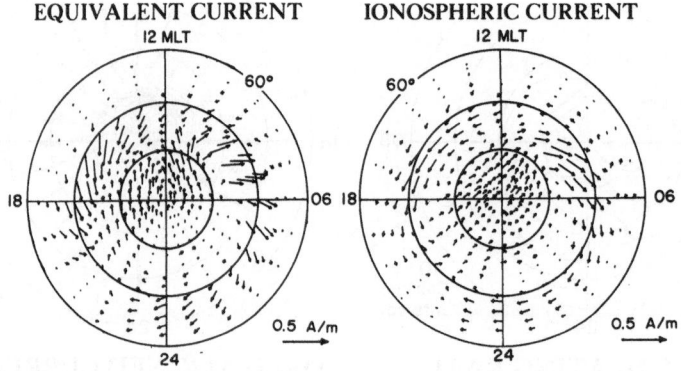

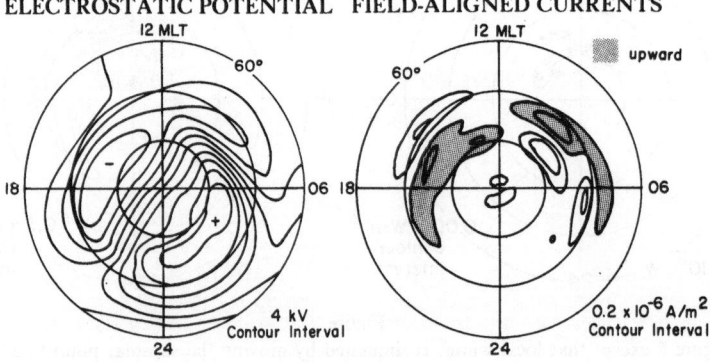

Figure 8
Application of the KRM method to averages of the geomagnetic variations at high latitudes (FRIIS-CHRISTENSEN, 1983) when $B_y = B_z = 0$), or relatively undisturbed conditions, without consideration for the solar wind velocity or density.

Perhaps the most pressing question deals with the existence of the NBZ current system proposed by MAEZAWA (1976) which is characterized by reversed convective flow in the polar cap, a four-celled electric potential pattern in the ionosphere, an additional pair of FAC's at latitudes higher than Region 1 and a bifurcation of the magnetotail. FRIIS-CHRISTENSEN (1984) presents the average ionospheric current, FAC and electric potential patterns derived from Greenland magnetometer chain data for different orientations of the IMF and reports no convection reversal in the polar cap and only a two-celled electric potential pattern during a northward B_z. AHN et al. (1988) shows a number of instantaneous, global maps of electric potential values and very few show the proposed four-celled pattern. Thus the NBZ current system may exist for selected times or it may have arisen in the statistical model from cusp FAC's, from FAC's associated with small-scale features or from FAC's when the auroral zone, and polar cap, was contracted toward the pole.

Thus, we conclude that a northward IMF does not yield a quiet polar cap. Rather our "undisturbed" polar cap must require that the IMF components $B_z = B_y = 0$. These conditions were mapped by FRIIS-CHRISTENSEN (1984) and are reproduced here as Figure 8. Because he used summertime data for 1972 and 1973, the resulting ionospheric currents and FAC's are relatively large as compared to those obtained from a wintertime analysis when the conductances are lower. You will also notice a nighttime current system in the region where we would expect substorm expansion currents which should not be present. These probably result from enhanced convection when the solar wind velocity was relatively large or from previous substorm activity. However, the important features in Figure 8 are the symmetry about the noon-midnight line in the current plots and the existence of the FAC's in the dayside only, even when significant electric fields and conductances (summer) exist in the "nightside."

6. Concluding Remarks

If we select our data to represent times when the polar ionosphere is not active as proposed and executed by CAMPBELL and SHIFFMACHER (1985), i.e., when solar wind-magnetosphere-ionosphere coupling is not enhanced due to the orientation of the IMF or the intensity of solar wind plasma pressure, we are able to define the high-latitude current systems resident in the quiet, instantaneous ionosphere. Such a schematic is presented in Figure 9. Note that magnetospheric convection always exists as manifested in auroral electrojet currents with vertical FAC's and ionospheric electric fields. These quiet-time currents are driven by the solar wind as opposed to the solar radiation-driven, Sq ionospheric dynamo. These current systems are symmetric about noon, exist in the sunlit hemisphere

SCHEMATIC OF ELECTRIC FIELDS AND CURRENTS IN THE
"UNDISTURBED POLAR IONOSPHERE"

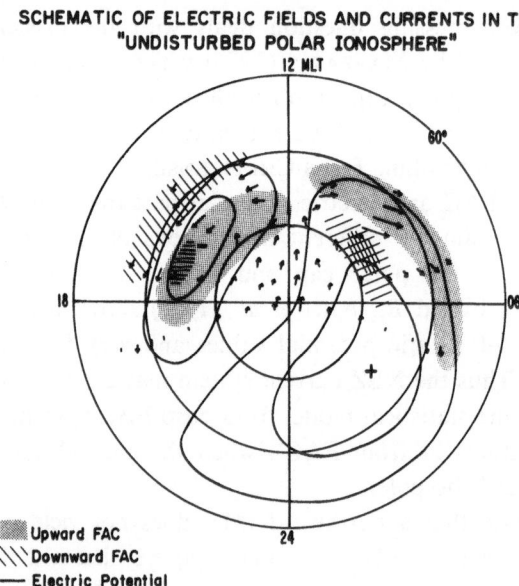

▨ Upward FAC
\\\ Downward FAC
— Electric Potential
→ Ionospheric Current

Figure 9

A schematic illustration of our definition of an undisturbed polar region dominated by very weak convection.

and are characterized as follows:

1. Region 1 FAC's are more intense than Region 2 FAC's and no vertical current exists above Region 1.
2. Anti-sunward plasma convection exists in the polar cap.
3. The cusp current system is not easily defined.
4. After noon, the eastward current maximum is colocated with the electric field maximum and between the Region 1 and 2 currents.
5. Before noon, the westward current maximum lies in the Region 2 current and is not colocated with the electric field maximum.
6. Significant electric fields exist in the dark hemisphere.

Finally, the effect of polar field-aligned currents will appear in mid-latitude magnetic recordings (LIU *et al.*, 1987) as variations in the D-component at higher latitudes and in the H-component at lower latitudes. When deriving quiet-time, *Sq* current systems, one must remove these effects.

Acknowledgements

I would like to express my appreciation to W. H. Campbell for his thoughtful comments, to B.-H. Ahn and Y. Kamide for our many fruitful discussions, and to Lorraine for her support, editing and manuscript preparation.

REFERENCES

AHN, B.-H., FRIIS-CHRISTENSEN, E., GORNEY, D. J., KAMIDE, Y., KROEHL, H. W., MIZERA, P. F., RICHMOND, A. D., SUCKSDORFF, C. G., and WELLS, C. D., *Numerical Modeling of Polar Ionospheric Electrodynamics for July 23–24, 1983 Utilizing Ionospheric Conductances Deduced from DMSP X-ray Images*, Report UAG-97 (World Data Center-A for STP) (Boulder, Co. 80303 1988) pp. 133.

AKASOFU, S.-I., ROMICK, G. J., and KROEHL, H. W. (1980), *The IMF By Effects Observed by the IMS Alaska Meridian Chain*, J. Geophys. Res. *85*, 2079.

ALLEN, J. H., and KROEHL, H. W. (1975), *Spatial and Temporal Distributions of Auroral Electrojets as Derived from AE Indices*, J. Geophys. Res. *80*, 3667.

ANGOT, A., *The Aurora Borealis* (Kegan Paul, Trench, Truber & Co. Ltd., Paternoster House, London 1896) pp. 135.

BERTHELIER, A., BERTHELIER, J. J., and GUERIN, C. (1974), *The Effect of the East-west Component of the Interplanetary Magnetic Field on Magnetospheric Convection as Deduced from Magnetic Perturbations at High Latitudes*, J. Geophys. Res. *87*, 3187.

BIRKELAND, K., *The Norwegian Aurora Polaris Expedition 1902–1903* (H. Aschehoug and Co., Christiania 1908) pp. 315.

BROSTROM, R., (1964), *A Model of the Auroral Electrojet*, J. Geophys. Res. *69*, 4983.

CAMPBELL, W. H., and SHIFFMACHER, E. R. (1985), *Quiet Ionospheric Currents of the Northern Hemisphere Derived from Geomagnetic Records*, J. Geophys. Res. *90*, 6475.

FELDSTEIN, Y. I., and STARKOV, G. V. (1967), *Dynamics of the Auroral Bolt and Polar Geomagnetic Disturbances*, Planet, Space Sci. *15*, 209.

FOSTER, J. C., (1983), *An Empirical Electric Field Model Derived from Chatanika Radar Data*, J. Geophys. Res. *88*, 981.

FRIIS-CHRISTENSEN, E., *Polar cap current systems*, In *Magnetospheric Currents* (ed. Potemra, T. A.) (Geophysical Monograph 28, American Geophysical Union, Washington D.C. 1984) pp. 86.

FULLER-ROWELL, T. J., and EVANS, D. S. (1987), *Height-integrated Pedersen and Hall Conductivity Patterns Inferred from the TIROS-NOAA Satellite Data*, J. Geophys. Res. *92*, 7606.

GUSSENHOVEN, M. S. (1982), *Extremely High Latitude Auroras*, J. Geophys. Res. *87*, 2401.

GUSTAFSSON, G. (1970), *A Revised Corrected Geomagnetic Coordinate System*, Aarkiv. fur Geofysik *5*, 595.

HARANG, L. (1946), *The Mean Field of Disturbance of Polar Geomagnetic Storms*, J. Geophys. Res. *51*, 353.

HARDY, D. A., GUSSENHOVEN, M. S., RAISTRICK, R., and MCNEIL, W. J. (1987), *Statistical and Functional Representations of the Pattern of Auroral Energy Flux, Number Flux and Conductivity*, J. Geophys. Res. *92*, 12,275.

HEPPNER, J. P. (1977), *Polar Cap Electric Field Distributions Related to the Interplanetary Magnetic Field Direction*, J. Geophys. Res. *82*, 1115.

HEPPNER, J. P., and MAYNARD, N. C. (1987), *Empirical High-latitude Electric Field Models*, J. Geophys. Res. *92*, 4467.

IIJIMA, T., and POTEMRA, T. A. (1978), *Large-scale Characteristics of Field-Aligned Currents Associated with Substorms*, J. Geophys. Res. *83*, 599.

IIJIMA, T., *Field-aligned currents during northward IMF*, In *Magnetospheric Currents* (ed. Potemra, T. A.) (Geophysical Monograph 28, American Geophysical Union, Washington, D.C. 1984) pp. 358.

KAMIDE, Y. (1979), *Recent Progress in Observational Studies of Electric Fields and Currents in the Polar Ionosphere: A Review*, Antartic Record *63*, 61.

KAMIDE, Y., RICHMOND, A. D., and MATSUSHITA, S. (1981), *Estimation of Ionospheric Electric Fields, Ionospheric Currents and Field-aligned Currents from Ground Magnetic Records*, J. Geophys. Res. *86*, 801.

KROEHL, H. W., and KAMIDE, Y. (1985), *High Latitude Indices of Electric and Magnetic Variability During the CDAW-6 Intervals*, J. Geophys. Res. *90*, 1367.

LEVITIN, A. E., AFONINA, R. G., BELOV, B. A., and FELDSTEIN, Y. I. (1982), *Geomagnetic Variations and Field-aligned Currents at Northern High-latitudes and their relations to solar wind parameters*, Phil. Trans., R. Soc. London *A304*, 253.

LIU, Z. X., WEI, C. Q., LEE, L. C., and AKASOFU, S.-I. (1987), *Magnetospheric Substorms: An Equivalent Circuit Approach*, EOS *68*, 1423.

MAEZAWA, K. (1976), *Magnetospheric Convection Induced by the Positive and Negative Z-Components of the Interplanetary Magnetic Field: Quantitative Analysis Using Polar Cap Magnetic Records*, J. Geophys. Res. *81*, 2289.

RICHMOND, A. D., KAMIDE, Y., AHN, B.-H., AKASOFU, S.-I., ALCAYDE, D., BLANC, M., DE LA BEAUJARDIERE, O., EVANS, D. S., FOSTER, J. C., FRIIS-CHRISTENSEN, E., FULLER-ROWELL, T. J., HOLT, J. M., KNIPP, D., KROEHL, H. W., LEPPING, R. P., PELLINEN, R. J., SENIOR, C., and ZAITSEV, A. N., (1988), *Mapping Electrodynamic Features of the High-latitude Ionosphere from Localized Observations: Combined Incoherent-scatter Radar and Magnetometer Measurements for January 18–19, 1984*, J. Geophys. Res. *93*, 5760.

ROSTOKER, G., *Recent Developments in the Area of Electric Fields, Magnetic Fields Including Ground-based Observations* (Univ. of Alberta, Edmonton, Canada 1977) pp. 75.

SPIRO, R. W., REIFF, P. H., and MAHER, L. J., Jr. (1982), *Precipitating Electron Energy Flux and Auroral Zone Conductances: An Empirical Model*, J. Geophys. Res. *87*, 8215.

VESTINE E. H. (1944), *The Geographic Incidence of Aurora and Magnetic Disturbance, Northern Hemisphere*, J. Terr. Magn. Atmos. Elect. *49*, 77.

WALLIS, D. D., and BUDZINSKI, E. E. (1981), *Empirical Models of Height-integrated Conductivities*, J. Geophysics. Res. *86*, 125.

ZMUDA, A. J., MARTIN, J. H., and HEURING, F. T. (1966), *Transverse Magnetic Disturbances at 1100 km in the Auroral Region*, J. Geophys. Res. *71*, 5033.

(Received June 6, 1988, accepted July 1, 1988)

PAGEOPH, Vol. 131, No. 3 (1989)

0033–4553/89/030413–23$1.50 + 0.20/0

Modeling the Ionosphere Wind Dynamo: A Review

A. D. Richmond[1]

Abstract—This paper reviews the current status of research concerned with modeling the ionospheric wind dynamo. Simulation models have been reasonably successful in reproducing the types of magnetic perturbations that are produced by the dynamo. Ionospheric electric fields are less well simulated, particularly at night. The primary areas of research needed to improve our ability to simulate realistically the ionospheric wind dynamo are in the modeling of nighttime conditions, hemispherical asymmetries of thermospheric tides, and mutual dynamic coupling among winds, conductivities, electric fields, and electric currents.

Key words: Ionosphere, thermosphere, dynamo current, *Sq*.

1. Introduction

Models of the ionospheric wind dynamo are useful for several purposes. They help elucidate the characteristics of the dynamo mechanism, by which thermospheric winds interact with the electrically conducting ionosphere to generate electric fields and currents. The dynamo models can be used to infer properties of the global-scale thermospheric winds and to test the validity of model wind distributions by comparison of calculated electric fields and magnetic variations with observed values. The dynamo models may also have potential value as predictive tools, giving quantitative information about upper atmospheric electrodynamic properties under conditions that have not yet been measured.

Some earlier reviews of ionospheric electrodynamics have been presented by MATSUSHITA (1977), BLANC (1979), RICHMOND (1979), KATO (1980), and WAGNER et al. (1980), while the equatorial electrojet has been reviewed by FORBES (1981). The emphasis of the current review is on global dynamo phenomena, with a discussion of the physical mechanisms, modeling techniques, comparisons with observations, and some untested predictions.

---

[1] High Altitude Observatory, National Center for Atmospheric Research,[2] P.O. Box 3000, Boulder, CO 80307, U.S.A.
[2] The National Center for Atmospheric Research is sponsored by the National Science Foundation.

2. Dynamo Theory

The central equation relating the winds, electric fields, and electric currents is a form of Ohm's Law:

$$\mathbf{J} = \sigma_0 \mathbf{E}_\parallel + \sigma_1 (\mathbf{E}_\perp + \mathbf{v} \times \mathbf{B}) + \sigma_2 \mathbf{b} \times (\mathbf{E}_\perp + \mathbf{v} \times \mathbf{B}) \tag{1}$$

where $\mathbf{J}$ is the current density; $\mathbf{E}_\parallel$ and $\mathbf{E}_\perp$ are the components of the electric field parallel and perpendicular to the geomagnetic field $\mathbf{B}$; $\mathbf{v}$ is the wind velocity; $\mathbf{b}$ is a unit vector in the direction of $\mathbf{B}$; σ_0 is the conductivity parallel to $\mathbf{B}$; σ_1 is the Pedersen conductivity; and σ_2 is the Hall conductivity. For time scales longer than a minute or so, the electric field can be assumed to be electrostatic,

$$\mathbf{E} = -\nabla \Phi \tag{2}$$

where Φ is the electrostatic potential. For time scales longer than a second the current can be considered to be divergenceless:

$$\nabla \cdot \mathbf{J} = 0. \tag{3}$$

Most models of the dynamo mechanism take specified distributions of the conductivities, winds, and geomagnetic field, and then solve (1)–(3) for the electric fields and currents. Magnetic perturbations can be directly calculated from the current

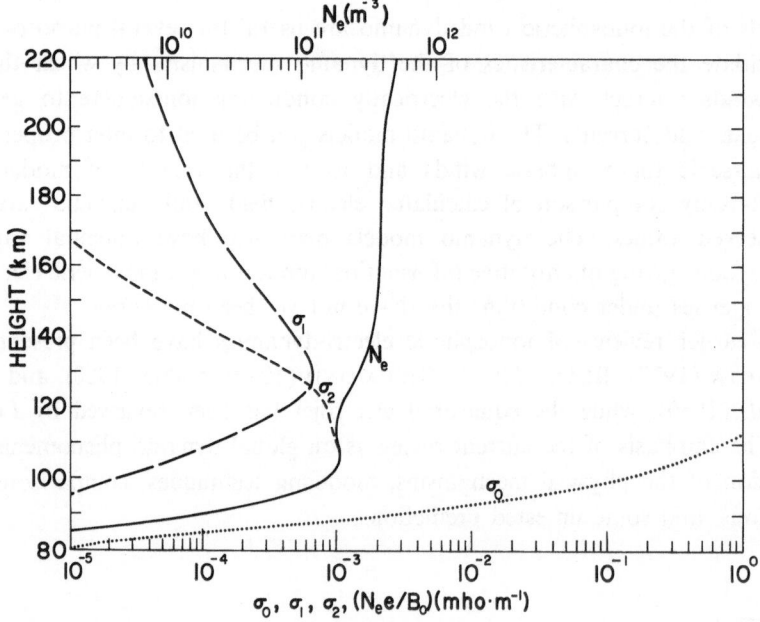

Figure 1
Electron density N_e, and parallel, Pedersen, and Hall conductivities σ_0, σ_1, and σ_2 for an overhead sun of moderate activity, with a magnetic field strength B_0 of 2.93×10^{-5} T. From RICHMOND (1973a).

distribution, if account is also taken of the perturbations associated with induced earth currents.

The selection of boundary conditions for the solution of (1)–(3) will influence the model. A lower boundary around 90 km altitude is generally chosen, where the vertical component of electric current is usually assumed to vanish due to the very small conductivity of lower altitude regions. (Current flow across this boundary has been considered, however, in studies such as those of HAYS and ROBLE (1979), ROBLE and HAYS (1979) and MAKINO and TAKEDA (1984).) The upper boundary is chosen with regard to the nature of electromagnetic coupling that one wishes to consider between the ionosphere and magnetosphere. Since Ohm's Law is no longer valid in the presence of energetic magnetospheric plasma, the upper boundary must

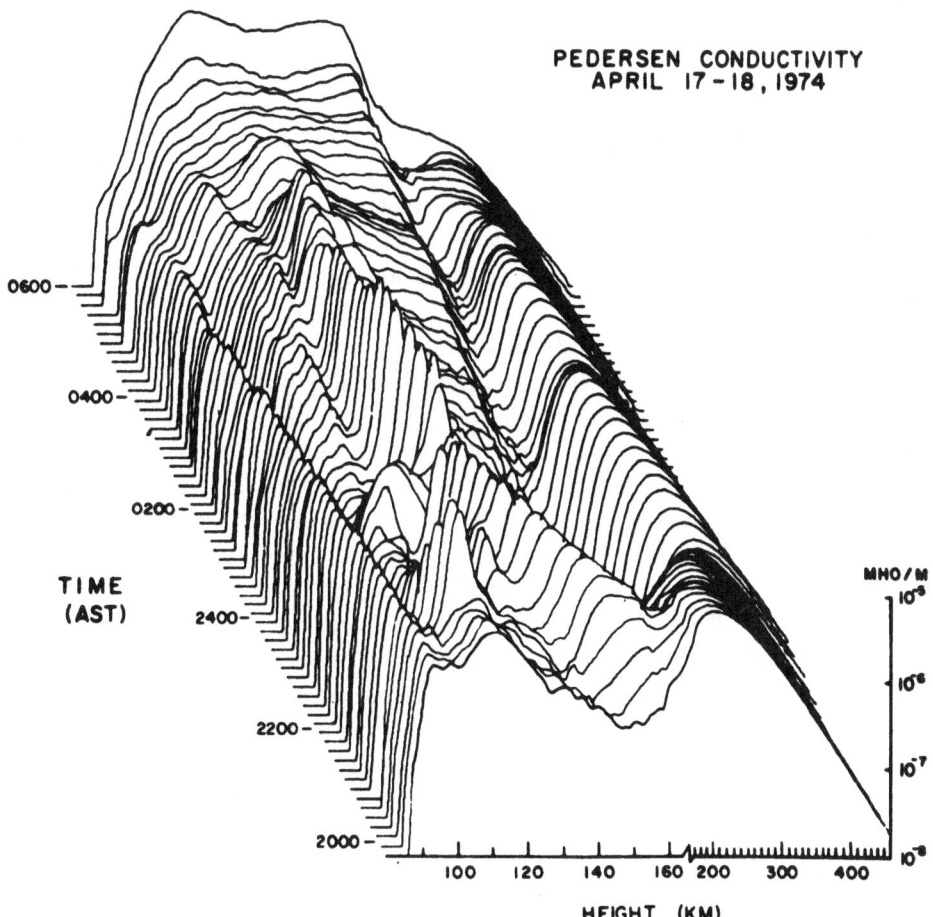

Figure 2

Pedersen conductivity over Arecibo for the night of 17–18 April 1974. Note change in altitude scale at 165 km. From HARPER and WALKER (1977).

either ignore energetic magnetospheric processes or else account for them in the form of specified electric currents or electric fields imposed on the ionosphere from the outer magnetosphere.

The electric conductivities $\sigma_0, \sigma_1, \sigma_2$ can be calculated from theoretical expressions involving the densities and temperatures of the ionized and neutral constituents of the upper atmosphere (for a recent treatment and discussion of assumptions, see RICHMOND (1983)). Figure 1 shows typical low-latitude noontime profiles of the conductivities, while Figure 2 shows representative nighttime profiles of the Pedersen conductivity. The nighttime conductivities are much smaller than those at day, and are also considerably more sensitive to plasma advection associated with winds and electric fields.

3. Feedback Mechanisms

The above description of the dynamo problem does not explicitly take into account the fact that a physical feedback occurs from the electric fields to the winds, conductivities, and magnetospheric field-aligned currents, as well as from the winds to the conductivities and magnetospheric currents. In reality, all of the electrodynamic parameters have mutual physical interactions. The full extent of these interactions has not yet been evaluated, but a number of studies have highlighted several of the effects.

One of the main feedback mechanisms is the effect that electric fields have on the neutral wind distribution. The electric currents driven by the electric fields exert an Ampere force on the medium, which can be comparable to other forces in the horizontal direction. There is a tendency for the neutrals to be driven in the direction of the $\mathbf{E} \times \mathbf{B}/B^2$ plasma drift, and to produce an induced $\mathbf{v} \times \mathbf{B}$ electric field that opposes the electrostatic field (BAKER and MARTYN, 1953; PIDDINGTON, 1954; AXFORD and HINES, 1961). This effect can be characterized as a reduction of the Pedersen conductivity (BAKER and MARTYN, 1953; AKASOFU and DEWITT, 1965; MATUURA, 1974; MAEDA, 1977). The consequent reduction in electric current means a reduced Ampere force on the medium. At low latitudes, where this force generally opposes the wind motion and is characterized by the concept of ion drag, its reduction allows stronger winds to blow. These stronger winds can in turn generate stronger electrostatic fields. RISHBETH (1971b) pointed out that the neutrals and ions at F-region heights can move rapidly together in the east-west direction at equatorial latitudes at night because of this effect.

This feedback from the electric fields and ion drifts to the neutral motions also causes a modification of the neutral pressure field that helps drive the winds. VOLLAND (1976a,b) and VOLLAND and GRELLMANN (1978) examined the characteristics and consequences of this mutual coupling between neutral and ion global dynamics with simplified models. VOLLAND (1976a) and VOLLAND and

GRELLMANN (1978) found that the horizontal and vertical structures of atmospheric tidal modes are altered when the atmosphere is electrically conducting. VOLLAND (1976a,b) showed that, in the limit of very large conductivity, an asymptotic value of electric current is attained.

The transport of plasma by winds and electric fields can affect the distribution of electrical conductivity (e.g., COLE, 1969). In the F-region the Pedersen conductivity is proportional to the product of electron and neutral densities. A raising (lowering) of the bottomside F-layer by equatorward (poleward) winds or by eastward (westward) electric fields will produce a reduction (enhancement) of the conductivities. In the lower E-region, electron density changes associated with the strong vertical electric fields in the equatorial electrojet can have a relatively small influence on the conductivities and on the electrojet current (RICHMOND, 1973a; STENING, 1986). Vertical shears in the winds also affect the electron densities (e.g., FOOTITT et al. 1983). The wind shears that affect the electron density are associated with preferred wind directions within conducting layers. Consequently, the conducting layers tend to have a systematic sense of dynamo current generation (RISHBETH and WALKER, 1982; RISHBETH, 1983; JALONEN et al., 1984).

The electrodynamics of the ionosphere and outer magnetosphere are closely coupled through electric fields and field-aligned currents, so that feedback processes occur between the ionospheric and magnetospheric dynamos. The magnetospheric dynamo is driven by energetic plasma, whose distribution is influenced by the electric fields of both magnetospheric and ionospheric origin. VASYLIUNAS (1972) presented a simplified picture showing how the magnetospheric plasma tends to react in a steady state, as though it were the source of large Hall conductivity. The ring of large effective Hall conductivity that maps into the auroral region of the ionosphere leads to a diminution of east-west electric fields at its equatorward edge, corresponding to an apparent electric field shielding region that tends to decouple the electric fields at auroral altitudes from those at middle and low latitudes. Although the primary interest of this effect is to understand how electric fields originating in the solar wind/magnetosphere dynamo are influenced, electric fields of ionospheric dynamo origin will be influenced to a comparable extent (WOLF et al., 1986).

The strong electric fields at high latitudes associated with magnetospheric convection tend to generate neutral winds in the same direction as the ion convection. When the magnetospheric dynamo weakens, the ionospheric dynamo action of these long-lasting winds can become relatively important, with a tendency to generate electric fields in the same sense as those previously maintained by the magnetospheric dynamo, though weaker. These ionospherically generated electric fields will thus tend to maintain a weakened ionospheric/magnetospheric convection, producing a type of "flywheel" effect that reduces the magnitudes of changes in convection that would otherwise occur (BANKS, 1972). RICHMOND and MATSUSHITA (1975) modeled this flywheel effect and found it to be moderately important for time scales of a few hours. LYONS et al. (1985) proposed that the thermospheric winds set up by magnetospheric convection during periods of southward interplanetary magnetic field (IMF) could persist when the IMF turns northward, and give rise to horizontal ionospheric currents similar to those inferred from ground magnetometers. SPIRO et al. (1988) suggested that these winds may extend equatorward of the shielding region as it contracts during the recovery phase of magnetospheric disturbances, and that the electric fields generated by these winds are no longer shielded from low latitudes, but become visible at the equator.

4. Modeling Techniques

The desire to model features of the dynamo process realistically while at the same time keeping the problem computationally tractable has led to a variety of approaches. This section examines the main elements of the various modeling techniques.

An important simplification common to almost all dynamo modeling approaches is the reduction in the number of spatial dimensions that are explicitly included in the computations, at least for some intermediate steps. There are reasons besides computational simplicity that make this dimensionality reduction desirable. For some purposes one is only interested in two-dimensional quantities, such as height-integrated ionospheric currents that are closely related to ground magnetic perturbations. Moreover, the electric potential turns out to be nearly constant along magnetic field lines, because of the high parallel conductivity, so that it is most conveniently expressed as a two-dimensional function varying perpendicularly to magnetic field lines.

BAKER and MARTYN (1953) used scale-length considerations to show that the high parallel conductivity of the ionosphere causes the horizontal electric field to be nearly constant with height over most of the earth. By assuming that the vertical current density is negligible, HIRONO (1952), BAKER and MARTYN (1952, 1953), and FEJER (1953) derived an effective horizontal conductivity tensor which interrelates the horizontal current with the horizontal electric field. The dynamo equations can thus be reduced to two dimensions by integration with respect to height. Many studies used this simplification to analyze the distributions of ionospheric winds, currents, and electric fields that are consistent with observed ground magnetic perturbations (HIRONO, 1952; BAKER, 1953; MAEDA, 1955, 1957; HIRONO and KITAMURA, 1956; KATO, 1956, 1957; MATSUSHITA, 1969; TARPLEY, 1970a,b; HEELIS et al., 1974; RICHMOND et al., 1976; FORBES and LINDZEN, 1976a; HANUISE et al., 1983).

One region where the effective horizontal conductivity tensor described above becomes invalid is close to the magnetic equator. Although this tensor qualitatively reproduces the effects of current enhancement due to Cowling polarization in the equatorial electrojet, it does not quantitatively reproduce the observed structure of the electrojet. UNTIEDT (1967) showed that vertical current densities, though small, are nonnegligible at equatorial latitudes. RICHMOND (1973a) showed that a much better approximation than $J_z = 0$ is to assume that the parallel conductivity is infinite, so that magnetic field lines are equipotentials. Only below about 100 km does this $\sigma_0 = \infty$ approximation start to break down. As pointed out by DOUGHERTY (1963), the approximation of equipotential field lines should be valid even along lines extending well into the magnetosphere and connecting conjugate points in the ionosphere. Since this $\sigma_0 = \infty$ approximation also allows a reduction from three to two dimensions in the solution of the global dynamo equations, it has generally replaced the $J_z = 0$ approximation (STENING, 1968; SCHIELDGE et al., 1973; JONES, 1974; MAEDA, 1974; MATUURA, 1974; MÖHLMANN, 1974, 1977; WALTON and BOWHILL, 1979; TAKEDA and MAEDA, 1980; TAKEDA, 1982; SINGH and COLE, 1987a,b; RICHMOND and ROBLE, 1987).

The equatorial region can give rise to numerical difficulties in global modeling studies because of the fact that the electric fields and currents change strongly over relatively small distances. One way to overcome this problem is to increase the number of grid points close to the magnetic equator (FEJER, 1953; HIRONO and KITAMURA, 1956; TARPLEY, 1970a; WALTON and BOWHILL, 1979; STENING, 1968, 1985; TAKEDA and MAEDA, 1980). Another way is to treat the whole equatorial electrojet as a single conducting strip along the boundary of the computation region, with the conductance of the strip determined from an infinitely long electrojet model (BLANC and RICHMOND, 1980; HANUISE et al., 1983; WOLF et al., 1986). Electric fields, currents, and conductances vary rapidly with latitude at auroral latitudes, too, but these have not yet been realistically simulated in a global dynamo model.

Dynamo action is in general asymmetric about the magnetic equator, because of seasonal variations in the winds and conductivities and because of the displacement between the magnetic and geographic equators. Several studies have examined consequences of this asymmetry (see review by WAGNER et al., 1980). It is necessary to account for the strong interhemispheric coupling associated with the very large parallel conductivity, which tends to keep conjugate points at nearly the same electric potential up to auroral latitudes. Studies that have examined this coupling have generally used a purely dipolar geomagnetic field to simplify the geometry (MAEDA, 1966; MAEDA and FUJIWARA, 1967; STENING, 1968, 1971, 1977; VAN SABBEN, 1969; SCHIELDGE et al., 1973; TAKEDA, 1982; SINGH and COLE, 1987a,c; RICHMOND and ROBLE, 1987), although WALTON and BOWHILL (1979) did examine the effects of a more realistic field geometry. The flow of electric current along magnetic field lines between hemispheres complicates the computation of ground magnetic perturbations, but computational procedures to do this have been developed (VAN SABBEN, 1966; SCHIELDGE et al., 1973; RICHMOND, 1974).

One of the goals of dynamo modeling has been to improve our knowledge of global thermospheric winds. The observed electrodynamic fields, especially the magnetic perturbations, can be used to constrain models of possible wind distributions. The early studies assumed, for want of better information, that the driving winds were constant in height. The nature of geomagnetic variations led also to the assumption that the wind pattern varies periodically on a daily basis, and migrates westward around the earth in synchronization with the apparent solar position, as do the dominant atmospheric tides. MAEDA (1955, 1956) and KATO (1956, 1957) used observed magnetic variations and a conductivity model to show that the dominant winds have a 24-hour period, with 12-hour winds of lesser importance. Later studies (STENING, 1969, 1981; TARPLEY, 1970b; VOLLAND, 1971; SCHIELDGE et al., 1973; MURATA, 1974; RICHMOND et al., 1976; MÖHLMANN, 1976, 1977; TAKEDA and MAEDA, 1980, 1981; TAKEDA et al., 1986) have tended to confirm this fact, but some have also shown that 12-hour winds play an important role in explaining observed ionospheric electric fields (RICHMOND et al., 1976; HANUISE et al., 1983; RICHMOND and ROBLE, 1987; TAKEDA and YAMADA, 1987).

VAN SABBEN (1962) was the first to use height-varying winds in a dynamo model, whereby he emphasized that even steady winds can be effective in generating currents. Observations and theoretical calculations of height-varying tidal winds in the atmosphere led to dynamo models incorporating more realistic winds (SCHIELDGE et al., 1973; MURATA, 1974; RICHMOND et al., 1976; STENING, 1981; TAKEDA and MAEDA, 1980, 1981; TAKEDA et al., 1986; RICHMOND and ROBLE, 1987; TAKEDA and YAMADA, 1987). A recent advance in the attempt to get physically realistic winds to use in dynamo modeling is the work by RICHMOND and ROBLE (1987). These authors used the three-dimensional thermospheric wind simulation of FESEN et al. (1986) coming from a general circulation mode of the thermosphere, with the inclusion of upward-propagating semidiurnal tidal winds.

5. Comparisons of Model Results with Observations

Ionospheric dynamo models have been able to reproduce observed features of ionospheric electrodynamics in several areas. Concerning the electrical conductivities, a number of observations support the basic characteristics of the highly anisotropic conductivity tensor employed in most dynamo models. The models predict that the very high parallel conductivity will make magnetic field lines nearly equipotentials all the way between conjugate points, which is supported by the observations of PETERSON et al. (1977). A consequence of this is that electric fields in the plasmasphere exhibit the effects of ionospheric dynamo action (CARPENTER, 1978; BAUMJOHANN et al., 1985; RASH et al., 1986). The large Hall-to-Pedersen conductivity ratio around 100 km altitude that is responsible for the existence of the equatorial electrojet is confirmed by comparisons of models of the electrojet with

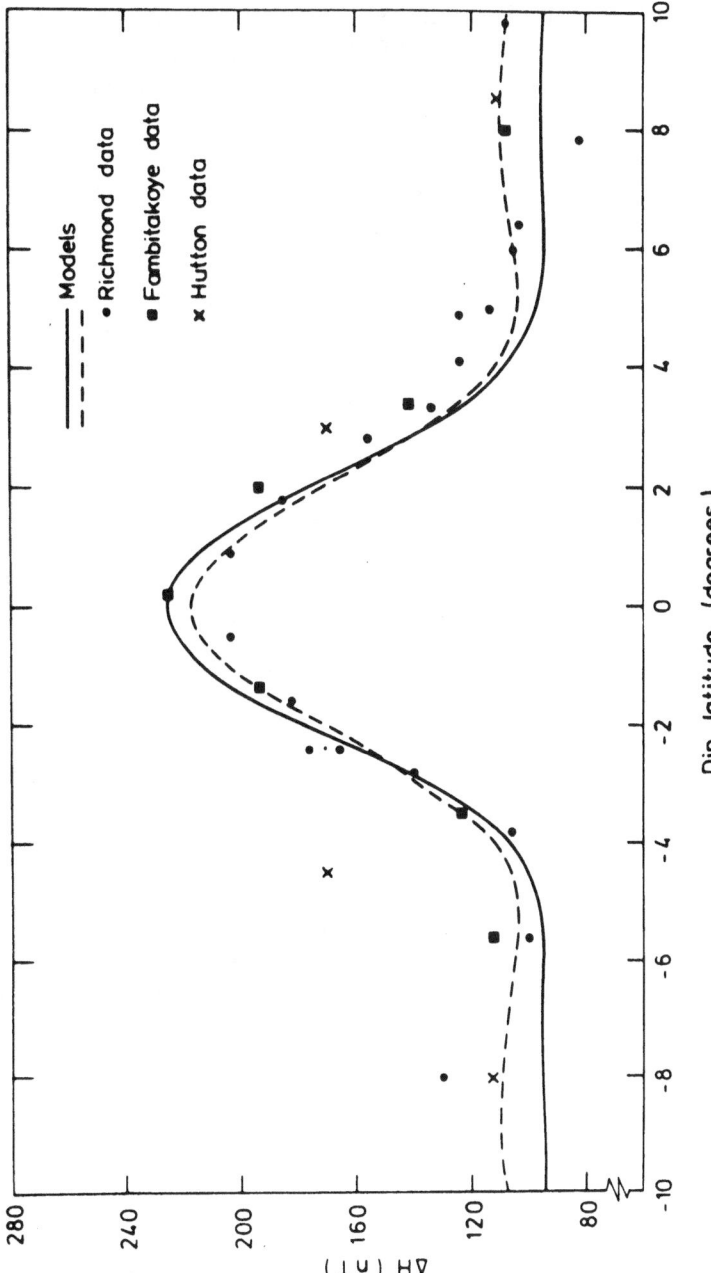

Figure 3

Variation of ΔH on the ground near noon. Full and dashed line models are simulations using different wind models. From STENING (1985).

observations (e.g., UNTIEDT, 1967; RICHMOND, 1973b; FAMBITAKOYE *et al.*, 1976; FORBES and LINDZEN, 1976b; STENING, 1985). Figure 3 (STENING, 1985) compares the latitudinal profile of modeled magnetic variations across the magnetic equator with various observations, showing that the width of the electrojet is rather well simulated. However, some quantitative discrepancies exist in the lower part of the electrojet (KRYLOV *et al.*, 1973; RICHMOND, 1973b; FORBES and LINDZEN, 1976b; GAGNEPAIN *et al.*, 1977; STENING, 1985; ANANDARAO and RAGHAVARAO, 1987; SINGH and COLE, 1987c) which UNTIEDT (1968) and GAGNEPAIN *et al.* (1977) were able to resolve by increasing the effective electron-neutral collision frequency, and hence the effective Pedersen conductivity, in an *ad hoc* fashion. Figure 4 (STENING, 1985) compares the vertical profile of current density observed at the magnetic equator with two models that do not make this *ad hoc* adjustment. Both models, which differ primarily in their electron density profiles, show current maxima

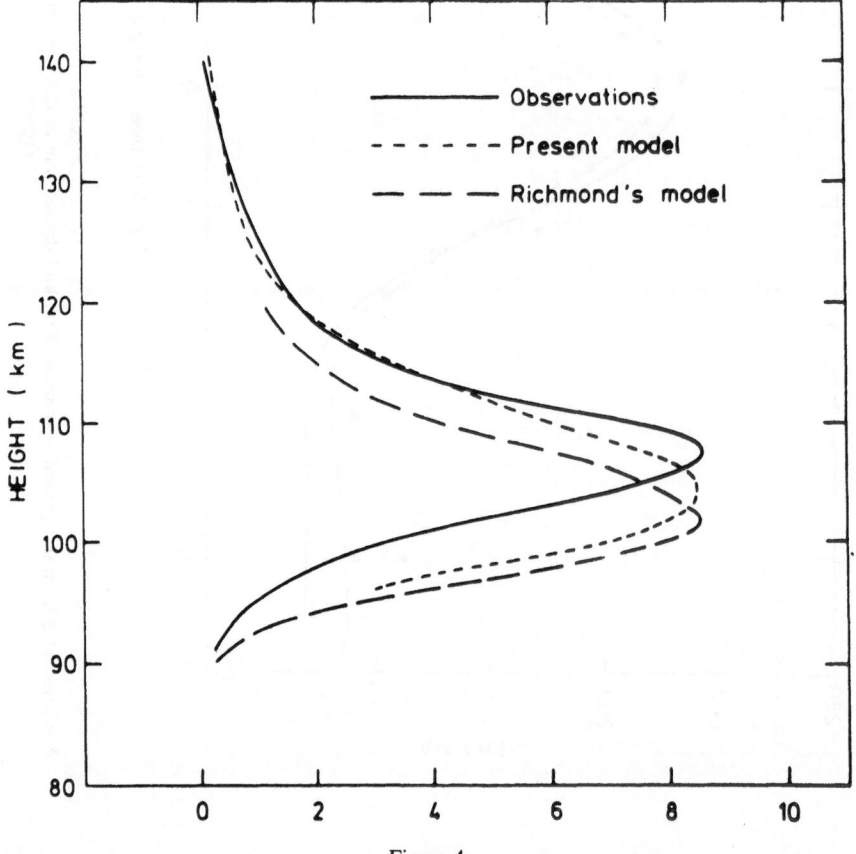

Figure 4

Current density profiles. Observations are the mean from rocket flights in Peru. The model of RICHMOND (1973b) has been scaled to give the same maximum current density. "Present Model" is from a two-layer model at 1200 LT. From STENING (1985).

located below the observed peak. STENING (1986) suggested that the discrepancies may be associated with errors in the neutral density models used.

Dynamo models have generally been successful in reproducing the basic features of observed Sq currents and magnetic variations using thermospheric wind systems that are compatible with observed winds, although some discrepancies remain. Figure 5 shows a comparison between observed and computed magnetic variations as presented by RICHMOND and ROBLE (1987). The simulation using winds driven by direct solar radiation absorption in the thermosphere has the basic pattern of the observations, although the computed amplitude is too small. The addition of upward propagating semidiurnal tides increases the amplitude, perhaps too strongly, and tends to give the daily variation a somewhat more rapid oscillatory character than the observations indicate. In general, dynamo simulations with semidiurnal tides modeled after observations have had some difficulty in reproducing observed electric fields and magnetic perturbations. Several studies have found that the magnitude of the semidiurnal dynamo effects tends to be too strong to be fully compatible with observations (FORBES and LINDZEN, 1976a; FORBES and GARRETT, 1979; TAKEDA and MAEDA, 1981; RICHMOND and ROBLE, 1987; TAKEDA and YAMADA, 1987). STENING (1969) and TAKEDA and MAEDA (1981)

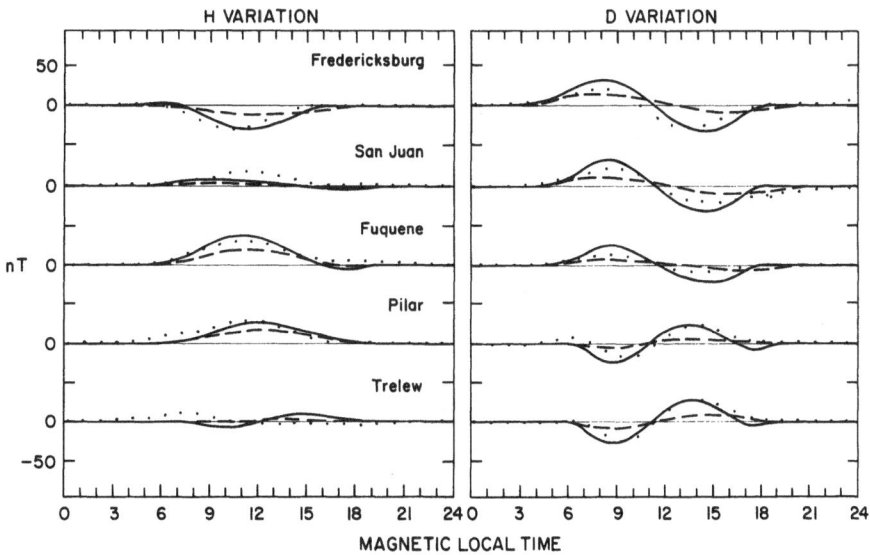

Figure 5

Comparisons of average equinoctial solar minimum northward (H) and eastward (D) magnetic perturbations measured at Fredericksburg, San Juan, Fuquene, Pilar, and Trelew (points) with perturbations computed from the two simulations: without tidal forcing (dashed lines) and with tidal forcing (solid lines). The computations show perturbations at 1700 UT as a function of magnetic local time (or magnetic longitude), rather than actual local-time variations for these stations. From RICHMOND and ROBLE (1987).

also found that the phase of semidiurnal dynamo effects in their simulations did not agree with observations.

Although dynamo simulations have had some measure of success in reproducing observed electric fields at middle and low latitudes (STENING, 1973; BEHNKE and HAGFORS, 1974; HEELIS *et al.*, 1974; RICHMOND *et al.*, 1976; FORBES and LINDZEN, 1977; RICHMOND and ROBLE, 1987; TAKEDA and YAMADA, 1987), large discrepancies remain, particularly at night. Figure 6 shows a comparison between equinox observations of $E \times B$ drifts at various locations and two model simulations of RICHMOND and ROBLE (1987). It is suspected that the discrepancies are related not only to uncertainties in the winds but also to the highly variable and uncertain nature of nightside conductivities and to magnetospheric effects that are not yet adequately treated in dynamo models (FORBES and LINDZEN, 1977; BURNSIDE *et al.*, 1983; RICHMOND and ROBLE, 1987).

The dynamo effects of storm disturbances in thermospheric winds have been modeled in a simplified manner by BLANC and RICHMOND (1980) and by MAZAUDIER *et al.* (1987). They note that there is a tendency for poleward electric fields to develop at midlatitudes, which is at least qualitatively in agreement with observations (Figure 7). At equatorial latitudes a westward electric field should develop during the day which would tend to reduce the strength of the normal

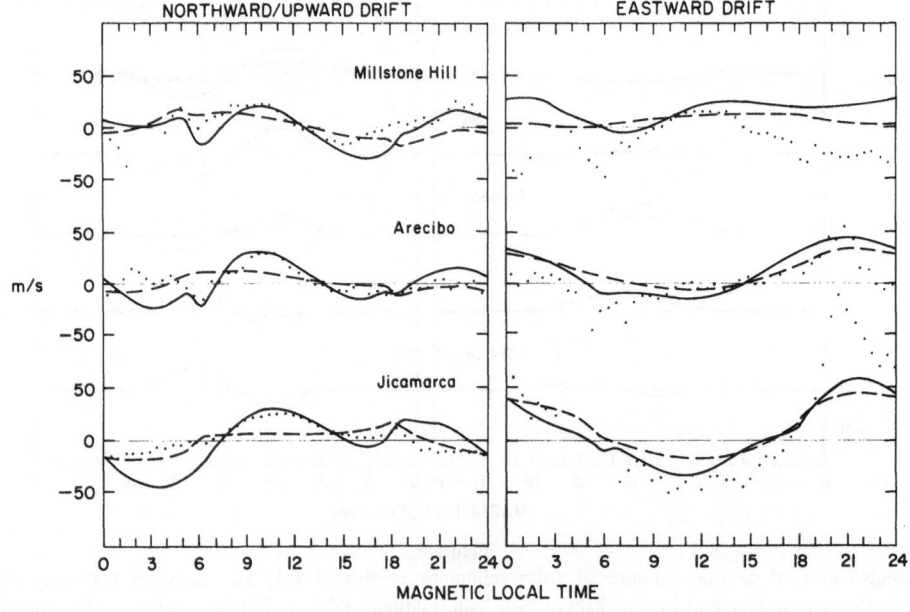

Figure 6

Comparisons of average equinoctial northward/upward and eastward plasma drifts measured at Millstone Hill, Arecibo, and Jicamarca (points) with drifts computed from the two simulations: without tidal forcing (dashed lines); and with tidal forcing (solid lines). From RICHMOND and ROBLE (1987).

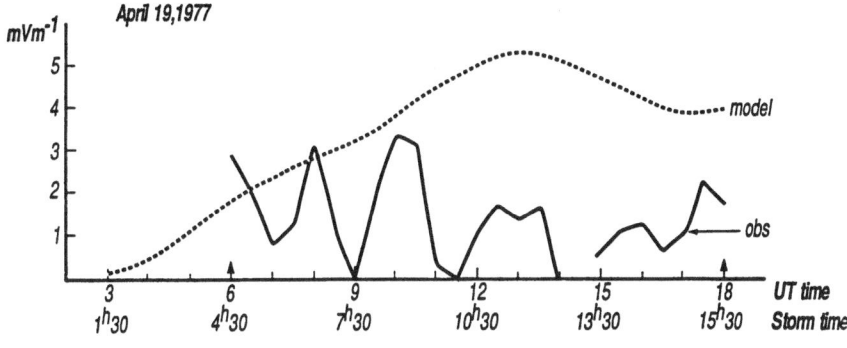

Figure 7

Comparison between observed northward electric field and model predicted electric fields over Saint-Santin on April 19, 1977. From MAZAUDIER et al. (1987).

eastward electric field and of the equatorial electrojet currents. BLANC and RICH-MOND (1980) and FEJER et al. (1983) noted that these effects do indeed tend to occur at equatorial latitudes in connection with magnetic storms, especially the recovery phase. The magnitude of the modeled effects is not large enough, however, to account for electric fields observed in the lower atmosphere during magnetic activity (e.g., OGAWA and KAWAMOTO, 1982).

At night the height-integrated conductivity of the lower F-region is comparable to that of the E-region during solar minimum (HARPER and WALKER, 1977; TAKEDA and ARAKI, 1985), and is much greater than the latter during solar maximum (BURNSIDE et al., 1983; TAKEDA and ARAKI, 1985). Since F-region winds tend to be stronger than E-region winds, at least when averaged over the thickness of the conducting layer, F-region winds tend to dominate the dynamo process at night (RISHBETH, 1971a,b; TAKEDA and MAEDA, 1983). HEELIS et al. (1974), MATUURA (1974), TAKEDA and MAEDA (1980, 1981), STENING (1981), and TAKEDA et al. (1986) modeled the coupled E- and F-region dynamos and showed that both are important. TAKEDA and MAEDA (1983) were able to reproduce low-latitude signatures of meridional currents observed by MAGSAT at dusk by simulating the relatively strong F-region dynamo effects around this time of day. Figure 8 compares the observed ranges of D-component variations with two simulations for different solar activity conditions, plotted as a function of altitude at 1800 LT.

Observations of the vertical structure of ionospheric currents have often shown strong variations with scale sizes of several kilometers and greater. These variations can be modeled by considering the effects of height-varying winds of similar scale sizes (RICHMOND, 1973a; FORBES and LINDZEN, 1976a,b; REDDY and DEVASIA, 1981; DEVASIA, 1986; ANANDARAO and RAGHAVARAO, 1987; SINGH and COLE, 1987c). Figure 9 shows examples of electric current profiles computed by FORBES and LINDZEN (1976a). Height-varying winds can also explain the latitudinal

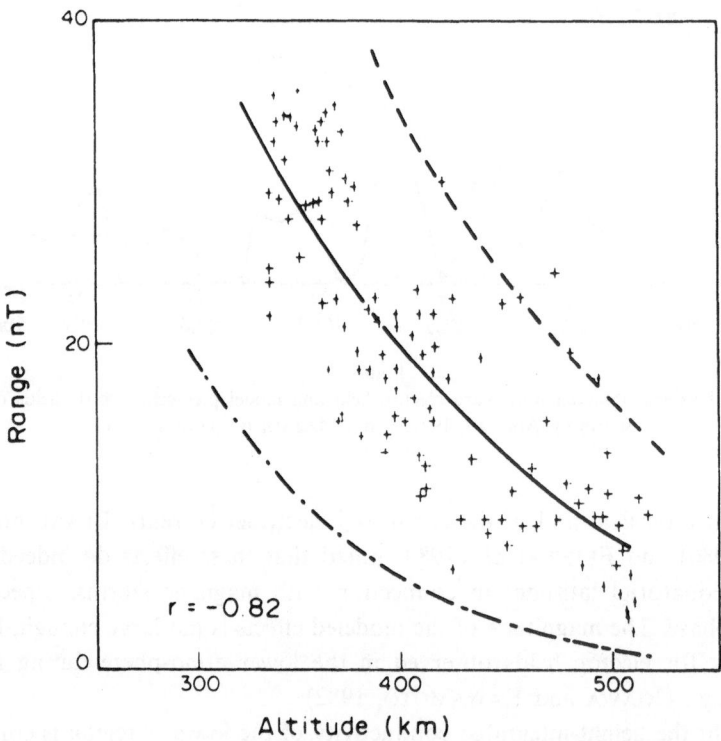

Figure 8

Height dependence of the maximum amplitude of the D-component observed by MAGSAT. The full line shows a least square fit and r is the correlation coefficient. Calculated ranges of D-component are shown by a dotted line ($R = 280$) and a dot-dash line ($R = 100$). From TAKEDA and MAEDA (1984).

structure of east-west currents at low latitudes (e.g., FAMBITAKOYE *et al.* 1976; ANANDARAO and RAGHAVARAO, 1979, 1987; REDDY and DEVASIA, 1981; DEVASIA, 1986).

6. Untested Predictions of Dynamo Theory

Dynamo models produce many types of results concerning quantities or variations that have not yet been measured. In a sense, such results can be considered to be untested predictions of the models. In this section only three of the many possible predictions of dynamo models are discussed.

TAKEDA *et al.* (1986) and TAKEDA and YAMADA (1987) found strong solar-cycle changes in the ionospheric electric fields modeled with a fixed wind system but variable conductivities. The strong electric fields they find for a sunspot number of 200 have not yet been tested against observations for this high level of solar activity.

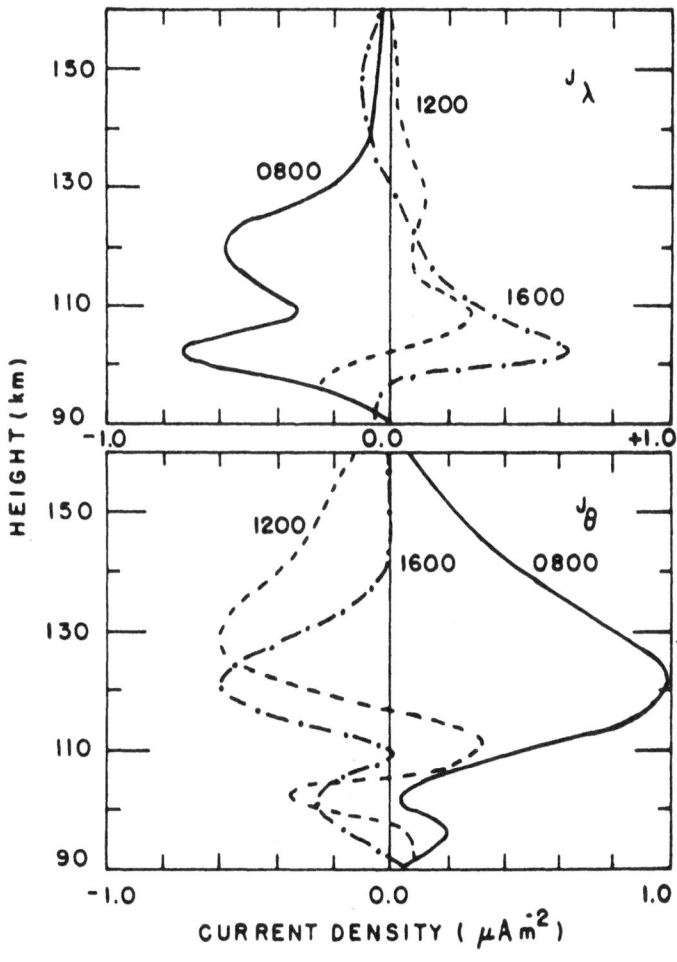

Figure 9

Vertical structures of eastward (J_λ) and southward (J_θ) current densities at 30° N 0800 LT, and 1600 LT. Southward currents are of opposite sign in the Southern Hemisphere. Current densities between 2000 LT and 0400 LT are less than 0.07 $\mu A m^{-2}$ at all heights. From FORBES and LINDZEN (1976a).

Figure 10 compares the modeled electrostatic potential distributions of TAKEDA *et al.* (1986) for two levels of solar activity.

A second prediction of dynamo theory is that substantial electrical currents should flow along geomagnetic field lines between hemispheres, with total amounts comparable to the total horizontal current flowing in the ionosphere (VAN SABBEN, 1966, 1968, 1970; MATVEEV, 1971; SCHIELDGE *et al.*, 1973; MAEDA, 1974; STEN-ING, 1977; SCHÄFER, 1978; WAGNER *et al.*, 1980; TAKEDA, 1982; RICHMOND and ROBLE, 1987). Figure 11 shows the distribution of vertical current density at the top of the ionosphere from a simulation of RICHMOND and ROBLE (1987). Although

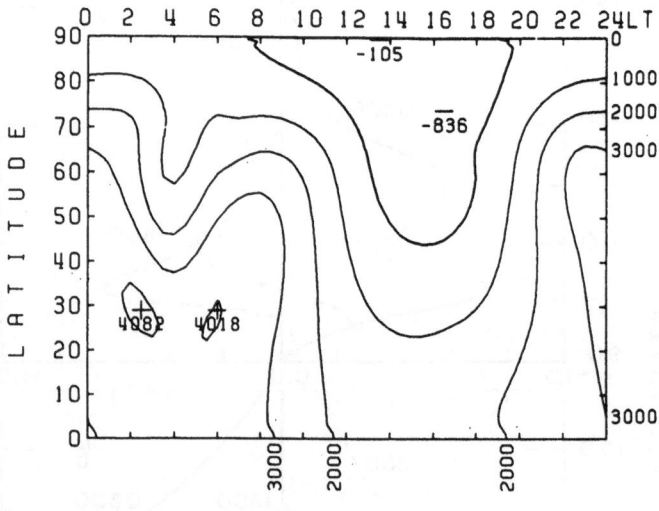

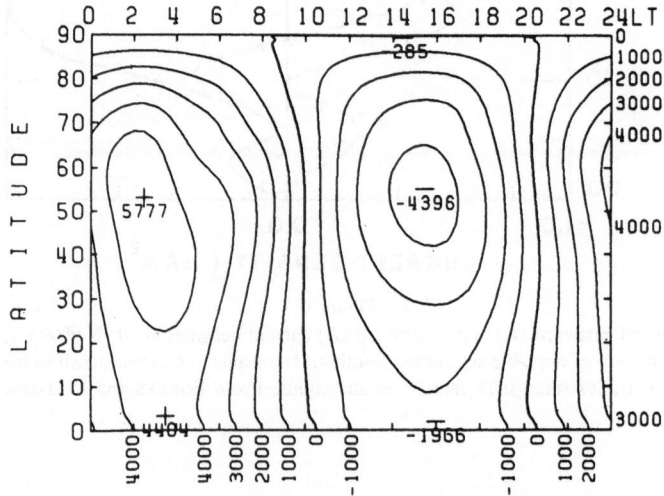

Figure 10

Electrostatic potential distribution at 90 km altitude for $R = 35$ (top) and 200 (bottom). Contours are drawn at every 1000 V. From TAKEDA *et al.* (1986).

the magnetic effects of these currents must affect satellite magnetometers, the magnitude of the effects is likely to be on the order of 10–20 nT and to vary smoothly over large spatial distances, so that its discrimination from main geomagnetic field variations would be difficult.

Another prediction concerns the effects of thunderstorm electric fields on the

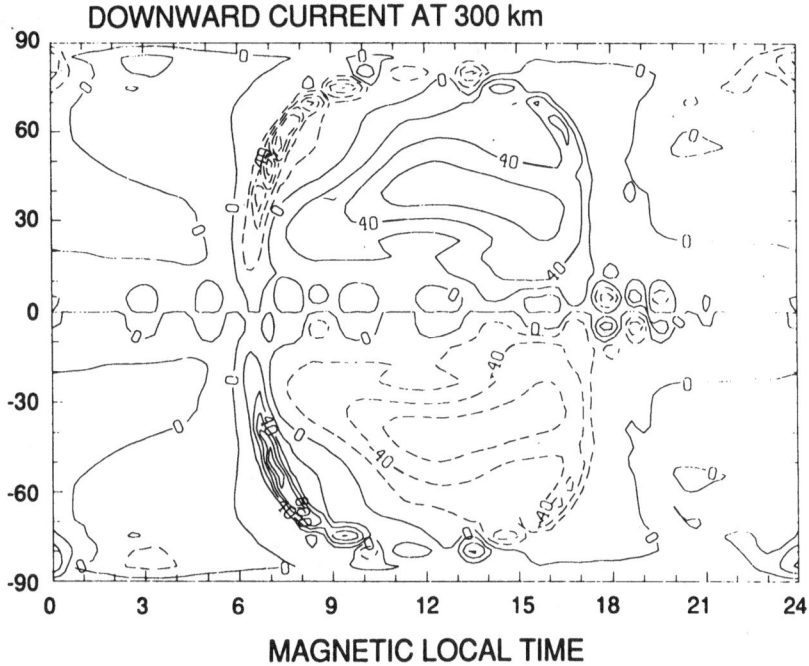

Figure 11
Downward component of field-aligned current at 300 km computed for 1700 UT. The contour interval
is 2×10^{-9} A/m². Downward currents have solid contours; upward currents have dashed contours.
From RICHMOND and ROBLE (1987).

large-scale ionospheric electric fields. HAYS and ROBLE (1979) modeled potential
differences of several thousand volts over the globe associated with thunderstorms
effects. However, they deliberately used very small ionospheric conductivities in
order to determine the maximum electric fields that might be produced over
thunderstorm areas, and so their global potential differences are overestimated.
MAKINO and TAKEDA (1984) used more realistic ionospheric conductivities and
found a total ionospheric potential difference of only 55 V associated with thunderstorm effects. This is two orders of magnitude less than the potential generated by
ionospheric winds, and would thus be practically impossible to observe. It may still
be possible, however, to observe regional ionospheric electric field effects near
thunderstorms at night.

7. Remaining Problems

In order to obtain improved simulations of ionospheric dynamo effects, better
global models of the electrical conductivities are required, particularly at night. It
will also be important to model the mutual interactions among conductivities,
winds, and electric fields at night in order to account properly for the relevant

feedback mechanisms. Another problem related to conductivities concerns the discrepancy between modeled and observed equatorial electrojet currents (Figure 4), which has not yet been fully resolved.

Most of the efforts to simulate dynamo effects with realistic global wind systems to date have concentrated on equinox conditions, with winds usually assumed to be symmetric about the equator. Most solstitial simulations that have been presented so far have examined only the effects of asymmetry in the conductivities, and not asymmetry in the winds (VAN SABBEN, 1970; STENING, 1971, 1977; RICHMOND and VENKATESWARAN, 1971; WALTON and BOWHILL, 1979). Simulations of solstice conditions with realistic distributions of asymmetric winds, including symmetric and antisymmetric tidal modes, have been attempted only on a few occasions (SCHIELDGE et al., 1973; MARRIOTT et al., 1979), primarily because of the difficulty of estimating realistic antisymmetric tides from the available observations. It should be noted, moreover, that observed hemispheric asymmetries in Sq current patterns at equinox (e.g., VAN SABBEN, 1964) probably indicate the presence of asymmetric winds at equinox as well as at solstice (VAN SABBEN, 1966, 1968).

There is substantial day-to-day variability in dynamo effects (e.g., MARRIOTT et al., 1979; CLAUER et al., 1980; BRIGGS, 1984) which is believed to be caused by a corresponding variability in the global winds. However, we do not yet understand the sources causing the variability in lower thermospheric winds at midlatitudes. MARRIOTT et al. (1979) and HANUISE et al. (1983) suggest that variability in the amplitudes and phases of tidal winds, particularly lunar or solar semidiurnal tides, could cause the global changes in currents that are associated with equatorial counterelectrojet events. However, simultaneous wind and current observations on a global scale during such an event have not yet been made.

The variability of electric fields at auroral latitudes is dramatic, and there is evidence that ionospheric dynamo effects play a significant role, at least during quiet times (BAUMJOHANN et al., 1985). Modeling studies of the interacting magnetospheric and ionospheric dynamos, as initiated by WOLF et al. (1986) and SPIRO et al. (1988), will be an important direction of future research.

Acknowledgments

This work was supported by the NASA Solar-Terrestrial Theory Program and by NASA Global Solar-Terrestrial Electrostatic Coupling Study, Contract # S–15,028.

REFERENCES

AKASOFU, S.-I., and DE WITT, R. N. (1965), *Dynamo Action in the Ionosphere and Motions of the Magnetospheric Plasma, 3. The Pedersen Conductivity, Generalized to Take Account of Acceleration of the Neutral Gas*, Planet. Space Sci. *13*, 737–744.

ANANDARAO, B. G., and RAGHAVARAO, R. (1979), *Effects of Vertical Shears in the Zonal Winds on the Electrojet*, Space Res. *19*, 283–286.

ANANDARAO, B. G., and RAGHAVARAO, R. (1987), *Structural Changes in the Currents and Fields of the Equatorial Electrojet due to Zonal and Meridional Winds*, J. Geophys. Res. *92*, 2514–2526.

AXFORD, W. I., and HINES, C. O. (1961), *A Unifying Theory of High-Latitude Geophysical Phenomena and Geomagnetic Storms*, Can. J. Phys. *39*, 1433–1464.

BAKER, W. G. (1953), *Electric Currents in the Ionosphere, II. The Atmospheric Dynamo*, Phil. Trans. Roy. Soc. *A246*, 295–305.

BAKER, W. G., and MARTYN, D. F. (1952), *Conductivity of the Ionosphere*, Nature *170*, 1090–1092.

BAKER, W. G., and MARTYN, D. F. (1953), *Electric Currents in the Ionosphere, I. The Conductivity*, Phil. Trans. Roy. Soc. *A246*, 281–294.

BANKS, P. M. (1972), *Magnetospheric Processes and the Behaviour of the Neutral Atmosphere*, Space Res. *12*, 1051–1067.

BAUMJOHANN, W., HAERENDEL, G., and MELZNER, F. (1985), *Magnetospheric Convection Observed Between 0600 and 2100 LT: Variations with Kp*, J. Geophys. Res. *90*, 393–398.

BEHNKE, R. A., and HAGFORS, T. (1974), *Evidence for the Existence of Nighttime F-Region Polarization Fields at Arecibo*, Radio Sci. *9*, 211–216.

BLANC, M. (1979), *Electrodynamics of the Ionosphere from Incoherent Scatter: A Review*, J. Geomag. Geoelectr. *31*, 137–164.

BLANC, M., and RICHMOND, A. D. (1980), *The Ionospheric Disturbance Dynamo*, J. Geophys. Res. *85*, 1669–1686.

BRIGGS, B. H. (1984), *The Variability of Ionospheric Dynamo Currents*, J. Atmos. Terr. Phys. *46*, 419–429.

BURNSIDE, R. G., WALKER, J. C. G., BEHNKE, R. A., and GONZALES, C. A. (1983), *Polarization Electric Fields in the Nighttime F-Layer at Arecibo*, J. Geophys. Res. *88*, 6259–6266.

CARPENTER, D. L. (1978), *New Whistler Evidence of a Dynamo Origin of Electric Fields in the Quiet Plasmasphere*, J. Geophys. Res. *83*, 1558–1564.

CLAUER, C. R., McPHERRON, R. L., and KIVELSON, M. G. (1980), *Uncertainty in Ring Current Parameters Due to the Quiet Magnetic Field Variability at Mid-Latitudes*, J. Geophys. Res. *85*, 633–643.

COLE, K. D. (1969), *Theory of Electric Currents in Ionospheric E-layers*, Planet. Space Sci. *17*, 1977–1992.

DEVASIA, C. V. (1986), *The Role of Local Action of Tidal Winds in the Generation of Counter Electrojet*, Ann. Geophysicae *4A*, 301–310.

DOUGHERTY, J. R. (1963), *Some Comments on Dynamo Theory*, J. Geophys. Res. *68*, 2383–2384.

FAMBITAKOYE, O., MAYAUD, P. N., and RICHMOND, A. D. (1976), *Equatorial Electrojet and Regular Daily Variation S_R*, J. Atmos. Terr. Phys. *38*, 113–121.

FEJER, B. G., LARSEN, M. F., and FARLEY, D. T. (1983), *Equatorial Disturbance Dynamo Electric Fields*, Geophys. Res. Letters *10*, 537–540.

FEJER, J. A. (1953), *Semidiurnal Currents and Electron Drifts in the Ionosphere*, J. Atmos. Terr. Phys. *4*, 184–203.

FESEN, C. G., DICKINSON, R. E., and ROBLE, R. G. (1986), *Simulation of the Thermospheric Tides at Equinox with the National Center for Atmospheric Research Thermospheric General Circulation Model*, J. Geophys. Res. *91*, 4471–4489.

FOOTITT, R. J., BAILEY, G. J., and MOFFETT, R. J. (1983), *Ion Transport in the Mid-Lattitude F1-Region*, Planet. Space Sci. *31*, 671–687.

FORBES, J. M. (1981), *The Equatorial Electrojet*, Rev. Geophys. Space Phys. *19*, 469–504.

FORBES, J. M., and GARRETT, H. B. (1979), *Solar Tidal Wind Structures and the E-Region Dynamo*, J. Geomag. Geoelectr. *31*, 173–182.

FORBES, J. M., and LINDZEN, R. S. (1976a), *Atmospheric Solar Tides and Their Electrodynamic Effects—I. The Global Sq Current System*, J. Atmos. Terr. Phys. *38*, 897–910.

FORBES, J. M., and LINDZEN, R. S. (1976b), *Atmospheric Solar Tides and Their Electrodynamic Effects—II. The Equatorial Electrojet*, J. Atmos. Terr. Phys. *38*, 911–920.

FORBES, J. M., and LINDZEN, R. S. (1977), *Atmospheric Solar Tides and Their Electrodynamic Effects—III. The Polarization Electric Field*, J. Atmos. Terr. Phys. *39*, 1369–1377.

GAGNEPAIN, J., CROCHET, M., and RICHMOND, A. D. (1977), *Comparison of Equatorial Electrojet Models*, J. Atmos. Terr. Phys. *39*, 1119–1124.

HANUISE, C., MAZAUDIER, C., VILA, P., BLANC, M., and CROCHET, M. (1983), *Global Dynamo Simulation of Ionospheric Currents and Their Connection with the Equatorial Electrojet and Counter Electrojet: A Case Study*, J. Geophys. Res. *88*, 253–270.

HARPER, R. M., and WALKER, J. C. G. (1977), *Comparison of Electrical Conductivities in the E- and F-Regions of the Nocturnal Ionosphere*, Planet. Space Sci. *25*, 197–199.

HAYS, P. B., and ROBLE, R. G. (1979), *A Quasi-Static Model of Global Atmospheric Electricity, I. The Lower Atmosphere*, J. Geophys. Res. *84*, 3291–3305.

HEELIS, R. A., KENDALL, P. C., MOFFETT, R. J., WINDLE, D. W., and RISHBETH, H. (1974), *Electrical Coupling of the E- and F-Regions and Its Effect on F-Region Drifts and Winds*, Planet. Space Sci. *22*, 743–756.

HIRONO, M. (1952), *A Theory of Diurnal Magnetic Variations in Equatorial Regions and Conductivity of the Ionospheric E-region*, J. Geomag. Geoelectr. *4*, 7–21.

HIRONO, M., and KITAMURA, T. (1956), *A Dynamo Theory in the Ionosphere*, J. Geomag. Geoelectr. *8*, 9–23.

JALONEN, L., NYGRÉN, T. OKSMAN, J., and TURUNEN, T. (1984), *On the Current-Carrying Properties of Mid-latitude Type Sporadic E-layers*, J. Atmos. Terr. Phys. *46*, 383–387.

JONES, M. N. (1974), *Mathematical Theory of the Ionospheric Dynamo*, Planet. Space Sci. *22*, 831–831.

KATO, S. (1956), *Horizontal Wind Systems in the Ionospheric E-Region Deduced from the Dynamo Theory of the Geomagnetic Sq Variations—Part II. Rotating Earth*, J. Geomag. Geoelectr. *8*, 24–37.

KATO, S. (1957), *Horizontal Wind Systems in the Ionospheric E-Region Deduced from the Dynamo Theory of the Geomagnetic Sq Variation—Part IV.* J. Geomag. Geoelectr. *9*, 107–115.

KATO, S. (1980), *Dynamics of the Upper Atmosphere* (Center for Academic Publications, Japan 1980) pp. 141–199.

KRYLOV, A. L., SOBOLEVA, T. N., FISHCHUK, D. I., TSEDILINA, Ye. Ye., and SHCHERBAKOV, V. P. (1973), *Structure of the Equatorial Electrojet*, Geomag. Aeron. *13*, 400–404.

LYONS, L. R., KILLEEN, T. L., and WATERSCHEID, R. L. (1985), *The Neutral Wind Flywheel as a Source of Quiet-time Polar-cap Currents*, Geophys. Res. Lett. *12*, 101–104.

MAEDA, H. (1955), *Horizontal Wind Systems in the Ionospheric E-Region Deduced from the Dynamo Theory of the Geomagnetic Sq Variations—Part I. Non-rotating Earth*, J. Geomag. Geoelectr. *7*, 121–132.

MAEDA, H. (1957), *Horizontal Wind Systems in the Ionospheric F-Region Deduced from the Dynamo Theory of the Geomagnetic Sq Variation—Part III.* J. Geomag. Geoelectr. *9*, 86–93.

MAEDA, H. (1966), *Note on the Ionospheric Sq Deduced Winds in Summer and in Winter*, J. Atmos. Sci. *23*, 363–369.

MAEDA, H. (1974), *Field-aligned Current Induced by Asymmetric Dynamo Action in the Ionosphere*, J. Atmos. Terr. Phys. *36*, 1395–1401.

MAEDA, H., and FUJIWARA, M. (1967), *Lunar Ionospheric Winds Deduced from the Dynamo Theory of Geomagnetic Variations*, J. Atmos. Terr. Phys. *29*, 917–936.

MAEDA, K. (1977), *Conductivity and Drifts in the Ionosphere*, J. Atmos. Terr. Phys. *39*, 1041–1053.

MAKINO, M., and TAKEDA, M. (1984), *Three-dimensional Ionospheric Currents and Fields Generated by the Atmospheric Global Circuit Current*, J. Atmos. Terr. Phys. *46*, 199–206.

MARRIOTT, R. T., RICHMOND, A. D., and VENKATESWARAN, S. V. (1979), *The Quiet-time Equatorial Electrojet and Counter-electrojet*, J. Geomag. Geoelectr. *31*, 311–340.

MATSUSHITA, S. (1969), *Dynamo Currents, Winds, and Electric Fields*, Radio Science *4*, 771–780.

MATSUSHITA, S. (1977), *Upper-atmospheric Tidal-interaction Effects on Geomagnetic and Ionospheric Variations—A Review*, Ann. Geophys. *33*, 115–126.

MATUURA, N. (1974), *Electric Fields Deduced from the Thermospheric Model*, J. Geophys. Res. *79*, 4679–4689.

MATVEEV, M. I. (1971), *Conductivity and E.M.F.-Dynamo Nonuniformities: Field-aligned Currents in the Magnetosphere*, Gerlands Beitr. Geophysik *80*, 155–170.

MAZAUDIER, C., RICHMOND, A. D., and BRINKMAN, D. (1987) *On Thermospheric Winds Produced by Auroral Heating during Magnetic Storms and Associated Dynamo Electric Fields*, Ann. Geophysicae *5A*, 443–448.

MÖHLMANN, D. (1974), *Ionospheric Dynamo-electric Fields*, Gerlands Beitr. Geophysik *83*, 101–112.

MÖHLMANN, D. (1976), *Two Wind Systems Causing the Global Ionospheric Dynamo-electric Field*, Gerlands Beitr. Geophysik *85*, 343–344.

MÖHLMANN, D. (1977), *Ionospheric Electrostatic Fields*, J. Atmos. Terr. Phys. *39*, 1325–1332.

MURATA, H. (1974), *An Estimation of the Electric Potential Field Generated by the Diurnal Atmospheric Tide with the First Negative Mode Excited in the Lower Ionosphere*, Planet. Space Sci. *22*, 569–582.

OGAWA, T., and KAWAMOTO, H. (1982), *Mid-latitude Horizontal Electric Fields in the Stratosphere During Magnetically Disturbed Periods*, Planet. Space Sci. *30*, 1013–1024.

PETERSON, W. K., DOERING, J. P., POTEMRA, T. A., BOSTROM, C. O., BRACE, L. H., HEELIS, R. A., and HANSON, W. B. (1977), *Measurement of Magnetic Field-aligned Potential Differences Using High Resolution Conjugate Photoelectron Energy Spectra*, Geophys. Res. Letters *4*, 373–376.

PIDDINGTON, J. H. (1954), *The Motion of Ionized Gas in Combined Magnetic, Electric, and Mechanical Fields of Force*, Mon. Not. R. Astron. Soc. *114*, 651–663.

RASH, J. P. S., HANSEN, H. J., and SCOURFIELD, M. W. J. (1986), *Electric Field Sources in the Quiet Plasmasphere from Whistler Observations*, J. Atmos. Terr. Phys. *48*, 399–414.

REDDY, C. A., and DEVASIA, D. V. (1981), *Height and Latitude Structure of Electric Fields and Currents due to Local East-West Winds in the Equatorial Electrojet*, J. Geophys. Res. *86*, 5751–5767.

RICHMOND, A. D. (1973a), *Equatorial Electrojet—Part I. Development of a Model Including Winds and Instabilities*, J. Atmos. Terr. Phys. *35*, 1082–1103.

RICHMOND, A. D. (1973b), *Equatorial Electrojet—Part II. Use of the Model to Study the Equatorial Ionosphere*, J. Atmos. Terr. Phys. *35*, 1105–1118.

RICHMOND, A. D. (1974), *The Computation of Magnetic Effects of Field-aligned Magnetospheric Currents*, J. Atmos. Terr. Phys. *36*, 245–252.

RICHMOND, A. D., and MATSUSHITA, S. (1975), *Thermospheric Response to a Magnetic Substorm*, J. Geophys. Res. *80*, 2839–2850.

RICHMOND, A. D. (1979), *Ionospheric Wind Dynamo Theory: A Review*, J. Geomag. Geoelectr. *31*, 297–310.

RICHMOND, A. D., *Thermospheric dynamics and electrodynamics*, In *Solar-Terrestrial Physics* (eds. Carovillano, R. L., and Forbes, J. M.) (D. Reidel, Dordrecht 1983) pp. 523–607.

RICHMOND, A. D., and ROBLE, R. G. (1987), *Electrodynamic Effects of Thermospheric Winds from the NCAR Thermospheric General Circulation Model*, J. Geophys. Res. *92*, 12365–12376.

RICHMOND, A. D., and VENKATESWARAN, S. V. (1971), *Geomagnetic Crochets and Associated Ionospheric Systems*, Radio Sci. *6*, 139–164.

RICHMOND, A. D., MATSUSHITA, S., and TARPLEY, J. D. (1976), *On the Production Mechanism of Electric Currents and Fields in the Ionosphere*, J. Geophys. Res. *81*, 547–555.

RISHBETH, H. (1971a), *The F-Layer Dynamo*, Planet. Space Sci. *19*, 263–267.

RISHBETH, H. (1971b), *Polarization Fields Produced by Winds in the Equatorial F-Region*, Planet. Space Sci. *19*, 357–369.

RISHBETH, H. (1983), *Further Studies of Directional F-Layer Currents*, Planet. Space Sci. *31*, 1177–1180.

RISHBETH, H., and WALKER, J. C. G. (1982), *Directional Currents in Nocturnal E-Region Layers*, Planet. Space Sci. *30*, 209–214.

ROBLE, R. G., and HAYS, P. B. (1979), *A Quasi-static Model of Global Atmospheric Electricity, 2. Electrical Coupling Between the Upper and Lower Atmosphere*, J. Geophys. Res. *84*, 7247–7256.

SCHÄFER, K. (1978), *Dynamo-electric Field-aligned Currents in the Plasmasphere*, J. Atmos. Terr. Phys. *40*, 755–760.

SCHIELDGE, J. P., VENKATESWARAN, S. V., and RICHMOND, A. D. (1973), *The Ionospheric Dynamo and Equatorial Magnetic Variations*, J. Atmos. Terr. Phys. *35*, 1045–1061.

SINGH, A., and COLE, K. D. (1987a), *A Numerical Model of the Ionospheric Dynamo—I. Formulation and Numerical Technique*, J. Atmos. Terr. Phys. *49*, 521–527.

SINGH, A., and COLE, K. D. (1987b), *A Numerical Model of the Ionospheric Dynamo—II. Electrostatic Field at Equatorial and Low Latitudes*, J. Atmos. Terr. Phys. *49*, 529–537.

SINGH, A., and COLE, K. D. (1987c), *A Numerical Model of the Ionospheric Dynamo—III. Electric Current at Equatorial and Low Latitudes*, J. Atmos. Terr. Phys. *49*, 539–547.

SPIRO, R. W., WOLF, R. A., and FEJER, B. G. (1988), *Penetration of High-latitude-electric-field Effects to Low Latitudes during SUNDIAL 1984*, Ann. Geophysicae *6*, 39–50

STENING, R. J. (1968), *Calculation of Electric Currents in the Ionosphere by an Equivalent Circuit Method*, Planet Space. Sci. *16*, 717–728.

STENING, R. J. (1969), *An Assessment of the Contributions of Various Tidal Winds to the Sq Current System*, Planet. Space Sci. *17*, 889–908.

STENING, R. J. (1971), *Longitude and Seasonal Variations of the Sq Current System*, Radio Sci. *6*, 133–137.

STENING, R. J. (1973), *The Electrostatic Field in the Ionosphere*, Planet. Space Sci. *21*, 1897–1910.

STENING, R. J. (1977), *Field-aligned Currents Driven by the Ionospheric Dynamo*, J. Atmos. Terr. Phys. *39*, 933–937.

STENING, R. J. (1981), *A Two-layer Ionospheric Dynamo Calculation*, J. Geophys. Res. *86*, 3543–3550.

STENING, R. J. (1985), *Modeling the Equatorial Electrojet*, J. Geophys. Res. *90*, 1705–1719.

STENING, R. J. (1986), *Inter-relations Between Current and Electron Density Profiles in the Equatorial Electrojet and Effects of Neutral Density Changes*, J. Atmos. Terr. Phys. *48*, 163–170.

TAKEDA, M. (1982), *Three Dimensional Ionospheric Currents and Field Aligned Currents Generated by Asymmetrical Dynamo Action in the Ionospheric*, J. Atmos. Terr. Phys. *44*, 187–193.

TAKEDA, M., and ARAKI, T. (1985), *Electric Conductivity of the Ionospherie and Nocturnal Currents*, J. Atmos. Terr. Phys. *47*, 601–609.

TAKEDA, M., and MAEDA, H. (1980), *Three-dimensional Structure of Ionospheric Currents—1. Currents Caused by Diurnal Tidal Winds*, J. Geophys. Res. *85*, 6895–6899.

TAKEDA, M., and MAEDA, H. (1981), *Three-dimensional Structure of Ionospheric Currents—2. Currents Caused by Semidiurnal Tidal Winds*, J. Geophys. Res. *86*, 5861–5867.

TAKEDA, M., and MAEDA, H. (1983), *F-region Dynamo in the Evening—Interpretation of Equatorial ΔD Anomaly Found by MAGSAT*, J. Atmos. Terr. Phys. *45*, 401–408.

TAKEDA, M., and YAMADA, Y. (1987), *Simulation of Ionospheric Electric Fields and Geomagnetic Field Variation by the Ionospheric Dynamo for Different Solar Activity*, Ann. Geophysicae *5A*, 429–434.

TAKEDA, M., YAMADA, Y., and ARAKI, T. (1986), *Simulation of Ionospheric Currents and Geomagnetic Field Variations of Sq for Different Solar Activity*, J. Atmos. Terr. Phys. *48*, 277–287.

TARPLEY, J. D. (1970a), *The Ionospheric Wind Dynamo—I. Lunar Tide*, Planet. Space Sci. *18*, 1075–1090.

TARPLEY, J. D. (1970b), *The Ionospheric Wind Dynamo—II. Solar Tides*, Planet. Space Sci. *18*, 1091–1103.

UNTIEDT, J. (1967), *A Model of the Equatorial Electrojet Involving Meridional Currents*, J. Geophys. Res. *72*, 5799–5810.

UNTIEDT, J. (1968), *Der äquatoriale Elektrojet—Stromsystem und Magnetfeld*, Habilitationsschrift, Technische Hochschule Carolo-Wilhelmina zu Braunschweig.

VAN SABBEN, D. (1962), *Ionospheric Current Systems Caused by Nonperiodic Winds*, J. Atmos. Terr. Phys. *24*, 959–974.

VAN SABBEN, D. (1964), *North-south Asymmetry of Sq*, J. Atmos. Terr. Phys. *26*, 1187–1195.

VAN SABBEN D. (1966), *Magnetospheric Currents, Associated with the N-S Asymmetry of Sq*, J. Atmos. Terr. Phys. *28*, 965–981.

VAN SABBEN, D. (1968), *Errata*, J. Atmos. Terr. Phys. *30*, 327–329.

VAN SABBEN, D. (1969), *The Computation of Magnetospheric Currents, Caused by Dynamo Action in the Ionosphere*, J. Atmos. Terr. Phys. *31*, 469–474.

VAN SABBEN D. (1970), *Solstitial Sq-Currents Through the Magnetosphere*, J. Atmos. Terr. Phys. *32*, 1331–1336.

VASYLIUNAS, V. M. (1972), *The interrelationship of magnetospheric processes*, In *Earth's Magnetospheric Processes* (ed. McCormac, B. M.) (D. Reidel Pub. Co. 1972) pp. 29–38.

VOLLAND, H. (1971), *A Simplified Model of the Geomagnetic Sq-Current System and the Electric Fields Within the Ionosphere*, Cosmic Electrodynamics *1*, 428–459.

VOLLAND, H. (1976a), *Coupling Between the Neutral Wind and the Ionospheric Dynamo Current*, J. Geophys. Res. *81*, 1621–1628.

VOLLAND, H. (1976b), *The Atmospheric Dynamo*, J. Atmos. Terr. Phys. *38*, 869–877.

VOLLAND, H., and GRELLMANN, L. (1978), *A Hydromagnetic Dynamo of the Atmosphere*, J. Geophys. Res. *83*, 3699–3708.

WAGNER, C.-U., MÖHLMANN, D., SCHÄFER, K., MISHIN, V. M., and MATVEEV, M. I. (1980), *Large-scale Electric Fields and Currents and Related Geomagnetic Variations in the Quiet Plasmasphere*, Space Sci. Rev. *26*, 391–446.

WALTON, E. K., and BOWHILL, S. A. (1979), *Seasonal Variations in the Low Latitude Dynamo Current System Near Sunspot Maximum*, J. Atmos. Terr. Phys. *41*, 937–949.

WOLF, R. A., MANTJOUKIS, G. A., and SPIRO, R. W. (1986), *Theoretical Comments on the Nature of the Plasmapause*, Adv. Space Res. *6*, 177–186.

(Received February 8, 1988, revised/accepted April 11, 1988)

PAGEOPH, Vol. 131, No. 3 (1989)

0033–4553/89/030437–10$1.50 + 0.20/0

Contribution to Geomagnetic *Sq*-Field and Equatorial Electrojet from the Day/Night Asymmetry of Ionospheric Current under Dawn-to-Dusk Electric Field of Magnetospheric Origin[1]

NAOSHI FUKUSHIMA[2]

Abstract—If the earth and its ionosphere are immersed in a large-scale dawn-to-dusk electric field (of the order of 0.5 mV/m), the resultant dawn-to-dusk ionospheric currents are much stronger on the dayside than on the nightside. These asymmetric currents over the earth produce a magnetic field detectable on the ground, which will contribute to a considerable extent to the *Sq*-field and equatorial electrojet.

Key words: Geomagnetic *Sq*-field, equatorial electrojet, ionospheric current, electric field of magnetospheric origin.

1. Introduction

This paper argues that some portion of the global geomagentic *Sq*-field and equatorial electrojet seems to result from an asymmetric current flow around the earth, due to a strong day/night contrast in the ionospheric conductivity, if the earth with its ionosphere is immersed in a large-scale dawn-to-dusk electric field of magnetospheric origin. Although the horizontal electric currents in the ionosphere (caused mainly by the atmospheric dynamo action) were considered a principal contributor to the geomagnetic *Sq*-field, a part of the ground magnetic field variations originates from other electric currents in the earth's environmental space, such as those discussed earlier by MEAD (1964), OLSON (1970) and MATSUSHITA (1971). This paper deals also with one such current which has not yet been considered, i.e., a current produced under a dawn-to-dusk electric field in the magnetosphere and the ionosphere.

---

[1] This paper was presented at the IAGA General Assembly meeting (Session 9.1) held in Vancouver, Canada, during August 1987.
[2] Geophysics Research Laboratory, University of Tokyo, Tokyo 113, Japan.

2. Assumed External Electric Field and Ionospheric Conductivity

It has been seriously discussed (e.g., NOPPER and CAROVILLANO, 1979; FOSTER, 1987; reviews by STERN, 1977; WAGNER et al., 1980; GREENWALD, 1983; MAUK and ZANETTI, 1987) whether or not a large-scale dawn-to-dusk electric field in the magnetosphere will penetrate down as far as the ionosphere at middle and low latitudes. However, it is simply assumed in this paper that a uniform electric field exists around the earth in the dawn-to-dusk direction with a magnitude of 0.5 mV/m, with an electric potential difference of 6.5 kV between the dawn and dusk equatorial regions. Figure 1 is an electrostatic potential pattern obtained by RICHMOND (1976) for the F-region level on quiet days. The potential difference between the values around the sunrise and sunset at the equator is assumed in this paper to be due mainly to the large-scale dawn-to-dusk electric field of magnetospheric origin.

The height-integrated electric conductivity of the ionospheric E region (the so-called dynamo region) is given by Σ_0 (for the current parallel to the lines of magnetic force), Σ_1 (for the Pedersen current) and Σ_2 (for the Hall current), and these values are assumed to depend on the location above the earth (given by the geomagnetic colatitude θ and longitude ϕ reckoned eastward from the noon meridian) as:

$$\Sigma_{i(i=0,1,2)} = \Sigma_{i0} \sin \theta \cos \phi \quad \text{for} \quad |\phi| < \pi/2, \tag{1}$$

i.e., the dayside conductivity is proportional to the cosine of the solar zenith angle, whereas

$$\Sigma_{i(i=0,1,2)} = 0 \quad \text{for} \quad |\phi| > \pi/2 \tag{2}$$

for simplicity, as schematically illustrated in Figure 2 for an idealized case when the geomagnetic and geographic axes of the earth coincide.

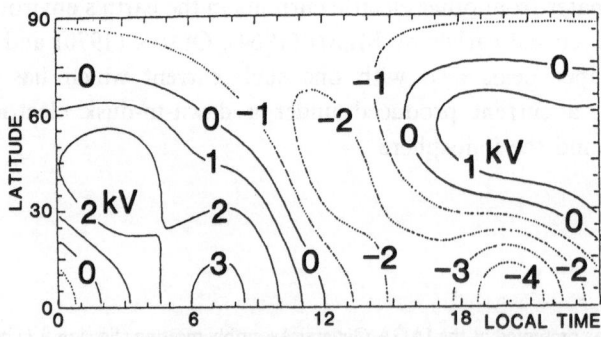

Figure 1
Quiet day F region electrostatic potential pattern illustrated by RICHMOND (1976).

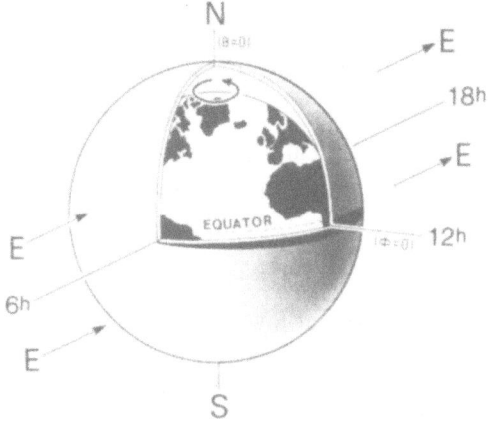

Figure 2

The earth and its ionosphere immersed in a dawn-to-dusk electric field. The electric conductivity of the ionosphere has a great day/night contrast and an enhancement along the magnetic equator.

The neglect of the nighttime conductivity is not at all unnatural, because of the great day/night contrast in the electric conductivity of the E region. Furthermore, it must be remembered that the ground magnetic field originates from the day/night difference in the electric current flow above the earth. Note here that any current of symmetric flow around the dawn-dusk axis of the earth produces no magnetic effect at all inside the spherical ionosphere (FUKUSHIMA, 1976).

3. Ohmic Currents in the Ionosphere and Associated Magnetospheric Currents

It is assumed in this paper that all equipotential planes in the magnetosphere are perpendicular to the dawn-dusk direction. Ohmic currents can flow only in the ionosphere horizontally surrounding the earth. The height-integrated current $i(i_\theta, i_\phi)$ in the ionosphere and the existing electric field $E(E_\theta, E_\phi)$ are connected with the following relations when the electric currents are confined to a thin horizontal layer with the effect of an additional vertical polarization electric field inside the conducting layer.

$$\left.\begin{aligned}
i_\theta &= \Sigma_{\theta\theta} E_\theta + \Sigma_{\theta\phi} E_\phi \\
i_\phi &= \Sigma_{\phi\theta} E_\theta + \Sigma_{\phi\phi} E_\phi
\end{aligned}\right\} \tag{3}$$

with

$$\left.\begin{aligned}
\Sigma_{\theta\theta} &= \Sigma_0 \Sigma_1 / [\Sigma_0 \sin^2 I + \Sigma_1 \cos^2 I] \\
\Sigma_{\theta\phi} &= -\Sigma_{\phi\theta} = \Sigma_0 \Sigma_2 \sin I / [\Sigma_0 \sin^2 I + \Sigma_1 \cos^2 I] \\
\Sigma_{\phi\phi} &= [\Sigma_0 \Sigma_1 \sin^2 I + (\Sigma_1^2 + \Sigma_2^2) \cos^2 I] / [\Sigma_0 \sin^2 I + \Sigma_1 \cos^2 I]
\end{aligned}\right\} \tag{4}$$

where I is the magnetic dip angle given by

$$I = \tan^{-1}(2 \cot \theta). \tag{5}$$

Equations (4) show an effective increase in the ionospheric conductivity at low latitudes due to the effect of a vertical polarization electric field to keep the electric current everywhere horizontal inside the conducting layer. Equations (4) can be reduced to

$$\Sigma_{\theta\theta} \to \Sigma_1, \quad \Sigma_{\theta\phi} = -\Sigma_{\phi\theta} \to \Sigma_2, \quad \Sigma_{\phi\phi} \to \Sigma_1 \tag{4a}$$

at high and middle latitudes (even at low latitudes except along a narrow belt of a few degrees width centering at the magnetic equator), and

$$\Sigma_{\theta\theta} \to \Sigma_0, \quad \Sigma_{\theta\phi} = -\Sigma_{\phi\theta} \to 0, \quad \Sigma_{\phi\phi} \to \Sigma_1 + \Sigma_2^2/\Sigma_1 \tag{4b}$$

at the magnetic equator, where $I = 0$.

The electric field E to be used in Equation (3) consists of the primary electric field E^p and the secondary field E^s, namely

$$E = E^p + E^s. \tag{6}$$

Here E^p at the ionospheric level is a dawn-to-dusk field of magnetospheric origin, which can be written as

$$E_\theta^p = 0 \quad \text{and} \quad E_\phi^p = E_0 \cos \phi, \tag{7}$$

with E_0 assumed to be 0.5 mV/m throughout this paper. The secondary field E^s originates from a heterogeneous distribution of the excess electric charge in the ionosphere, produced in association with the space gradients of electric field and conductivity.

It is also assumed that an electric current continuity is preserved as schematically illustrated in Figure 3 for the dawn and dusk terminals of the spherical ionosphere. Between the two equipotential planes A and B, electric currents in the magnetosphere always flow towards the ionosphere on the morning side and away from the ionosphere on the afternoon side.

The magnetic field produced inside the ionosphere by the converging currents on plane A is the same as that produced by an incoming line-current perpendicular to plane A, while the magnetic field from the diverging currents on plane B can be replaced by that from an outgoing line-current. The resultant magnetic field inside the ionosphere produced by the electric currents (thick lines) in Figure 3 is exactly the same as that produced by the two equivalent line-currents of semi-infinite length (white arrows) and the actual horizontal currents in the ionosphere, with the aid of a theorem referred to by FUKUSHIMA (1969, 1971). If the ionospheric conductivity were uniform, no magnetic field would be produced inside the spherical ionosphere (FUKUSHIMA, 1976). Although the accurate calculations have not been carried out yet for the ground magnetic effect of various model currents in the magnetosphere, it has been generally accepted that the ground magnetic effect comes mostly from the currents flowing horizontally in the ionosphere.

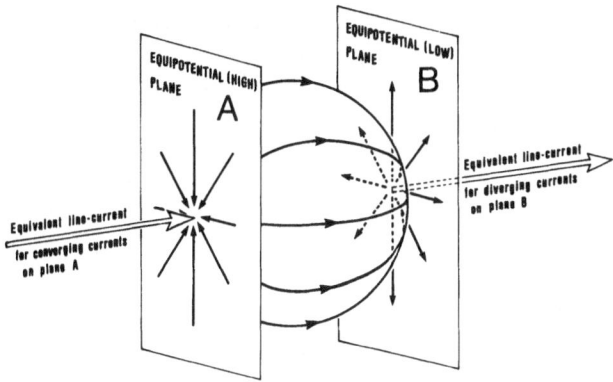

Figure 3

Schematic illustration for the ionospheric and associated magnetospheric currents. The magnetic effect inside the ionosphere from diverging or converging currents in plane A or B is the same as that from the infinite line-currents shown by white arrows.

4. Equivalent Current-system for Primary Ionospheric Pedersen Current

According to assumptions (1) and (2) concerning the ionospheric conductivity, horizontal electric currents flow only in the sunlit ionosphere. Such an asymmetric current flow around the earth produces a magnetic field detectable on the ground. Figure 4 illustrates the resultant ground magnetic effect by two equivalent overhead current-systems. The actual ionospheric current, whose noon-midnight meridian cross-section is shown by (A) in Figure 4, produces a northward magnetic field everywhere on the earth; its magnitude is greatest at the subsolar point and smallest at the midnight equator. The zonal current-system (B) is responsible for the northward magnetic field averaged for all longitudes, while the longitudinal inequality of the magnetic field is given by current-system (C). The current vortices on the dayside ionosphere in current-system (C) are very similar in their form to the *Sq* current vortices.

A simple quantitative check shows that the total amount of Pedersen currents on the dayside is 65 kA, if the average height-integrated Pedersen conductivity in the dayside ionosphere is 10 mho. Hence the total intensity of zonal currents in (B) will be about 30 kA, which produce a northward magnetic field of about 3 nT all over the earth. The total amount of eastward currents across the noon meridian between the northern and southern foci of vortex currents in (C) will also be approximately 30 kA, which will result in a northward magnetic field of about 5 nT at low latitudes.

Though current-system (C) in Figure 4 is several times weaker than the actual *Sq* current-vortex, there is a possibility for an intensification of the ionospheric currents by the Hall currents under the secondary electric field, as discussed in the next section.

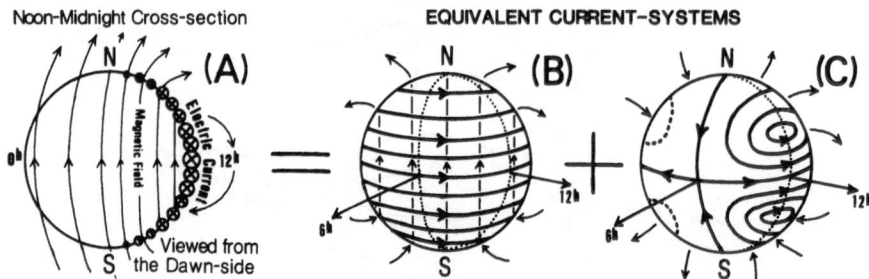

Figure 4

The magnetic effect from the actual current flow in the ionosphere surrounding the earth, whose noon-midnight cross-section is given by (A), is the same as the superimposition of the effects from two equivalent current-systems (B) and (C), respectively for zonal and remaining components.

5. Secondary Electric Field and Hall Currents

In the two-dimensional dynamo theory for the geomagnetic Sq-field, the dayside vortex-currents are mostly the secondary Hall currents produced by the secondary electric field originating from the negative charge accumulation at middle latitudes on the dayside. It must be carefully examined, whether or not the charge excess will also take place in the ionosphere so as to produce the secondary electric field, even when electric currents can flow from the ionosphere to the magnetosphere or *vice versa* easily along the magnetic field-lines. If the current production is symmetric with respect to the equator, the field-aligned current from one hemisphere will be cancelled out by the same kind of current from the opposite hemisphere, except for the polar-cap regions where the magnetic field-lines extend to the magnetotail and not to the opposite hemisphere. Hence we may say that the two-dimensional current pattern will be a good approximation for the electric current flow in the earth's environmental space for equinoctial seasons.

In the present model, electric currents of 65 kA are fed from the magnetosphere to the ionosphere, and this amount is a few times smaller in comparison with the vortex-currents for the observed Sq-field. The average density of incoming and outgoing current at the ionospheric level is $65 \text{ kA}/25 \cdot 10^{13} \text{ m}^2 = 2.5 \cdot 10^{-10} \text{ A/m}^2$, and this current density is much smaller than that of field-aligned currents in the polar region. Although we must carefully check whether we may or may not apply the two-dimensional confinement of electric currents flowing in the ionosphere, the divergence-free approximation seems to hold rather well for horizontal currents in the ionosphere.

A rough estimation is made below for the secondary electric field at the noon meridian ($\phi = 0$ in Figure 2), where there is no vertical current between the ionosphere and magnetosphere, and the horizontal divergence-free condition is satisfied for the ionospheric currents, i.e.,

$$\text{div } \boldsymbol{i} = \text{div}(\boldsymbol{i}^p + \boldsymbol{i}^s) = 0, \tag{8}$$

where i^p and i^s denote the horizontal currents in the ionosphere driven by the primary and secondary electric fields, respectively. The primary electric field E^p is given in (7), whereas the secondary electric field E^s is given from an electrostatic potential S, which satisfies

$$\left.\begin{aligned} E_\theta^s &= -\partial S/r_0\partial\theta, \\ E_\phi^s &= -\partial S/r_0 \sin\theta\partial\phi, \end{aligned}\right\} \tag{9}$$

where r_0 denotes the radius of the ionosphere. An elaborate numerical computation is generally required to obtain a world map of the electrostatic potential $S(\theta, \phi)$ using a realistic model of the ionospheric conductivity dependent on latitude and local time, but a simple analytical calculation is shown below which enables us to obtain an approximate latitude profile of $S(\theta, 0)$ along the noon meridian, using the $\partial/\partial\phi = 0$ condition for S and the ionospheric conductivity, along with a simple replacement of $\Sigma_{\theta\theta}$ and $\Sigma_{\phi\phi}$ by Σ_1, and $\Sigma_{\theta\phi}(=-\Sigma_{\phi\theta})$ by Σ_2.

A calculation by means of Equations (1), (3), (6), (7) and (9) shows that Equation (8) can be satisfied at least along the noon meridian, if S is given as

$$\left.\begin{aligned} S(\theta, \phi) &= \pm 2r_0 E_0(\Sigma_{20}/\Sigma_{10}) \sum_n a_n \sin^n\theta \cos\theta \cos\phi \\ &\text{according to whether } \theta \lessgtr \pi/2, \\ a_{n+2}/a_n &= (n+1)(n+3)/\{(n+2)(n+3)-1\} \text{ for } n = 1, 3, 5, 7, \ldots, \\ &\text{with } a_1 = 1, \quad\text{and}\quad a_2 = a_4 = a_6 = \cdots = 0. \end{aligned}\right\} \tag{10}$$

From this calculation, we see that the secondary electric field (poleward at low latitudes) intensity is as strong as E_0 of the primary field. Then the ground magnetic effect due to vortex current (C) of Figure 4 will be intensified approximately $(\Sigma_2/\Sigma_1)^2$-times by the secondary Hall current in the ionosphere, which is shown in

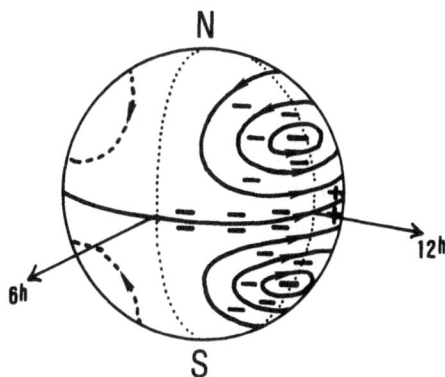

Figure 5

Schematic sketch for the secondary charge distribution and Hall currents in the dayside ionosphere, which intensify the ground magnetic effect of vortex-current (C) of Figure 4.

Figure 5 schematically (without an accurate computation) along with the excess charge pattern (including that associated with the electrojet, see the next section) in the dayside ionosphere.

6. Equatorial Electrojet

At the magnetic equator, $\Sigma_{\theta\theta}$ tends to Σ_0, $\Sigma_{\phi\phi}$ to Σ_3, but $\Sigma_{\theta\phi}$ and $\Sigma_{\phi\theta}$ vanish. We assume here that an additional electrostatic potential S_E exists along a narrow belt of the magnetic equator. The divergence-free condition of the ionosphere currents (8), which is approximately satisfied, holds if the additional electrostatic potential $S_E(\theta, \phi)$ along the equator has the following form:

$$S_E(\theta, \phi) = [r_0 E_0/(\alpha^2 k + 1)] \cdot \exp[-\alpha^2(\theta - \pi/2)^2]\sin\phi \qquad (11)$$

with $k = \Sigma_0/\Sigma_3$; α is a positive number much greater than 1, r_0 denotes the radius of the ionosphere, E_0 is assumed to be 0.5 mV/m. This electrostatic potential produces a weak dusk-dawn electric field along the equator, and the resultant eastward electrojet will have the density of

$$i_\phi(\text{at } \theta = 90°) = E_0 \Sigma_{30}[1 - 1/(\alpha^2 k + 1)]\cos^2\phi. \qquad (12)$$

Putting Σ_{30} (the subsolar-point Σ_3 value) = 600 mho (referring to Figure 5 of FEJER, 1964), the maximum current density of the eastward equatorial electrojet around noon is calculated to be 300 mA/m. This current produces a northward magnetic field of approximately 150–200 nT on the ground, which is of the same magnitude as the actual electrojet.

7. Concluding Remarks

This paper points out that the geomagnetic Sq-field and equatorial electrojet will contain a considerable contribution from the asymmetric current flow in the ionosphere surrounding the earth, due to a strong day/night contrast in the electric conductivity of the lower ionosphere, if a dawn-to-dusk electric field (of the order of 0.5 mV/m) permeates down to the ionosphere from the magnetosphere even at middle and low latitudes. According to a preliminary semi-quantitative estimation, the contribution to Sq seems to be slightly less than the observed magnitude, whereas the contribution to the equatorial electrojet will be very significant. However, the author does not intend to ignore the importance of the dynamo action in the lower ionosphere, but he wishes to attract attention to geomagnetists to a possible effect from the electric field of magnetospheric origin. We do need an eastward electric field at low latitudes to explain the geomagnetic Sq field and equatorial electrojet, whatever its origin is. The traditional dynamo theory attributes the electric field to a dynamo action in the lower ionosphere, whereas it is assumed in this paper that the earth and its ionosphere are immersed in a large-scale dawn-to-dusk electric field in the magnetosphere.

We have no convincing experimental evidence yet for the global dawn-to-dusk electric field around the earth (except in the polar-cap regions), but future experiments will clarify a large-scale electric field in the earth's environmental space. According to the present model current-system, the Pedersen current at high latitudes (on the poleward side of the *Sq* current vortex) is eastward on the dayside, but its equivalent overhead current is westward. The detection of the real ionospheric current at the dayside high latitudes is desirable to evaluate the new idea in this paper.

Acknowledgements

The author is very grateful to Dr. W. H. Campbell for his kind permission to present this work as a late paper in the session he convened (Quiet-time external current systems including *Sq*, *L*, electrojet and magnetospheric origin) for the IAGA General Assembly held in Vancouver, Canada, in August 1987. The writer wishes to extend his thanks to Mr. D. Hesse for explaining the valuable data of his earlier observation of the equatorial electrojet in Brazil; the idea for the present work originated during the discussion with him. The author thanks Prof. B. E. Brunelli for his kind comment regarding the need for a careful check on the validity of divergence-free approximation for the ionospheric currents.

Note Added in Proof

§7 should have contained the following remark. The validity of the new idea described in this paper can be simply checked by a direct experimental detection of the net dawn-to-dusk electric current in the ionosphere. We could know the existence of a net sunward or antisunward space current below the satellite altitude, by applying Ampère's theorem to the analysis of MAGSAT (in a dawn-dusk orbit from November 1979 to June 1980) data, as reported by SUZUKI and FUKUSHIMA (J. Geomag Geoelectr., *36*, 493–506, 1984) and SUZUKI *et al* (J. Geophys. Res., *90*, 2465–2471; 1985). If a low-altitude satellite with three-axis magnetometers is launched in the future in the noon-midnight meridian, the calculated dawn-to-dusk net current in the ionosphere will verify or deny the new idea of this paper.

REFERENCES

FEJER, J. A. (1964), *Atmospheric Tides and Associated Magnetic Effects*, Rev. Geophys. *2*, 275–309.
FOSTER, J. C., *Radar-deduced Models of the Convection Electric Field*, Extended abstract presented to the International Symposium on Quantitative Modeling of Magnetosphere-Ionosphere Coupling Processes, at Kyoto Sangyo University, March 9–13, 1987.

FUKUSHIMA, N. (1969), *Equivalence in Ground Geomagnetic Effect of Chapman-Vestine's and Birkeland-Alfvén's Electric Current-systems for Polar Magnetic Storms*, Rept. Ionos. Space Res. Japan *23*, 219–227.

FUKUSHIMA, N. (1971), *Electric Current-systems for Polar Substorms and their Magnetic Effect Below and Above the Ionosphere*, Radio Science *6*, 269–275.

FUKUSHIMA, N. (1976), *Generalized Theorem for no Ground Magnetic Effect of Vertical Currents Connected with Pedersen Currents in the Uniform-conductivity Ionosphere*, Rept. Ionos. Space Res. Japan *30*, 35–40.

GREENWALD, R. A. (1983), *Electric Fields in the Ionosphere and Magnetosphere*, Space Sci. Rev. *34*, 305–315.

MATSUSHITA, S. (1971), *Interactions Between the Ionosphere and the Magnetosphere for Sq and L Variations*, Radio Science *6*, 279–294.

MAUK, B. H., and ZANETTI, L. J. (1987), *Magnetospheric Electric Fields and Currents*, Rev. Geophys. *25*, 541–554.

MEAD, D. G. (1964), *Deformation of the Geomagnetic Field by the Solar Wind*, J. Geophys. Res. *69*, 1181–1195.

NOPPER, R. W. Jr., and CAROVILLANO, R. L., *Ionospheric electric fields driven by field-aligned currents*, In *Quantitative Modeling of Magnetospheric Processes* (ed. Olson, W. P.) (Geophysical Monograph 21, Amer. Geophys. Union, Washington, D.C. 1979) pp. 557–568.

OLSON, W. P. (1970), *Contributions of Non-ionospheric Currents to the Quiet Day Magnetic Variations at the Earth's Surface*, Planet. Space Sci. *18*, 1471–1484.

RICHMOND, A. D. (1976), *Electric Field in the Ionosphere and Plasmasphere on Quiet Days*, J. Geophys. Res. *81*, 1447–1450.

STERN, D. P. (1977), *Large-scale Electric Fields in the Earth's Magnetosphere*, Rev. Geophys. Space Phys. *15*, 156–194.

WAGNER, C.-U., MÖHLMANN, D., SCHÄFER, K., MISHIN, V. M., and MATVEEV, M. I. (1980), *Large-scale Electric Fields and Currents and Related Geomagnetic Variations in the Quiet Plasmasphere*, Space Sci. Rev. *26*, 391–446.

(Received/accepted May 9, 1988)

PAGEOPH, Vol. 131, No. 3 (1989)

0033–4553/89/030447–16$1.50 + 0.20/0

The Contribution of Magnetospheric Currents to *Sq*

W. P. OLSON[1]

Abstract—Since the discovery of the magnetosphere, it has been known that the currents flowing in the magnetosphere contribute to *Sq*, the regular daily variation in the earth's surface magnetic field. The early models, however, were not very accurate in the vicinity of the earth. The magnetospheric contribution to *Sq* has therefore been recalculated by direct integration over the three major magnetospheric current systems; magnetopause, tail and ring. The finite electrical conductivity of the earth, which increases the horizontal and decreases the vertical components of the magnetospheric field at the earth's surface, has been taken into account. The magnetospheric currents are found to contribute 12 nanotesla to the day to night difference in the mid-latitude *Sq* pattern for steady, quiet magnetospheric conditions. They also contribute to the annual variation in the surface field and must be considered an important source of the observed day to day variation in the *Sq* pattern.

Key words: Magnetospheric currents, *Sq*.

1. Introduction

It has been known for several hundred years that the earth possesses a planetary magnetic field (GILBERT, 1600). It has also been known for over 200 years (CANTON, 1759) that there is a daily variation in the strength of the magnetic field at the earth's surface. The earth's surface magnetic field also exhibits changes associated with solar activity (ADAMS, 1892). Even when the sun is "quiet," however, there is a day to night variation in the earth's surface magnetic field. This daily variation at mid-latitudes is designated by "*Sq*," the Solar Quiet daily variation (CHAPMAN, 1919). Typically, at mid-latitude, the magnitude of this daily variation ranges from 15 to 30 nanotesla (VESTINE *et al.*, 1947). This compares with a total surface magnetic field of about 50,000 nanotesla.

Near the turn of the century, the existence of the ionosphere was inferred from the presence of these magnetic variations by STEWART (1882). He suggested that the *Sq* pattern was caused by ionization in the upper regions of the atmosphere being moved by neutral particles through the earth's magnetic field. This led to the

[1] McDonnell Douglas Astronautics Company, 5301 Bolsa Avenue, Huntington Beach, California 92647, U.S.A.

ionospheric dynamo theory which explained how the diurnal solar heating of the upper atmosphere could produce a global system of electric current responsible for the *Sq* pattern. Later, when the existence of the ionosphere was proven by experiments which reflected radio waves off of it (BREIT and TUVE, 1925), the ionospheric dynamo theory became the accepted explantation for *Sq*.

With the observational discovery of the magnetosphere in the early 1960's, however, it was recognized that other currents flowing above the earth must also contribute to the surface magnetic field variations. In early work it was suggested that the magnetospheric currents produce a magnetic field structure at the earth's surface that is similar to the observed *Sq* pattern but considerably smaller in magnitude (MEAD, 1964; OLSON, 1970a,b).

Since that time, much has been learned concerning the nature of the magneto-spheric current systems. The purpose of this paper is to quantitatively reexamine the contribution of the magnetospheric currents to *Sq*. An accurate determination of the contribution of these currents to the *Sq* pattern can serve to place an upper bound on the neutral winds in the upper atmosphere required by the ionospheric dynamo theory. Such a determination is also necessary for the quantitative specifi-cation of the earth's main magnetic field. An accurate representation of the main field (to within several nanoteslas) is required for the study of the secular variation in the main field, and for the quantitative specification of magnetic anomalies in the earth's crust. The accurate representation of the magnetospheric contribution to the earth's surface magnetic field may also be of use in the study of magnetospheric dynamics since indices constructed from ground based magnetometer data are routinely used to describe magnetospheric events.

2. The Major Magnetospheric Current Systems

The magnetopause, tail and ring current systems are the three largest magneto-spheric current systems. They have been described in detail elsewhere. See, for example, the review by WALKER (1979). A brief description of each current system follows.

2.1 The Magnetopause Current System

The largest magnetospheric current system, and the first to be quantitatively modeled, flows on the outer boundary of the magnetosphere, the magnetopause. The magnetopause currents are produced by the interaction of the solar wind with the earth's magnetic field. The electrons and protons that comprise most of the solar wind are deflected from the geomagnetic field in opposite directions, giving rise to a large current system that flows around two foci. The magnetopause currents have the proper shape and sense to produce an *Sq*-like pattern at the earth's surface as was first shown by MEAD (1964).

2.2 The Tail Current System

Examination of early satellite magnetometer data indicated that the magnetosphere possessed an extended "tail" region directed away from the sun (WILLIAMS and MEAD, 1965). The most prominent feature of the magnetospheric tail is its magnetic lobe structure. Each lobe is characterized by its magnetic field, directed away from the earth in the magnetic Southern Hemisphere and toward the earth in the Northern Hemisphere. The strength and direction of the lobe fields could not be accounted for by the magnetopause currents but required an additional current system that flowed across the center of the tail, carried by the plasma sheet particles that are present across the center of the tail. (The plasma sheet region separates the two lobes of the tail.) The tail currents were first modeled by WILLIAMS and MEAD (1965) simply as a sheet current that flowed from dawn to dusk across the center of the tail of the magnetosphere. Later, more realistic models of the cross tail currents were developed which contained a return current system flowing at the magnetopause (OLSON and CUMMINGS, 1970). The tail current system also produces an *Sq*-like magnetic field pattern at the earth's surface. This is true because although this current system produces a magnetic field opposite in direction to that of the magnetopause currents, it is also located in the opposite direction ("behind" instead of "in front of" the earth).

Note that both the magnetopause and tail current systems are constantly present and are thus expected to continuously contribute to the variations in the surface magnetic field as the earth rotates under them.

2.3 The Ring Current System

Long before the discovery of the magnetosphere, it was understood that a ring of current at great altitudes must be formed during magnetic storms (STORMER, 1907; BIRKELAND, 1913). This ring current was responsible for the depression of the mid-latitude surface magnetic field during the main phase of the magnetic storm. It was understood that the ring current was a transient phenomenon and therefore not considered a source of the *Sq* pattern. Well after the discovery of the magnetosphere, however, careful analysis of satellite magnetometer data showed that a "quiet time ring current" appears to always be present in the magnetosphere (SUGIURA *et al.*, 1971). If the quiet time ring current is symmetric and geocentric, it will not produce a diurnal variation in the earth's surface magnetic field but rather only a constant contribution. However, Sugiura's data suggest that the quiet time ring current is not symmetric (although his analysis was restricted to geocentric distances greater than 3.5 earth radii $\{R_E\}$). Also, SISCOE (1966) argued that the ring must be positioned such that the forces acting on it from the earth's main field and the other magnetospheric current systems cancel. This causes the center of the ring to be displaced toward the sun which is in the direction required for the ring current to produce an *Sq*-like variation pattern.

3. The Magnetic Fields of the Magnetospheric Current Systems

Models of the magnetic fields produced by these magnetospheric current systems have been developed by two methods:
1) empirically, form satellite magnetometer data sets, and
2) from quantitative representations of the various magnetospheric current systems.

3.1 Empirical Models

In models developed using the first method, the data sets have typically been collected at geocentric distances 3.5 R_E and beyond. There have been few "near earth" data points used to fit the coefficients of the series expansion representation of the observed magnetospheric magnetic field. This method is also limited by the fact that only the total magnetic field can be measured at a given time and location. The total measured field contains the earth's main magnetic field which is not known with complete accuracy (in part because all measurements of the main field are "contaminated" by the contributions of other magnetic fields produced by currents that flow in the magnetosphere and ionosphere).

3.2 Models Based on Representations of Magnetospheric Current Systems

In the second method, the vector magnetic field from a quantitative representation of one of the magnetospheric currents is determined by using the Biot-Savart law to integrate over the current system at a set of points located throughout the magnetosphere. A least squares best fit is then performed on the field at these points to provide a functional representation of the magnetic field. The resulting "global" analytical model of the magnetic field is valid over the (large) region of the magnetosphere that contains these points. Typically, the spacing of this grid of points is at even intervals of 1–2 R_E. Thus, in this method of model development, the field in the vicinity of the earth is typically represented by only one point out of several thousand. It was therefore expected that these early global models would not accurately represent the contributions of the magnetospheric current systems to the earth's surface magnetic field variations. We have therefore directly integrated over model representations of the magnetopause, ring and tail currents at a set of points on the earth's surface. This was done individually for the magnetopause, ring and tail current systems.

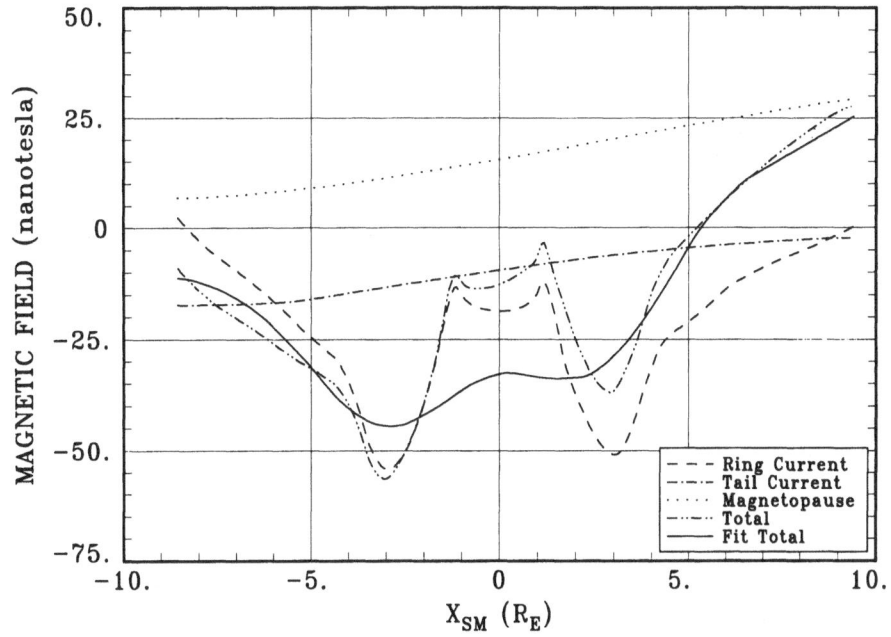

Figure 1

The field from the magnetopause, ring, and tail currents along the sun-earth line. Solar magnetospheric, SM, coordinates are used. (X_{SM} is in the direction of the sun, Z_{SM} is northward, perpendicular to the ecliptic plane, and Y_{SM} is in the dusk direction forming a right handed coordinate system.) All currents except the "fit total" are obtained by direct integration over the magnetopause, ring and tail current systems. Note that at the earth (noon-midnight) there is a significant difference between the total fields obtained by direct integration and from a magnetospheric model magnetic field.

The results of this direct integration are compared with those obtained from the early global models along the earth-sun line in Figure 1. All of the curves except the "fit total" were obtained by direct integration over the currents. The "fit total" curve is based on a global analytical model of the sum of the three major magnetospheric current systems. Note that, as expected, near the earth there is a large difference between the field obtained by direct integration (total) and that of the general magnetospheric model. Also, as expected, this difference is small throughout most of the magnetosphere. When the magnetospheric magnetic field *models* were used, the total contribution from the magnetospheric currents at the earth's surface produced a day to night variation of about 4 nanotesla (OLSON, 1970b). However, when a *direct integration* is performed over the current systems, this variation is actually about 7 nanotesla. The north, east, and vertical components of the ring, magnetopause, and tail currents and their sums are shown in Figures 2 through 5. In each figure, the daily variation in the earth's surface magnetic field is shown from −60° to 60° magnetic latitude in 15° intervals.

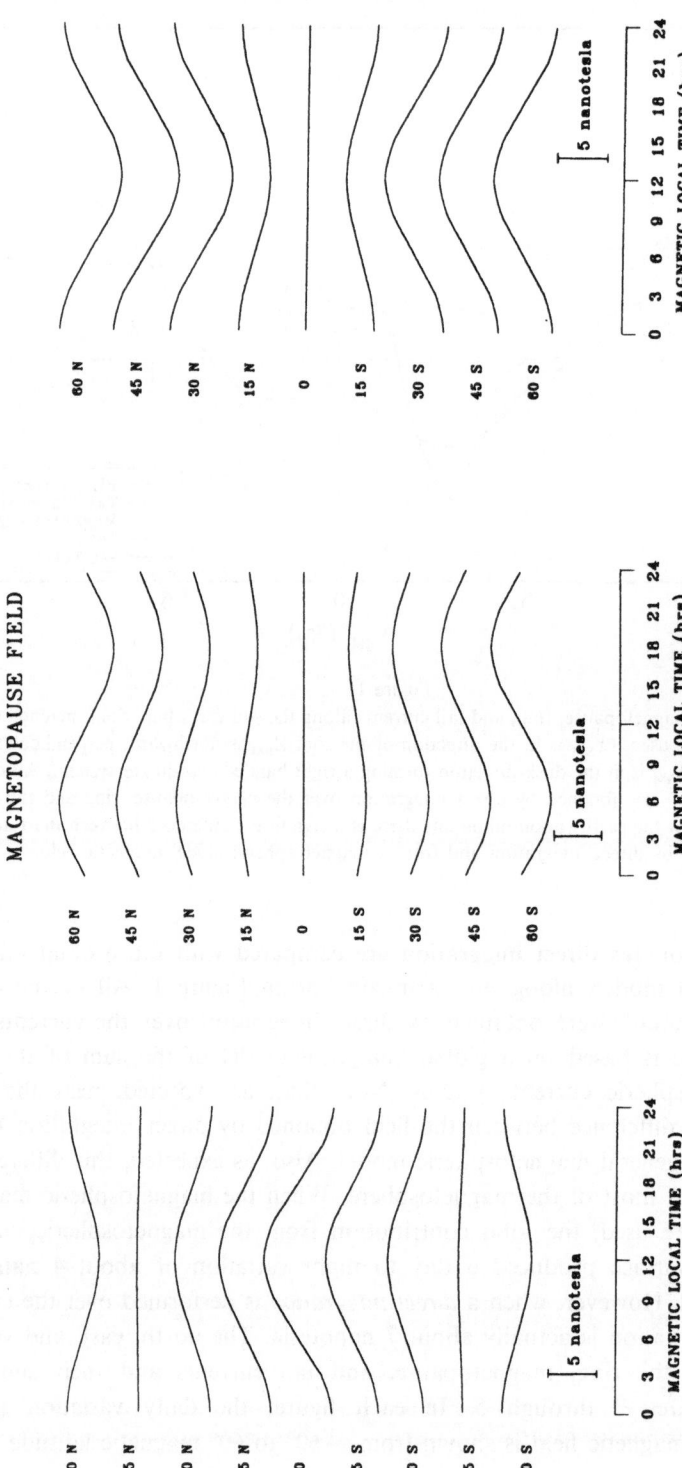

Figure 2

Magnetic field at the earth's surface from the magnetopause currents (by direct integration over the currents). The north, east, and vertical components of the field are shown.

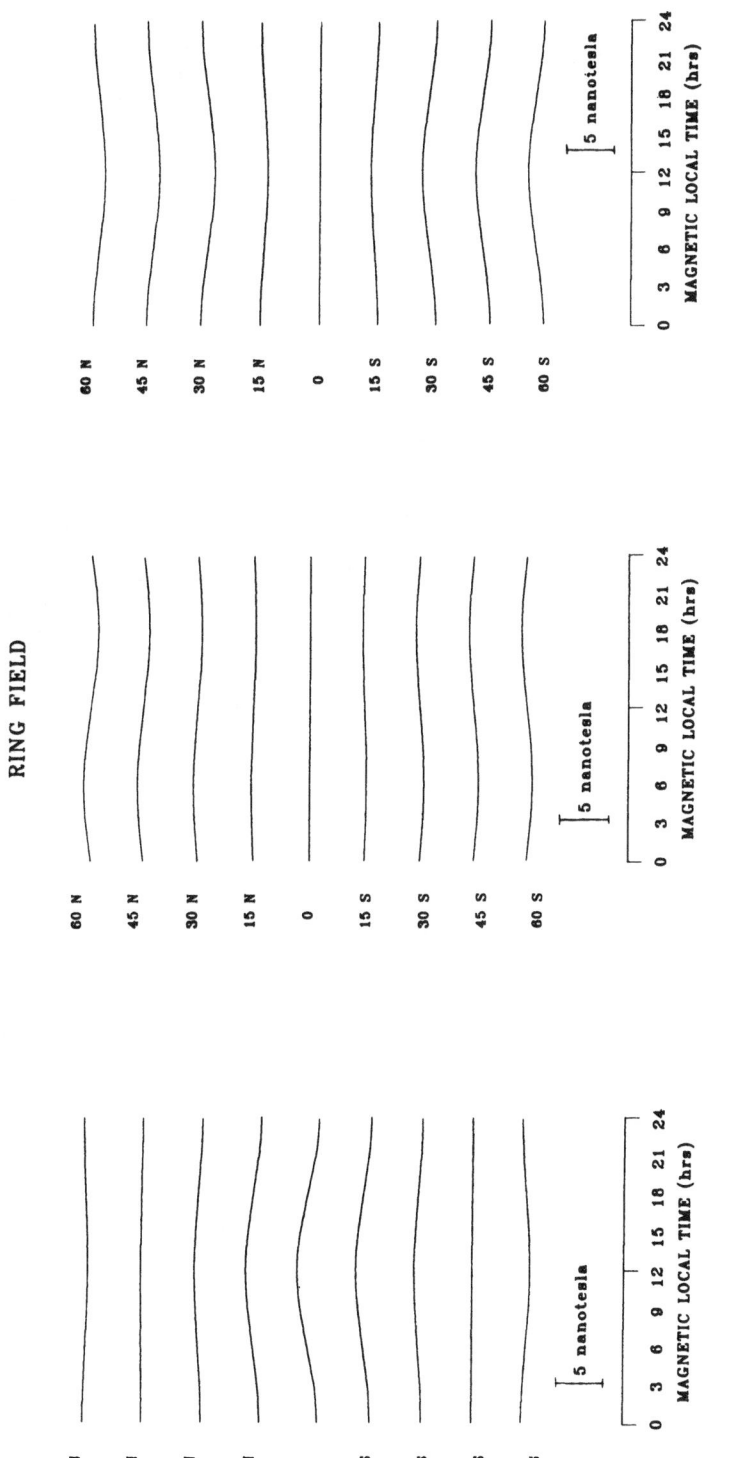

Figure 3

Magnetic field at the earth's surface from the ring currents (by direct integration over the currents).

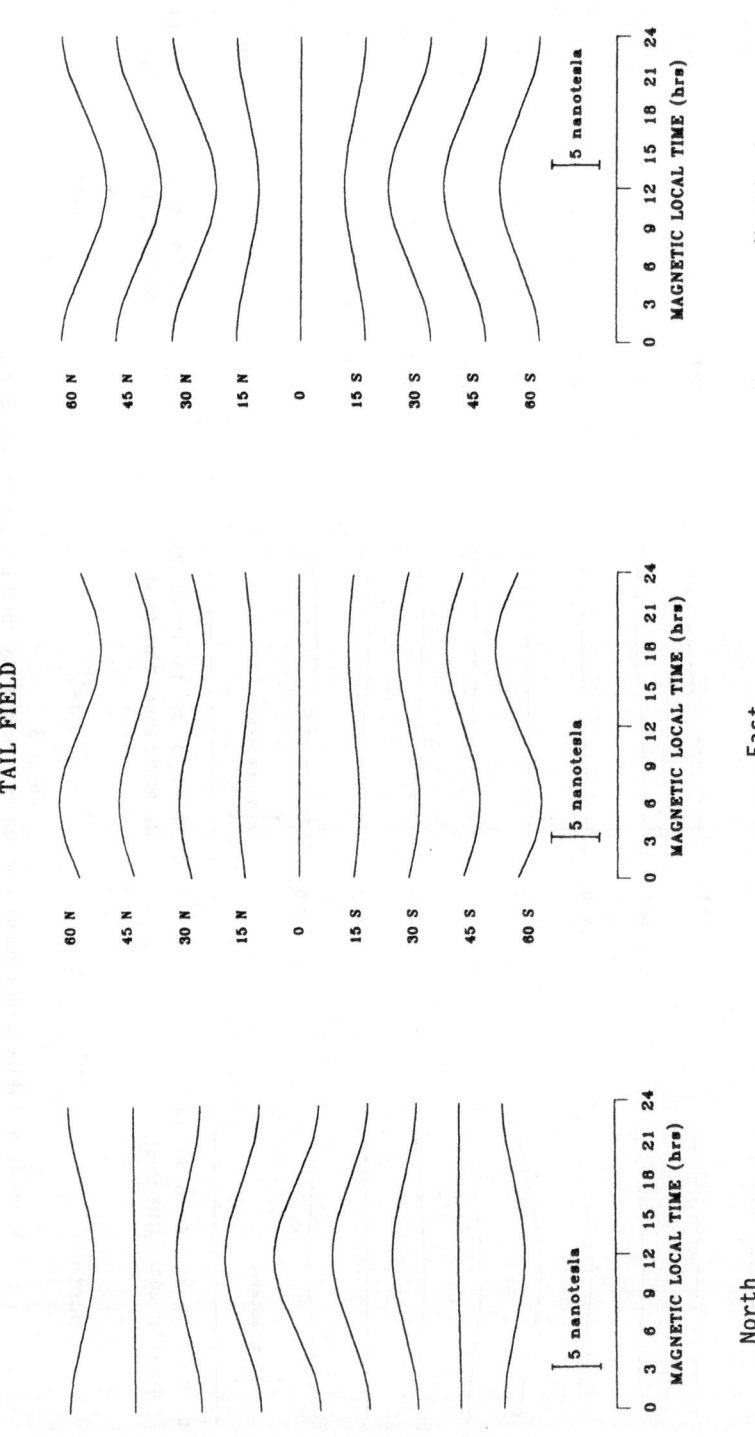

Figure 4

Magnetic field at the earth's surface from the tail currents (by direct integration over the currents).

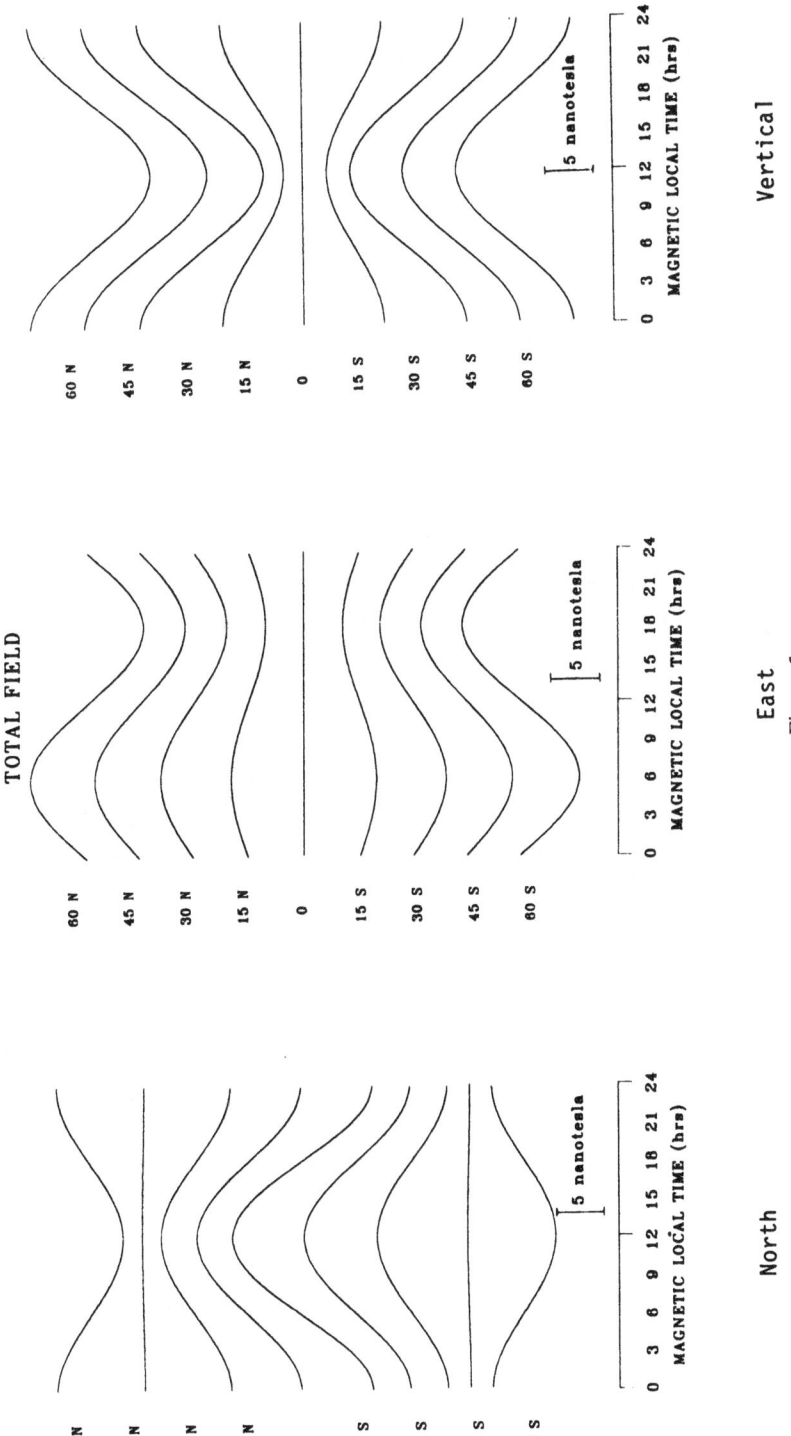

Figure 5

Combined field at earth's surface from the major magnetospheric current systems (magnetopause, ring and tail).

3.3 Model Accuracy

The accuracy of the magnetospheric contributions to the earth's surface magnetic field, as determined from such direct integration over models of the magnetospheric currents, can be assessed by comparing the agreement between these models of the magnetospheric magnetic field (based on the same currents) and the total magnetic field observed in the magnetosphere (taking into account the contribution from the earth's main magnetic field). "Event models" have been constructed for the magnetospheric events studied in NASA's CDAWs (Coordinated Data Analysis Workshops). In them, the magnetopause, ring and tail currents are allowed to vary in strength and size in response to the interplanetary forces that drive them. During magnetically quiet times, there is excellent agreement between the observed magnetic field and the models of the magnetic fields associated with these currents (OLSON and PFITZER, 1982). Even during magnetically disturbed periods, there is good agreement between model and observation on the dayside magnetosphere. Thus we are confident that the magnetic field at the earth's surface, determined by direct integration over models of the magnetospheric currents, provides an accurate description of the contribution of the magnetospheric magnetic fields to the earth's surface magnetic field variations during quiet times.

4. The Influence of the Earth's Electrical Conductivity

The most important remaining problem with the accurate determination of the contribution of the magnetospheric currents to the earth's surface magnetic field is the influence of the earth's finite electrical conductivity. To this point it has been implicitly assumed that the earth has zero electric conductivity and permeability, i.e., that it has the same electrical properties as free space. However, it has been known through this century that the earth's finite electrical properties cause the magnetic field from external sources to be significantly modified above and at the earth's surface (e.g., see SCHUSTER, 1889). The conducting properties of the earth act to prohibit the penetration of time varying magnetic and electric fields by the induction of electric currents in the earth's oceans and crust. These currents cancel out the external source magnetic field in the earth's interior but produce a field that adds to the external magnetic field at and above the earth's surface. More specifically, they will add to the horizontal and tend to cancel the vertical component of the source field just above the surface of the electrically conducting body. The electrical conductivity of the ionosphere, however, is such that the ionosphere cannot appreciably shield the earth's surface from the magnetic fields produced in the magnetosphere (OLSON, 1970a).

It is therefore necessary to quantitatively study the electromagnetic currents induced in the earth as it rotates under these magnetospheric current systems. Using the magnetospheric magnetic fields shown in Figure 5 as the forcing function, the daily variation in the *Sq* pattern was found for various values of conductivity for the earth. The value .0065 mho is most commonly used to represent the spatial average conductivity of the earth's crust. Since the earth's conductivity increases with depth, this value may somewhat underestimate the ability of the earth to induce currents and shield its interior from the magnetospheric magnetic field.

The total contribution to the *Sq* pattern produced by the three magnetospheric current systems and their associated induced currents is shown in Figure 6 for the north, east, and vertical components of the surface field at −60° to 60° magnetic latitude in 15° intervals (with the earth's conductivity set at .0065 mho). The amplitude of the day to night magnetospheric contribution to *Sq* at mid-latitudes is seen to be increased from 7 to 12 nanoteslas when the earth's electrical conductivity is taken into account (as represented by the dashed lines). The "free space earth" values are represented by the solid lines.

5. Discussion

5.1 Remaining Inaccuracies in the Determination of the Magnetospheric Contribution to Sq

There are two primary sources for any remaining inaccuracies in the determination of the magnetospheric contribution to *Sq*. First, the present work has not taken the Birkeland currents into account. These field-aligned currents (BIRKELAND, 1911) are also expected to contribute to the daily variation in the earth's surface magnetic field at high latitudes (XU, this issue). Quantitative models of these currents are not available since their extension into the magnetosphere and their closure through the magnetosphere are not at present well understood. However, since the classical *Sq* pattern is typically described at mid-latitudes, the magnetic field associated with the Birkeland currents should have only a minor effect since they intersect the ionosphere at auroral latitudes and above. The other possible inaccuracy stems from the ring current and the extent to which it is asymmetric with respect to the earth's center. Changes in the position of the ring current can produce large effects on the variation pattern of its magnetic field at the earth's surface while having almost no effect at geosynchronous orbit where much of the data on the quiet time ring current has been gathered. We believe, however, that this inaccuracy has also been minimized since our model of the ring current agrees well with the calculations of SISCOE (1966) and with the data of SUGIURA et al. (1971) on the location of the quiet time ring current.

5.2 The Effect of the Earth's Conductivity on the Time of Maximum Magnetospheric Contribution to Sq

The daily variation in the north component of the magnetic field at the equator is shown in relative units in Figure 7 for several values of conductivity. As the earth's

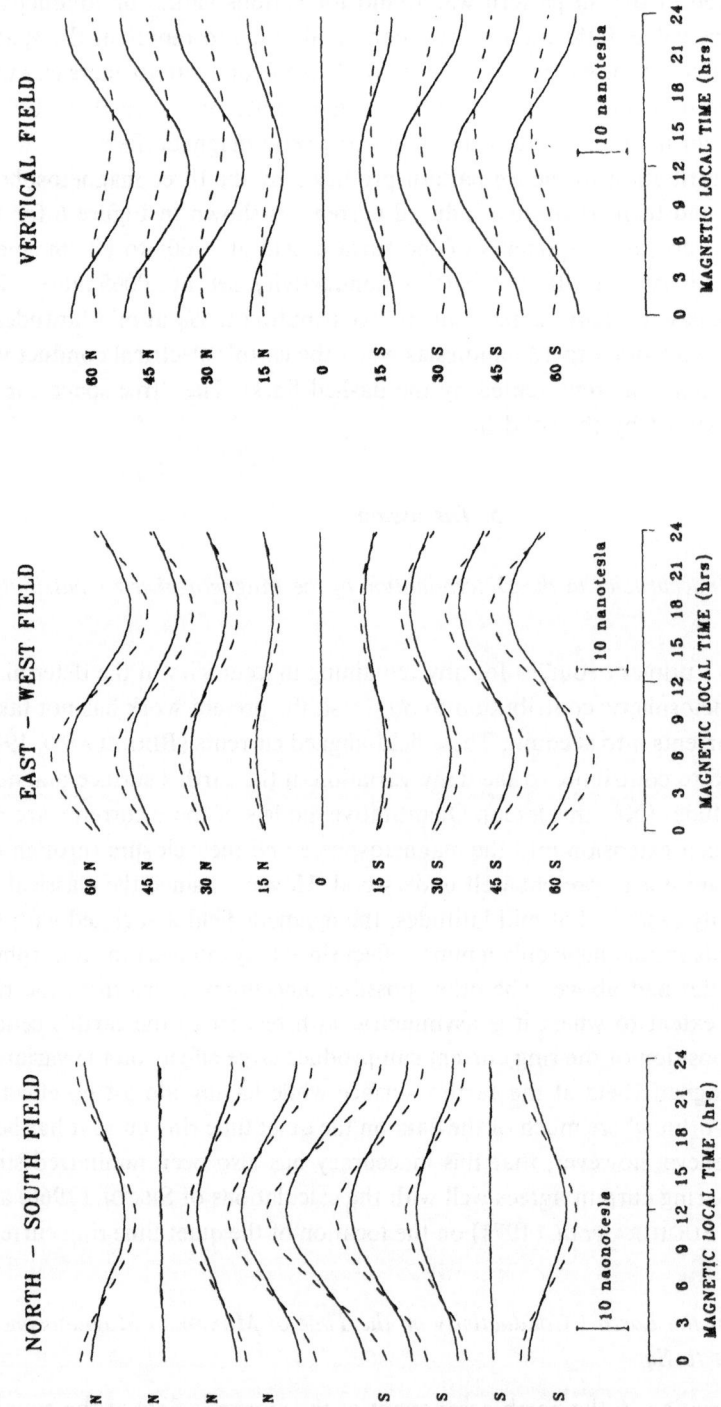

Figure 6

Total surface field produced by magnetospheric currents (magnetopause, ring and tail) including contribution from currents flowing in the electrically conducting earth (shown as dashed lines). Field for free space earth shown as solid lines.

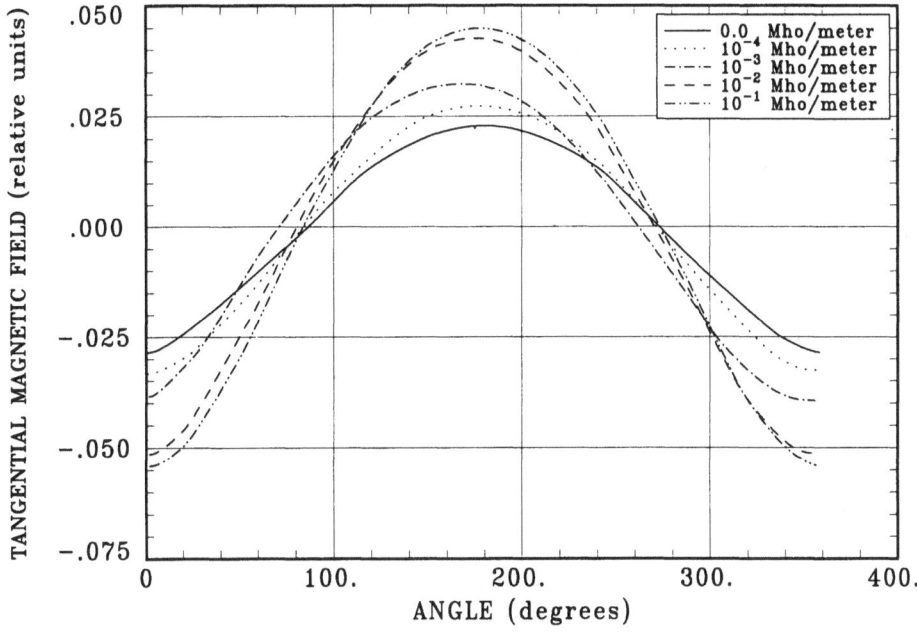

Figure 7

Relative amplitude and phase of total magnetospheric magnetic field at the equator for several values of conductivity for the earth. (Phase angle is shown instead of magnetic local time.)

electrical conductivity goes from zero to infinite, the phasing of the variation pattern changes such that it attains its maximum value before noon for finite conductivity, but occurs exactly at noon for both zero and infinite conductivity. The observed *Sq* pattern exhibits this feature (VESTINE *et al.*, 1947). With the earth's conductivity modeled at .0065 mho, the north-south component attains its maximum just before 1100 UT. The Walters effect (the solar wind does not come from the optical direction of the sun but rather along the "garden hose" angle because of the rotational velocity of the sun) adds another 20 minutes to this "phase delay." Thus it is expected that the magnetospheric contribution to *Sq* will reach its maximum value at about 1040 UT.

5.3 *The Magnetospheric Contribution to the Annual Variation in Sq*

It is also possible to estimate the contribution of the magnetospheric currents to the annual and semiannual variation observed in the *Sq* pattern. This is done without direct calculation since the present models are symmetric about the noon-midnight meridian plane and cannot be used to examine the "tilt effects" which contribute to the long-term variation in the *Sq* pattern. OLSON (1970a) determined that the magnetospheric currents produced about 1.5 nanoteslas for the annual and semiannual variation in Sq. Since the calculated day to night range for these currents is approximately three times larger than reported earlier, this is also true for the annual and semiannual variations produced by the magnetospheric currents since the diurnal and annual variations scale with the size of the magnetosphere (OLSON, 1970a). This suggests that the magnetospheric currents produce about 4.5 nanoteslas, about half of the observed 8–10 nanotesla seasonal variation in the Sq pattern.

5.4 The Day to Day Variability of Sq

Work done over the past two years as part of the Coordinated Data Analysis Workshops CDAWs (see Section 3) has shown that although the nominal position for the magnetopause is between 10 and 11 earth radii along the sun-earth line, it is sometimes compressed to well within geosynchronous orbit. It is this compression of the magnetosphere that is responsible for the well studied "sudden commencement phase" of the geomagnetic storm. Using pressure balance formalism it is easily shown that an inward motion of the magnetopause of 1 earth radius from its nominal 10.5 R_E standoff location will increase the magnetopause contribution to Sq by approximately 35 percent. Studies of the solar wind suggest that there are extended periods where the solar wind pressure is considerably enhanced. Yet, because the solar wind flow is steady, such intervals are characterized as magnetically quiet. Thus increases in the surface field from the magnetospheric currents over the nominal values shown in Figure 6 may be expected to occur on some days that have been chosen for the measurement of Sq.

To make a more precise determination of what fraction of Sq is contributed by the magnetospheric currents, it will be necessary to use such event models to determine the strength of the magnetospheric current systems (as a dependent variable of solar wind pressure, etc.) for the same days that were used to derive the observed Sq patterns. Clearly the magnetospheric values presented here are baseline and a detailed day to day investigation (where solar wind and interplanetary field data are available) should yield somewhat larger values for the magnetospheric contribution. It is anticipated that such an analysis will show that the magnetospheric currents make a sizable contribution to the day to day variability in Sq.

5.5 The Magnitude of the Magnetospheric Contribution to Sq

It is difficult to assess precisely what fraction of the observed Sq pattern is produced by these magnetospheric currents because there remains some uncertainty as to what value should be used for the standoff distance of the magnetosphere (which provides a measure of the size of the magnetosphere and of the strength of the magnetospheric current systems). A value of 10.5 R_C has been used in this study.

This size (its magnetopause intersecting the sun-earth line at a geocentric distance of 10.5 R_E) is traditionally used to represent the magnetosphere's "ground state" and occurs only during quiet magnetic conditions. The results shown in Figure 6 should therefore be compared only with Sq measurements made on very quiet days ($K_p = 0$ or 0^-). A day to night difference of 12 nanoteslas should therefore be considered the minimum contribution of the magnetospheric currents to the mid-latitude Sq pattern. Likewise, an annual variation of 4.5 nanoteslas should be considered the minimum contribution of the magnetospheric currents to the annual variation in Sq.

4. Summary

The purpose of this work has been to reexamine the role that the magnetosphere plays in producing magnetic variations at the earth's surface. Prior to the discovery of the magnetosphere, it had been assumed for half a century that the magnetic variations at the earth's surface during quiet times (the Sq pattern) were caused by currents flowing in the ionosphere. Early work on the magnetosphere suggested that it produced an inconsequential portion of this Sq pattern. However, the early magnetospheric models were developed to represent the magnetic field valid throughout much of the magnetosphere and were not constructed for accuracy at the earth's surface. Because of this and an increased understanding of the magnetospheric currents fifteen years later, the contribution of the magnetospheric currents to Sq has been reexamined. The minimum magnetospheric contribution to Sq, as represented in terms of a day to night range, is approximately 12 nanoteslas at mid-latitudes when the earth's finite conductivity is taken into account. This is significantly larger than the values reported earlier by MEAD (1964) and (OLSON (1970b). The magnetospheric currents may also be responsible for a significant portion of the day to day variability in the observed Sq pattern.

The fraction of the observed Sq pattern produced by the magnetospheric currents is still not known precisely because minimum values for the magnetospheric currents were used in this study. However, on some of the days for which the Sq pattern has been determined, the scale of the magnetosphere (as indicated by its standoff distance along the sun-earth line) is smaller and the strengths of the magnetospheric currents are therefore larger than those shown in Figure 6. The actual contribution of the magnetopause currents to the observed mid-latitude Sq pattern can therefore be accurately specified only by first using interplanetary data to calculate the strengths of the magnetospheric currents for those days on which the Sq pattern has been determined. The contribution of these magnetospheric currents to Sq, averaged over many days, is therefore expected to be larger than shown in Figure 6.

Acknowledgements

I gratefully acknowledge S. J. Scotti's work on computing the earth currents induced by the rotation of earth in the presence of the magnetospheric currents. I also thank Dr. K. A. Pfitzer for his help with the figures, his constructive comments and continual encouragement. This work was supported by the Air Force Office of Scientific Research Contract F49620–87–C–0039.

REFERENCES

ADAMS, W. G. (1892), *Comparison of Simultaneous Magnetic Disturbances at Several Observatories*, Phil. Trans. London (A) *183*, 131.

BIRKELAND, K. (1911), *Sur la lumiere zodiacale*, Comptes Rendes Acad. Sci. Paris *152*, 345.

BIRKELAND, K. (1913), *Sur la conservation à l'origine du magnetisme terrestre*, Comptes Rendes Acad. Sci. Paris *157*.

BREIT, G., and TUVE, M. A. (1925), *A Test of the Existence of the Conducting Layer*, Nature *116*, 357.

CANTON, (1759), *An Attempt to Account for the Regular Diurnal Variation of the Horizontal Magnetic Needle*, Phil. Trans. London *398*.

CHAPMAN, S. (1919), *The Solar and Lunar Diurnal Variation of the Earth's Magnetism*, Phil. Trans. R. Soc. *A218*, 1.

GILBERT, W. (1600), *De Magnete*, London, English translation by P. F. Mattely (J. Wiley and Sons, New York 1893).

MEAD, G. D. (1964), *Deformation of the Geomagnetic Field by the Solar Wind*, J. Geophys. Res. *69*, 1181.

OLSON, W. P. (1970a), *Variations in the Earth's Surface Magnetic Field from the Magnetopause Current System*, Planet. Spa. Sci. *18*, 1471.

OLSON, W. P. (1970b), *Contribution of Nonionospheric Currents to the Quiet Daily Magnetic Variations at the Earth's Surface*, J. Geophys. Res. *75*, 7244.

OLSON, W. P., and CUMMINGS, W. D. (1970), *Comparison of the Predicted and Observed Field at AB-1*, J. Geophys. Res. *75*, 7117.

OLSON, W. P., and PFITZER, K. A. (1982), *A Dynamic Model of the Magnetospheric Magnetic and Electric Fields for July 29, 1977*, J. Geophys. Res. *87*, 5943.

SCHUSTER, A. (1889), *On the Currents Induced in a Spherical Conductor by Variation of an External Magnetic Potential*, Phil. Trans. London (A) *180*, 467.

SISCOE, G. L. (1966), *A Unified Treatment of Magnetospheric Dynamics*, Planet. Spa. Sci. *14*, 947.

STEWART, B., *Terrestrial Magnetism* (Encyc. Brit., 9th ed., p. 36, 1882).

STORMER, C. (1907), *Sur les trajectoires des corpuscules electrisees dans l'espace sans l'action du magnetisme terrestre avec application, aux aurores boreales*, Arch. Sci. Phys. *24*, Geneve.

SUGIURA, M., LEDLEY, B. G., SKILLMAN, T. L., and HEPNER, J. P. (1971), *Magnetospheric Field Distortions Observed by 0603 and 5*, J. Geophys. Res. *76*, 7552.

VESTINE, E. H., LEPORTE, L., LANGE, I., and SCOTT, W. E., *The Geomagnetic Field, Its Description and Analysis* (Carnegie Institute of Washington, Publication 580, 1947).

WALKER, R. J., *Quantitative modeling of planetary magnetospheric magnetic fields*, In *Quantitative Modeling of Magnetospheric Processes*, Geophys. Monogr. Ser. Vol. 21 (ed. Olson, W. P.) (AGU, Washington, DC 1979).

WILLIAMS, D. J., and MEAD, G. D. (1965), *Night-side Magnetospheric Configuration as Obtained from Trapped Electrons at 1100 kilometers*, J. Geophys. Res. *70*, 3017.

XU, Wen-Yao (1989), *Polar Region Sq*, Pure Appl. Geophys. *131*, 371–393.

(Received April 4, 1988, accepted April 18, 1988)

PAGEOPH, Vol. 131, No. 3 (1989)

0033–4553/89/030463–21$1.50 + 0.20/0

Abnormal Sq Behaviour

E. C. BUTCHER[1]

Abstract—It is well-known that the amplitude and phase of the $Sq(H)$ variation show considerable variability from day to day. In this paper we consider one aspect of the phase variability—that associated with AQDs. AQDs (or "abnormal quiet days") are defined as magnetically quiet days where the maximum excursion of H at a mid-latitude station on the poleward side of the focus occurs outside the "normal" time range 0830–1330 LST. Such days exhibit properties, many of which appear quite distinct from the properties of the "normal" $Sq(H)$ variation. The properties of AQDs, and the proposals that have been made to explain them, are considered in detail. The consequences of these proposals and some problems which need to be addressed in order to obtain a fuller understanding of the dynamics of the ionosphere on AQDs are also discussed.

Key words: Abnormal quiet days, $Sq(H)$.

1. Introduction

It has been known for many years that there is considerable variability in the phase of the horizontal component H, of the quiet day geomagnetic variation. Although the maximum excursion of H may occur at any time, the most probable time of occurrence is around the hours centred on 1100 LST. Thus BROWN and WILLIAMS (1969) and BROWN (1974), using International quiet days over a period of nearly 90 years for the mid-latitude station Hartland/Abinger/Greenwich, on the poleward side of the focus, found that for approximately 80% of these days the minimum in H occurred between 0830 and 1330 LST. Such days were termed "normal" quiet days (NQDs) by BROWN and WILLIAMS (1969) and the significant minority of days whose minimum occurred outside this time range they termed "abnormal" quiet days (AQDs). An example of the form of the variation of H on an AQD compared to that on an NQD is shown in Figure 1.

Although the selection of the 0830–1330 LST time range was quite arbitrary, all subsequent work on AQDs using stations on the poleward side of the focus have used this time range distinction. Analysis of the variation of the time of maximum

[1] Department of Physics, La Trobe University, Bundoora, Vic 3083, Australia.

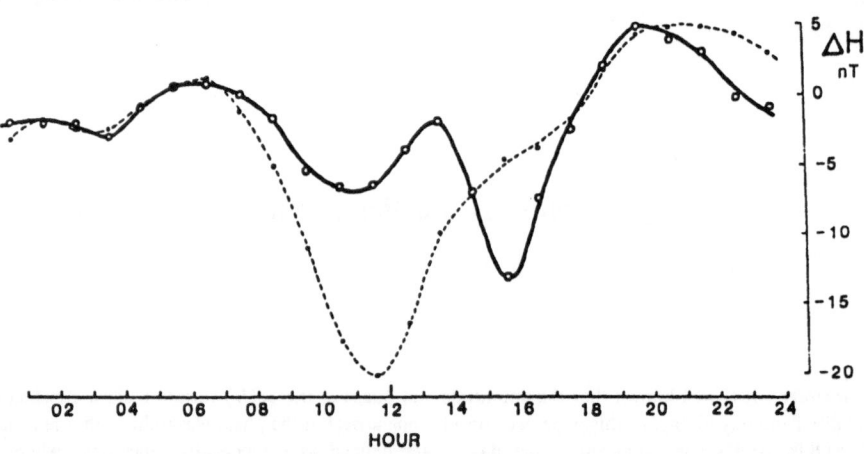

Figure 1
Magnetograms for Hartland for an NQD (dashed curve) and neighbouring AQD (solid curve) normalised
to the morning mean value. Both curves are averages of five days of each type (from BROWN, 1981).

H at stations on the equatorward side of the focus shows less variability and in such
cases the NQD range has been taken as 0930–1230.

Even though AQDs are separated from other quiet days by using these
arbitrarily selected time ranges, the study of AQDs has indicated many properties
which are not found in NQDs. The problem is then to determine the ultimate causes
of AQDs and the conditions that occur on AQDs as opposed to those which occur
on NQDs. Whether AQDs are different from NQDs in some fundamental way or
just part of the overall general variability of all quiet days, is a question which is
basic to our understanding of the dynamics associated with the production of the
currents associated with the daily geomagnetic variations. In this paper we shall
therefore examine the properties of AQDs and the proposals that have been made
to explain them.

2. *The Properties of AQDs*

The original work of BROWN and WILLIAMS (1969) and BROWN (1974) using
the Hartland (51°N 4°29′W) (Abinger/Greenwich) data indicated some interesting
properties of AQDs, particularly the occurrence distribution with time of day,
season and solar epoch. The properties include:

1. There is a greater number of AQDs in winter than in summer. This seasonal
   distribution was confirmed by considering the occurrence of AQDs at a Southern
   Hemisphere station.
2. There is a maximum in the occurrence of AQDs having minimum in *H* at
   midnight, although the midnight minimum appears to be exclusively confined to
   winter.

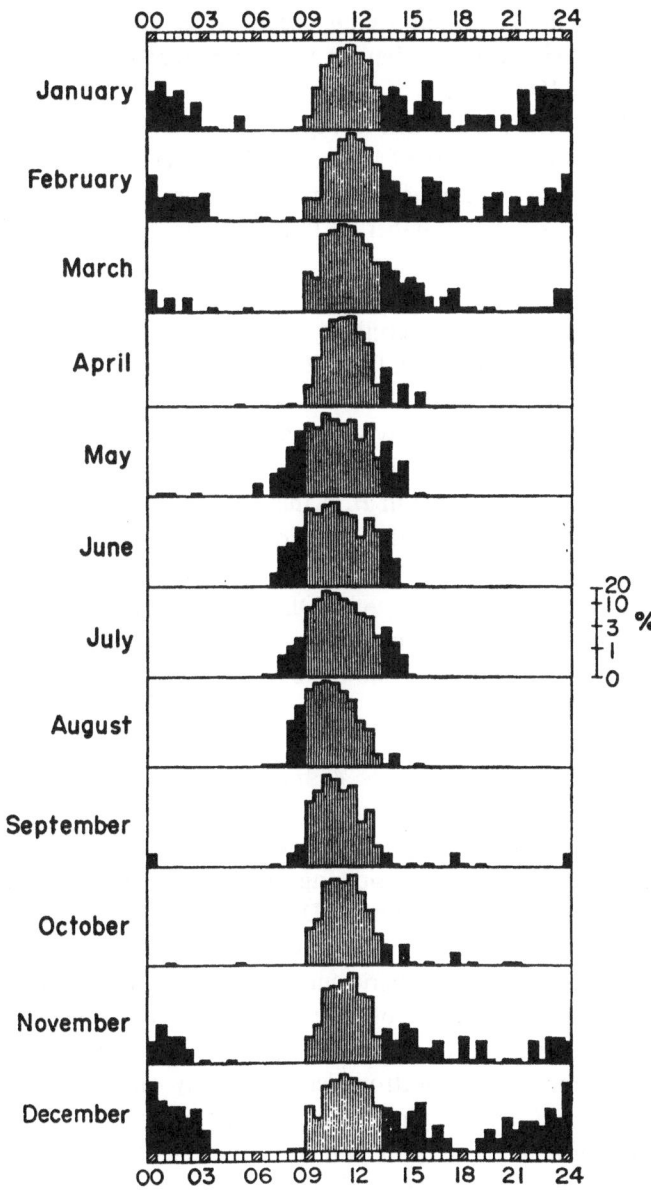

Figure 2
Monthly histogram of percentage number of days for which the diurnal minimum of *H* occurred during each half hour interval. Greenwich/Abinger/Hartland, 1884–1972. Hatched areas: NDQ; black areas: AQD (from BROWN, 1975).

3. Of those with minimum in *H* occurring near the NQD interval, in winter more in the afternoon period (> 1330 LST) (pm AQDs) whereas in the summer, more in the morning period (< 0830 LST) (am AQDs).

4. The minimum in H on most AQDs occurs either around midnight or near the NQD period, with very few having their minimum between these two periods near 0600 and 1800 LST.

These properties are illustrated in Figure 2.

5. More AQDs occur in years of low sunspot number, and the maximum number occurring in a given year near sunspot minimum is related to the size of the average annual sunspot number of the following sunspot maximum. Over the whole cycle the percentage occurrence of AQDs is in approximate anti-phase to the sunspot number variation.

6. The occurrence of AQDs determined at Wingst (53°45′N, 9°04′E), a station at a similar latitude but ~1000 km from Hartland, showed similar seasonal and solar cycle variations at Hartland. However, in winter 80% of AQDs at Hartland were AQDs at Wingst, whereas only 40% were common in summer.

7. Although the main minimum in H on AQDs occurs outside the 0830–1330 range, the normal minimum in H in this time range is still observable, giving a semi-diurnal appearance to the variation. This normal minimum, however, is reduced (see Figure 1).

8. The same seasonal variation of AQDs also occurred for a station on the equatorward side of the focus (where AQDs were defined as days where the maximum of H occurred outside the 0930–1230 LST range).

3. Explanations of the Occurrence Properties of AQDs

BROWN and WILLIAMS (1969) suggested that an explanation of the properties of AQDs might be found by considering the variations in the dynamo-induced currents that flow in the ionosphere. During the daytime on AQDs the normal $Sq(H)$ variation is still present, but somewhat reduced, and this reduction could be related to changes in the ionospheric currents. By considering the critical frequency of the E-region as an indicator of E-region conductivity, they concluded that the reduction in $Sq(H)$ amplitude was unlikely to be caused by a reduction in the conductivity (although the electron density is only one parameter on which the conductivity depends and may not be a good indicator of conductivity, BUTCHER, 1980). Wind variability was proposed as a possible cause since there was some evidence that the winds were more variable in the winter than the summer. However, the phenomena of AQDs with a minimum in H near midnight, particularly in winter, did not seem to be explained using dynamo theory since the nighttime E-region is not able to support the required current. They therefore suggested that the cause of these AQDs might be extra-terrestrial in origin. It was noted that magnetic bays occur most frequently around local midnight in mid-latitudes but it was found that the occurrence of (listed) bays near the time of the minimum in H, was very small. However, by considering five stations of approximately the same latitude, but extended over a longitude range of ~115°, MIZZI and

SCHLAPP (1971) showed, that for the AQDs where the minimum occurred near midnight, the minimum in *H* was caused by an "event" which occurred at all stations at the same UT, thus confirming that the actual minimum is probably caused by an event of extra-terrestrial origin. For AQDs where the minimum in *H* occurs near the NQD period, BROWN (1975) proposed that changes to the dynamo currents may be caused by upward propagation of planetary waves into the dynamo region. It had been found previously (BROWN and WILLIAMS, 1971) that changes in the heights of the 10 mbar level in winter months were correlated with the changes of the heights of E-region constant electron density contours. By dividing the AQDs into "am" and "pm" AQDs, BROWN (1975) found that on many occasions "am" AQDs had a tendency to occur when the isobaric heights were decreasing, whereas "pm" AQDs tended to occur when the isobaric heights were increasing. Since planetary scale waves are more likely to propagate to the dynamo region in winter than summer, this mechanism may be a possible cause of some AQDs in winter, although very few AQDs of the "am" type are found in winter (see Figure 2).

BUTCHER and BROWN (1981a) carried out an extensive analysis of *Sq(H)* in order to understand the nature and occurrence properties of AQDs. By considering eight stations along a line of approximately constant longitude, from 60°N to 14°N (geographic), spanning both sides of the *Sq(H)* focus they established that (1) on AQDs there is a reduction in the *Sq(H)* amplitude in the NQD period for all stations on the poleward side of the focus and an increase in the amplitude at stations on the equatorward side of the focus (see Figure 3). (2) The minimum in *H* on AQDs is formed by a substorm-(negative bay-)like event. These substorm events were very small (typically < 10 nT on average at Hartland) and were too small to be listed as substorm events. However, their correlation with the auroral indices *AE* and *AL* and the *z*-component of the IMF (B_z) (which changed from positive to negative with its maximum excursion occurring about one hour prior to the maximum excursion of the substorm event) shown in Figure 4 strongly suggested that they are magnetospheric substorm type events. The observed changes in the amplitude of the normal *Sq(H)* variation on AQDs suggests that a West-East current flows producing a superposed northward field (SPNF) at all latitudes. A comparison of the *H*-variation on NQDs and AQDs indicates that this implied additional current flows in the ionosphere (see Figure 3).

Thus in order for a quiet day at say Hartland to become an AQD we require that some additional current flows which reduces the normal *Sq(H)* amplitude such that a mini-substorm (which may occur randomly in time) of sufficient amplitude may form the daily minimum in *H*. Because the amplitude and "shape" of the daily variation changes with season and solar epoch (minimum amplitude in winter and sunspot minimum years, maximum in equinoxes and sunspot maximum years, with maximum in the daily value of *H* near 0600 and 1800 hours) most of the properties of the previous section may be explained.

The magnitude of the SPNF was also found to depend on both latitude and on

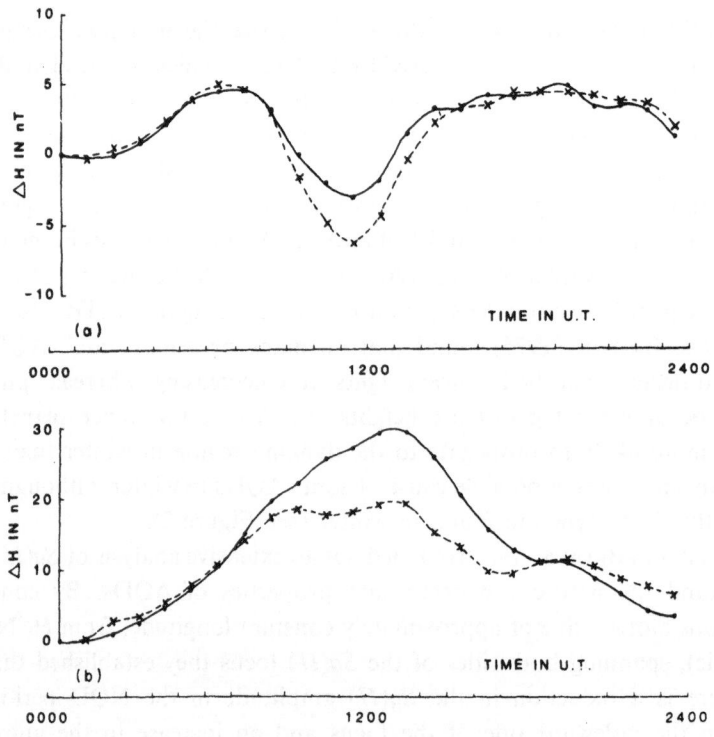

Figure 3
The average $Sq(H)$ variation, winter months for the years 1963–1965, for NDQs (x-----x) and ADQs
(●———●). (a) Lerwick (poleward side of focus), (b) Tenerife (equatorward side of focus). The
magnetograms have been normalized to zero at 0030 UT (from BUTCHER and BROWN, 1981a).

IMF polarity, being larger on A-days than T-days. Thus it was found that more
AQDs occurred on A-days (B_y positive) than T-days (B_y negative). The magnitude
of the SPNF was largest near 55°N in summer and 35°N in winter, tending to zero
near 20°N and 70°N (BUTCHER, 1982a).

4. AQDs and the Apparent Position of the Sq(H) Focus

It is well-known that the position of the $Sq(H)$ focus may vary considerably
from day to day. In one analysis HASEGAWA (1960), using data from a large
number of observatories, determined the internal and external equivalent current
system to obtain the position of the focus. He found changes up to 15° in the focus
latitude over a period of two days. He also found that the focus occurred at a higher
latitude in winter than summer for the European region, which is surprising as the
$Sq(H)$ current system is assumed to remain (approximately) fixed with respect to
the sun, as the earth rotates below it. PATIL et al. (1985) also found a similar

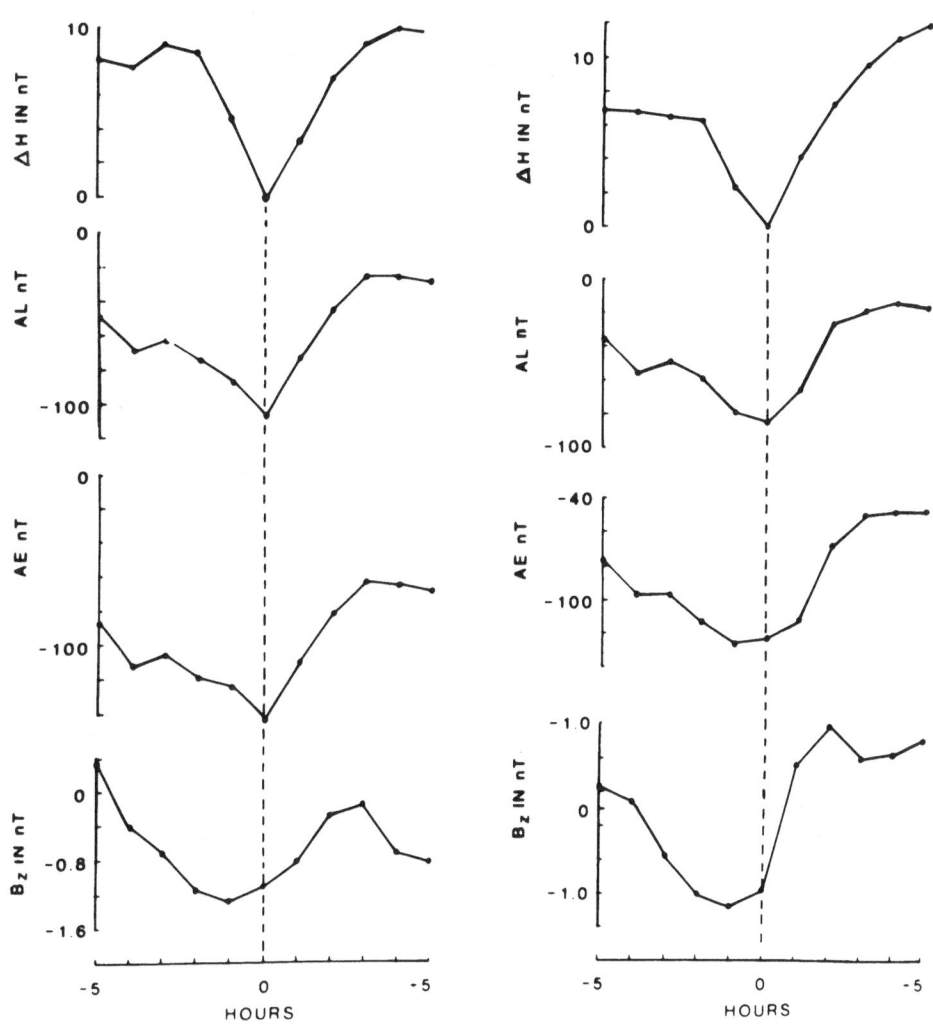

Figure 4

Superposed epoch analysis of the *H* variation of the AQD event, the auroral indices *AL* and *AE*, and the z-component, B_z, of the IMF for the period 1966–1973. The left panel is for A-days, the right panel for T-days. Zero time corresponds to the time of minimum for the AQD events (from BUTCHER and BROWN, 1981a).

seasonal effect for the Indian region, using a line of stations along an approximately constant longitude to determine the focus from the latitude where the amplitude of *H* tended to zero. However, CAMPBELL and SCHIFFMACHER (1985,1987a,b) using a selection of extremely quiet days for 1965 to determine the equivalent external current systems, found that the focus was most poleward in the summer season in the North American, European, Australian, Central and East Asian regions—the amount of movement with season varying considerably between the regions—but most poleward in winter in the Central Asian region.

MATSUSHITA *et al.* (1973) and MATSUSHITA (1975) found that the latitude of the $Sq(H)$ focus was dependent on the direction of the IMF. They found that the focus was more poleward on A-days than on T-days. Along the 75°W and 135°E meridians they found an average poleward shift of about 4° in summer months from T-days to A-days for the years 1965–1967.

None of these analyses noted the phase of the $Sq(H)$ variation. BUTCHER and BROWN (1980) separated the quiet days according to phase (i.e., NQDs and AQDs) and according to IMF polarity. It was found that for the 0° longitude meridian, for the sunspot minimum years 1963–64, the focus for NQDs showed no IMF dependence but that all the IMF dependence was attributable to the AQDs. There appeared to be no dependence on the magnitude of the IMF, only the polarity. The poleward motion was larger on A-days than T-days (where the effect was very small) and larger in winter than in summer. However, in winter the reduction of the amplitude was greater on T-days than A-days at latitudes $\gtrsim 50°N$. The sense of the IMF effect is consistent with a current system having a high latitude focus with current flowing anti-clockwise on A-days and clockwise on T-days in mid-latitudes similar to that found by TAKEDA and ARAKI (1984) although in winter the focus appears to be at a lower latitude than it is in summer. On A-days the IMF associated current therefore augments the West-East current associated with the SPNF on AQDs, whereas on T-days it opposes it. However, it is not clear why no effect was found on NQDs as the IMF related currents are not expected to be restricted to AQDs.

MANN (1986) carried out a detailed analysis of the IMF effects on $Sq(H)$ over the period 1957–1967. Although his analysis confirmed that of BUTCHER and BROWN (1980) for 1963–64, he found opposite (i.e., smaller amplitude of $Sq(H)$ on AQDS on T-days) or insignificant results for other years. However, a similar analysis for the sunspot minimum years 1974–77 tended to agree with the 1963–64 results.

Since the SPNF that occurs at most latitudes produces a large poleward motion of the focus, particularly in winter, removal of AQDs from the analysis should reduce the day-to-day variability in the focus latitude. In order to increase the number of days in the analysis, BUTCHER (1982b) considered quiet days as defined by MAYAUD (1973) using the aa indices. Such indices give an absolute measure of quietness rather than a relative one and in sunspot minimum years include most of the 5 monthly international quiet days anyway. When the AQDs were removed from these quiet days the focus variability was reduced significantly. In the summer and equinoctial months, the whole range of focus latitudes along the 0° meridian was contained within 12°, the mean latitude for the three years being 41.5° with a rms error of $\pm 2.3°$. In winter, the range was slightly larger, but significantly less than when the AQDs were included, and had a mean latitude of 36.7° with a rms error of $\pm 3.4°$. It is seen that the focus is nearer to the equator in winter than in summer and it is the inclusion of AQDs that gives a high latitude bias to the focus latitude in winter. Whether one is justified in removing the AQDs from the sample, and considering them as a separate phenomenon, will be considered later.

5. AQDs at Low Latitudes

In their original work on AQDs, BROWN and WILLIAMS (1969) also looked at the properties of AQDs defined at a station on the equatorward side of the focus (e.g., Apia, 13°48′S, 171°46′W). They found that the variability in the time of occurrence in the maximum of $Sq(H)$ was much less than for stations on the poleward side of the focus and so they considered AQDs to be days where the maximum in H occurred outside the range 0930–1230 LST. For the seven-year period considered, they again found that more AQDs occurred in winter and in sunspot minimum years. This work was extended by ARORA (1972) who considered the occurrence of AQDs at Alibag (18°38′N, 72°52′E) over three solar cycles. The percentage occurrence of AQDs appears predominantly semi-annual (i.e., similar to that found at Hartland) with the main maximum in winter and a subsidiary maximum in summer, and there was an inverse relation between the annual percentage occurrence and sunspot number. This semi-annual variation was also apparent in a smaller sample (ten years) of data at Trivandrum (8°29′, 76°57′E) near the magnetic equator. LAST et al. (1976) studied the occurrence of AQDs at Addis Ababa (9°N, 38°7′E), Trivandrum and Alibag, and found that no AQDs occurred having H-min between 2100 and 0800 LST but that they tended to occur for times bordering the NQD period. This observation appears to be confirmed by SASTRI and MURTHY (1978) for the electrojet station at Kodaikanal (10°14′N, 77°28′E) and shown in Figure 5. At all these stations, the semi-annual variation in the occurrence frequency of AQDs is observed as well as the inverse relation to the sunspot number (although it is not as clear as found at stations on the poleward side of the focus). These differences in the diurnal and sunspot number variation in the percentage occurrence of AQDs, compared to that observed in mid-latitudes, cause the authors to question whether the AQD phenomenon is the same in low and mid-latitudes, and whether the low latitude AQDs may be effected by the equatorial electrojet.

PENG FENGLIN and TSCHU (1987) have investigated AQD occurrence in H, D and Z at five Chinese stations on both sides of the focus. They also found a semi-annual variation in the H-data (with a maximum occurrence in winter) particularly for pm AQDs and also in the Z-data. There is an antiphase relationship between the percentage occurrence of AQDs in the H- and D-data with sunspot number, but it is not present in the Z-data. THAKUR and SONTAKKE (1981) also report an observation of AQDs in the Z-component at stations on the equatorward side of the focus (Alibag, Kodaikanal and Annamalainagar (11°22′N, 78°41′E)). They found that even on normal days in $Sq(H)$ at these stations the Z-component showed a maximum during the day rather than a minimum, and they termed these days AQDs in Z. Although such a definition of an AQD is rather different from

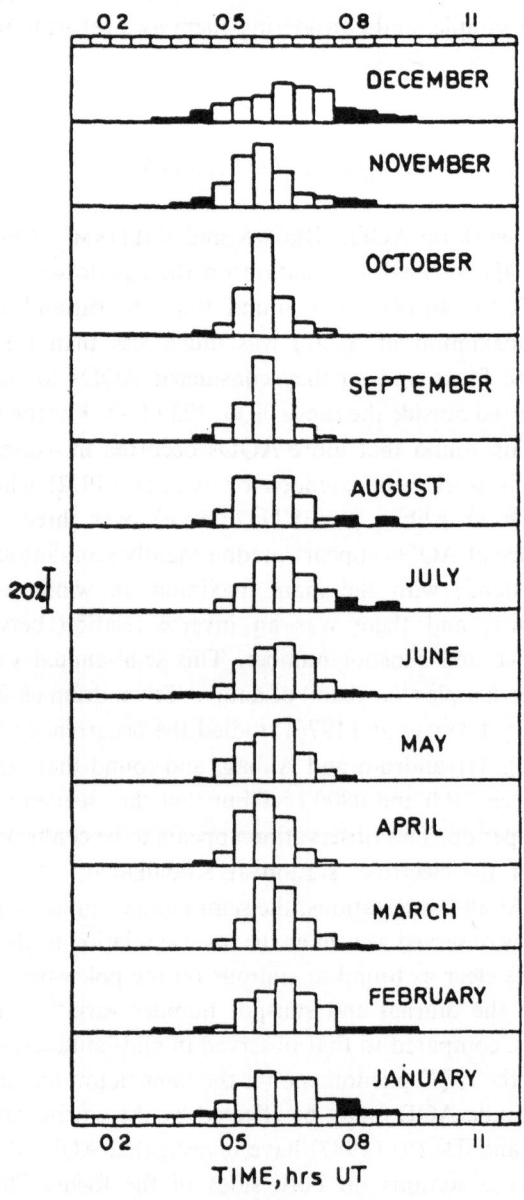

Figure 5
Monthly histograms of percentage occurrence of international quiet days on which diurnal maximum in
the *H*-component occurred during each half-hour interval at Kodaikanal over the period 1950–1975
(black areas represent AQDs) (from SASTRI and MURTHY, 1978).

that used for *H*, the percentage occurrence of these AQDs in $Sq(Z)$ is quite high
(from ~7% at Kodaikanal to ~22% at Alibag) and they also have a maximum
percentage occurrence in winter and an inverse relation with sunspot number which
is strong at Alibag but weak at the other two stations.

6. AQDs and Sunspot Number Prediction

As stated previously, BROWN and WILLIAMS (1969) found that the occurrence of AQDs was larger in sunspot minimum years than in sunspot maximum years. This observation is explained in Section 3 since the amplitude of $Sq(H)$ in sunspot minimum years is smaller than that in sunspot maximum years and so on a given day it is easier for the mini-substorm to form the minimum. They also found that the occurrence of AQDs was in anti-phase with the average annual sunspot number R_a (over $\sim 2\frac{1}{2}$ solar cycles) and that there was a noticeable variation between the percentage occurrence of AQDs at sunspot minima but not at sunspot maxima such that the maximum occurrence near sunspot minimum seemed to be related to the magnitude of the following sunspot maximum. BROWN (1974) extended this work to include 89 years (~ 8 solar cycles) of data and found a linear relationship between the average percentage occurrence of AQDs at a given minimum (using a mean over 3 years) and R_a at the following maximum as shown in Figure 6. BROWN (1979) has successfully used this method for predicting values of R_a for sunspot maximum.

In order for an AQD to occur, the normal $Sq(H)$ amplitude must be reduced sufficiently for a mini-substorm to form the minimum in H. Both phenomena must occur since, on quiet days, these mini-substorms are not, in general, large enough to form the minimum unless the normal $Sq(H)$ amplitude is reduced. Thus the percentage occurrence of AQDs has to be related to these two effects. BROWN and

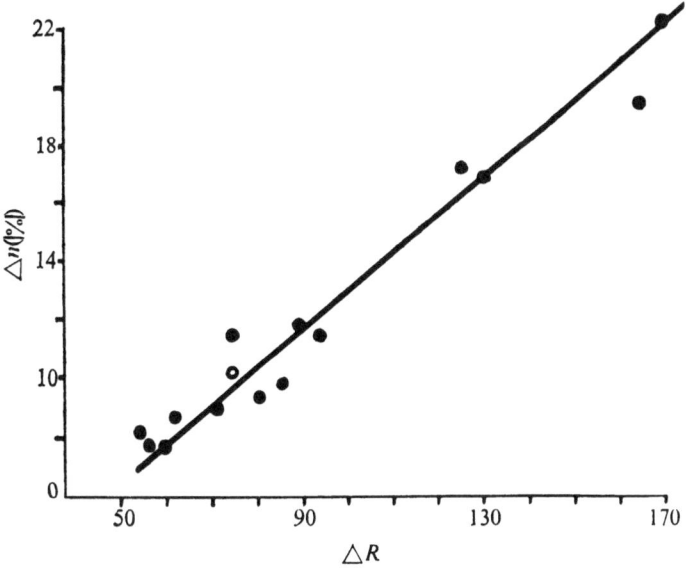

Figure 6

Amplitude of the cycles in AQD occurrence (Δn) and sunspot number (ΔR) one half-cycle ahead. (Open circle is for the ascending part of cycle 13 and involves an estimate of the value of Δn, since no magnetic data are available for the time of the AQD minimum) (from BROWN, 1974).

BUTCHER (1981) considered both effects separately and found that there was a weak inverse dependence of the $Sq(H)$ amplitude on AQDs at sunspot minimum with subsequent R_a-max, but a strong linear dependence on the amplitude of the substorm event. When the two were combined the correlation was very high (see Figure 7).

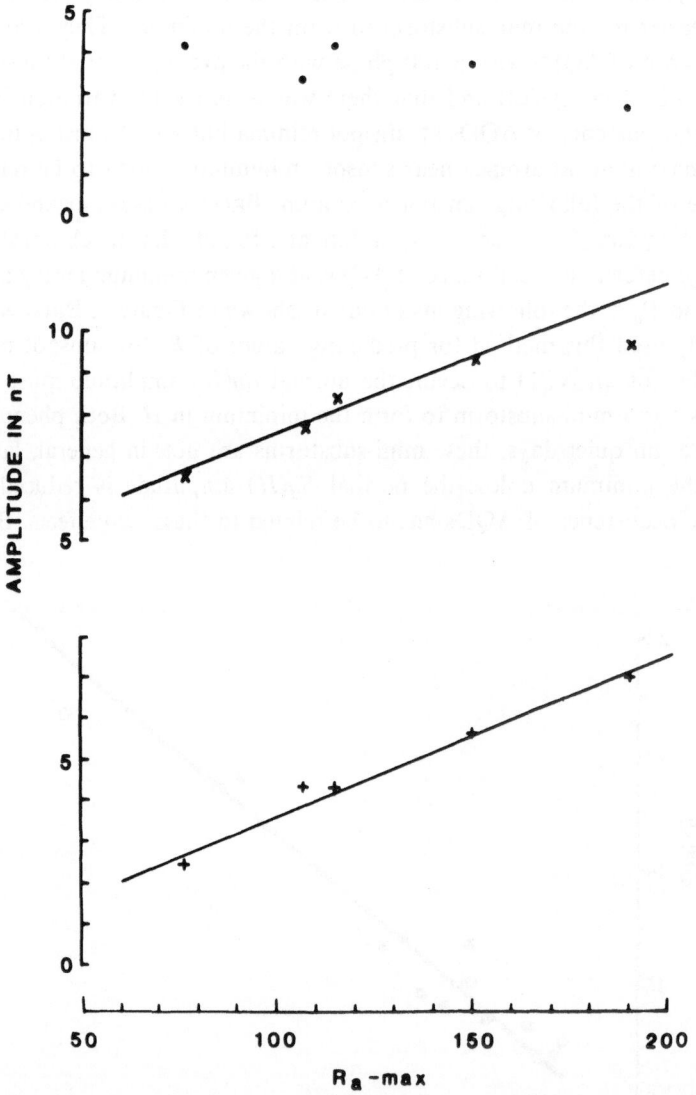

Figure 7

Variation with annual mean maximum sunspot number (R_a-max) for cycles 16 to 20 of (a) Top: Amplitude of $Sq(H)$ on AQDs during preceding sunspot minimum; (b) Centre: Amplitude of southward H perturbation on AQDs during preceding sunspot minimum; (c) Bottom: Differences between AQD-event and $Sq(H)$ amplitudes on AQDs during preceding sunspot minimum. All magnetic data refer to Hartland (from BROWN and BUTCHER, 1981).

The linear relationship between R_a and A, ((the substorm event amplitude) $- (Sq(H)$ amp on AQDs)) was found to be of the form

$$R_a = (26.1 \pm 5)A.$$

Only A-days were considered in view of the uncertainty of the IMF dependence mentioned previously. However, for the period considered (1922–1965) $\sim 80\%$ of AQDs occurred on A-days and inclusion of AQDs that were T-days would not be expected to affect the relationship in any significant way.

Thus, it is the magnitude of the mini-substorm at a given sunspot minimum which is the major influence as a predictor of R_a at the next sunspot maximum. It was mentioned earlier that the substorm event was well correlated with southward (negative) changes in B_z of the IMF (the maximum southward excursion occurring one hour earlier than the maximum excursion of the mini-substorm). It was also found that the larger mini-substorms were associated with larger excursions in B_z. Thus it would appear that the size of R_a at a given sunspot maximum is determined at the previous minimum (as might be expected on the dynamo theory of solar cycle generation) and is reflected in the magnitude of the southward excursion of B_z associated with the mini-substorms which produce the minimum in H on AQDs. Presumably the magnitude of the southward B_z excursion is related to the energy transferred into the geomagnetic tail and hence into field-aligned currents which are associated with the mini-substorm.

7. Longitudinal Effects

A comparison of the variation of H at stations on approximately the same latitude, but separated by $\sim 13°$ in longitude (Hartland and Wingst) shows that not all days designated AQDs at one station are AQDs at the other (BROWN and WILLIAMS, 1969). In fact the degree of correspondence depended on season and solar epoch, being $\sim 80\%$ for winter at sunspot minimum and only $\sim 35\%$ in summer at sunspot maximum. This suggests that either the magnitude of the SPNF or the amplitude of the mini-substorm, or both, are longitude dependent such that the mini-substorm which forms the minimum at one station is unable to form the minimum at the adjacent station. This problem was pursued by BUTCHER (1982a) who found that the amplitude of the SPNF fell off slowly with distance from the station where the AQD was determined and appeared to become negative (i.e., a superposed southward field, SPSF) after $\sim 145°$ of longitude. The mini-substorm amplitude was also found to fall off slowly with longitude and appeared consistent with the event being caused by a negative mini-substorm associated with a high latitude field-aligned current. Thus it would appear that the correspondence of AQDs at Hartland and Wingst is caused by a combination of both effects—variation of SPNF and substorm amplitude.

BROWN (1986) chose two pairs of stations (Eskdalemuir 55°19′N, 3°12′/Almeria 36°51′N, 2°22′W and Guam 13°27′N, 144°45′E/Juzhno-Sakhalinsk 47°N, 143°E), each pair along a given meridian with the meridians separated by ∼145°. Each station of the pair was on opposite sides of the $Sq(H)$ focus. A superposed epoch analysis was performed on the data for the four stations using 24 days centred on an AQD at Hartland. At Eskdalemuir and Almeria there was a SPNF as observed by BUTCHER and BROWN (1981a) and at Guam a SPSF. At Juzkno-Sakhalinsk, a statistically insignificant result was found, due to the proximity of the station to the $Sq(H)$ focus. However, the observation tends to confirm that the SPNF reverses to a SPSF at some distant longitude. Brown also observed that the SPNF (and SPSF) takes 4–5 days to build up and decay and the maximum SPNF occurs within a day of the AQD. Some evidence was also produced to suggest that B_y builds up and decays in a similar way.

8. The Causes of AQDs

From the analyses on AQDs that have been presented, it would appear that for a station on the poleward side of the $Sq(H)$ focus the actual minimum in H is caused by a mini-substorm. HIBBERD (1981) suggested that since there is some disturbance present on all quiet days AQDs are just the result of disturbance on the $Sq(H)$ variation. However, it was shown by BUTCHER and BROWN (1981a) that the normal $Sq(H)$ variation on AQDs is reduced by the presence of a SPNF thus facilitating (and on many occasions being a precondition for) the formation of the minimum by the mini-substorm. BROWN (1975) has also presented evidence to suggest that at least some AQDs (those whose minimum in H occurs near the time of the normal $Sq(H)$ minimum) are caused by changes in the dynamo region associated with an upward propagation of planetary waves from the lower atmosphere. At the lower latitudes, the evidence would suggest that the cause of the AQD phenomenon may be different. Of course, the mini-substorm causing the midlatitude minimum is a negative depression in H, whereas at low latitudes we are looking for changes in the occurrence of the time of maximum positive excursion. However, it should be noted that a negative substorm event is able to change the time of attainment of maximum excursion in H at stations on the equatorward side of the focus (BUTCHER and BROWN, 1981b). For this to occur, the substorm event needs to occur in, or very close to, the NQD range and would only produce AQDs with maxima near the NQD period. Although ARORA (1972) does give examples where the maximum in H occurs well away from the NQD period, SASTRI and MURTHY (1978) and LAST et al. (1976) find AQDs only adjacent to the NQD period.

The presence of the SPNF on AQDs in the Northern Hemisphere occurs at all latitudes and it is important to ask whether it is just part of the day-to-day variability in the dynamo currents or some additional process unrelated to what we consider to be the normal $Sq(H)$ current mechanism. Zonal currents are also required to explain the relationship of the normal $Sq(H)$ variation on AQDs to the IMF. BUTCHER and BROWN (1981a) suggested that these IMF dependent currents are driven by the penetration of the Z-component of the induced interplanetary electric field, E_z, into the high latitude ionosphere, which then drives zonal Hall currents.

BROWN (1986) points out that on A-days, E_z is positive and penetration of this field into the high latitude ionosphere drives a westward current. Return currents flow eastward at lower latitudes as required for the SPNF. These return currents would be westward in the opposite hemisphere as is observed. A current system of the form suggested by Brown has been identified by TAKEDA and ARAKI (1984).

TAKEDA (1982) extended the model of TAKEDA and MAEDA (1980), used to calculate ionospheric currents, to include the effect of field-aligned currents (FAC), which enter at high latitudes in the pre-noon period and exit in the post-noon period. Such currents produce a West-East current in mid-latitudes during the daytime which Takeda suggested could be the source of the SPNF. TAKEDA (1982) and TAKEDA and ARAKI (1984) have also identified several current systems which may contribute to the day-to-day variability of $Sq(H)$ and which may be related to the AQD problem, especially an IMF current which has a high latitude focus and flows to mid-latitudes. Their analysis suggests, however, that at low latitudes any IMF effect in $Sq(H)$ is unlikely to be due to currents of ionospheric origin.

The mid-latitude current associated with the high latitude FAC which enters in the pre-noon period, flows eastward and exits in the post-noon period, as suggested by Takeda, flows equatorward in the morning and poleward in the afternoon periods as shown in Figure 8. If this current is associated with the SPNF on AQDs, it should produce changes in the declination D on AQDs consistent with these directions of flow. BUTCHER (1987) analysed the D-data at six stations (four in the Northern Hemisphere and two in the Southern Hemisphere) near the $0°$ meridian for the years 1963–65 and found a well defined variation on AQDs at all stations. However, the sense of the variation on the AQDs (which were determined at Hartland) was consistent with a current flowing northward in the morning period and southward in the afternoon period in both hemispheres. This was opposite to the direction expected on the Takeda proposal. Also the two Southern Hemisphere stations were on opposite sides if the focus, and analysis of the H-data at these two stations indicated that the current was East-West on AQDs (i.e., a SPSF was present). The magnitude of the effect in D was found to be IMF dependent, being larger on A-days than on T-days in local summer and *vice versa* in local winter. By assuming that the effects of the IMF currents on A- and T-days were equal and opposite, the IMF dependent and independent parts may be separated. The form of the two resulting variations is shown in Figure 9 and it is seen that the former variation is consistent with the form of the IMF-dependent variation found by

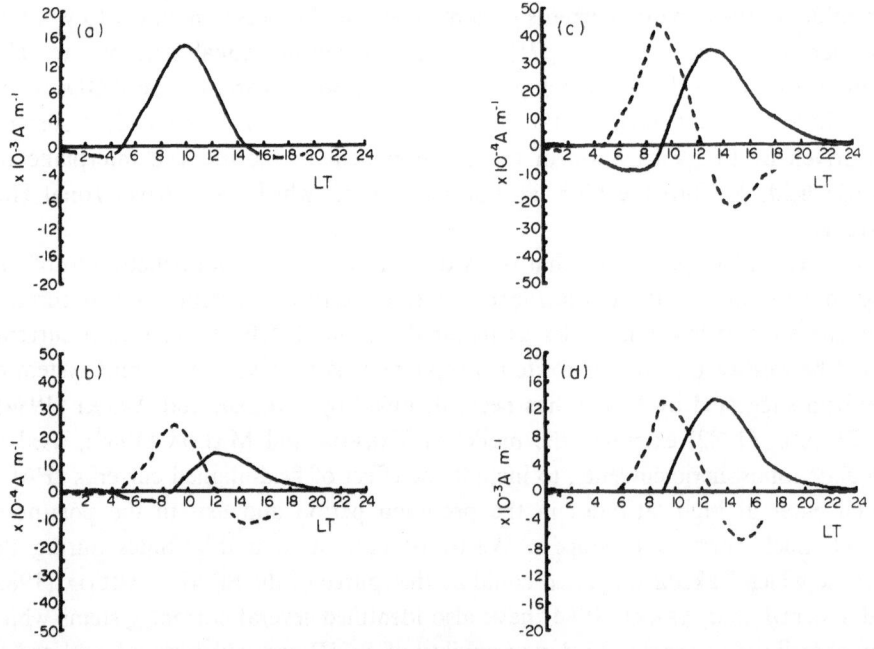

Figure 8

Daily variation of height integrated currents at latitudes of (a) 0°, (b) 40° and (c) 60°, respectively. Solid and dotted lines represent eastward and southward (equatorward) current, respectively (from TAKEDA, 1982).

TAKEDA and ARAKI (1984). Both the IMF dependent and independent parts were found to be of larger magnitude in local summer than local winter. In order to explain the IMF-independent variation it was suggested that the additional current system may be due to a large single current vortex (SCV) flowing clockwise on AQDs, extending over both hemispheres and centred near the equator. It was proposed that such a current could be driven by equator-asymmetric diurnal wind modes or equator-symmetric semidiurnal modes since interhemispheric current flow occurs along the highly conducting field lines.

On this proposed model, three (apparently) independent ionospheric current systems would be present on AQDs (besides that associated with the mini-substorm).

1. The "normal" two-cell current system (one in either hemisphere, with a focus near 35–40°) associated with NQDs.
2. The SCV (flowing clockwise on AQDs and with a focus near the equator).
3. The IMF associated current (with a focus at a high latitude, and which flows west-east on A-days in mid-latitudes in both hemispheres).

The proposed current systems are shown schematically in Figure 10.

Since the SCV is assumed to be produced by tidal winds, the question arises as to what is meant by "normal" in (1) above, since the "normal" Sq variation is assumed to be caused by currents driven by tidal winds.

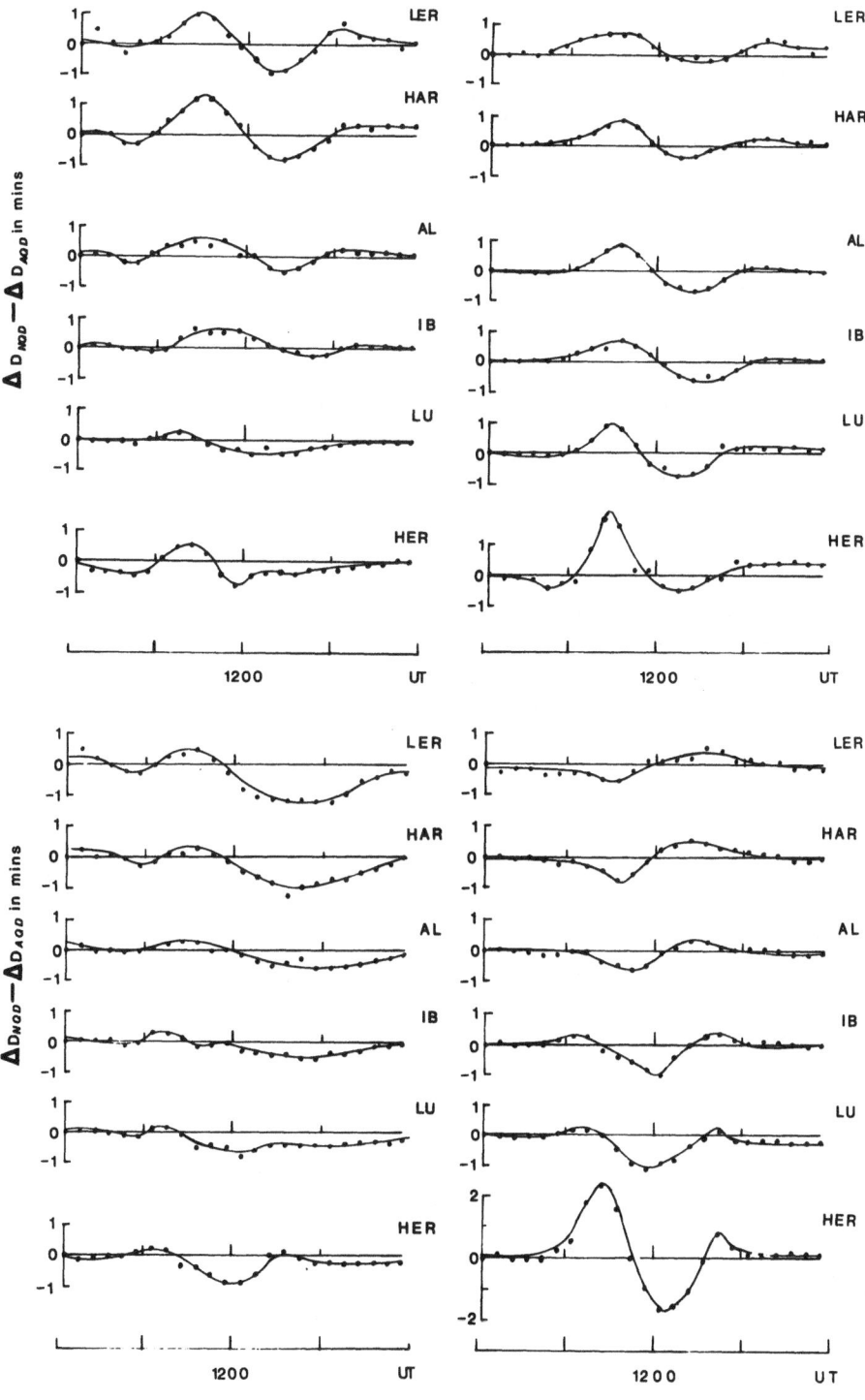

Figure 9

(a) Top: IMF-independent and (b) Bottom: IMF dependent part of $(\Delta D_{\mathrm{NQD}} - \Delta_{\mathrm{AQD}})$ for the stations Lerwick (LER), Hartland (HAR), Almeria (AL), Ibadan (IB), Luanda (LU) and Hermanus (HER); left (Northern Hemisphere) summer, right (Northern Hemisphere) winter. For (b) the variation represents the contribution of A-days (from BUTCHER, 1987).

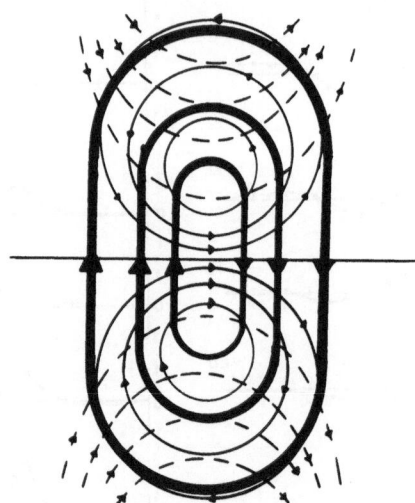

Figure 10
Schematic of the equivalent ionospheric current systems assumed to be present of AQDs. ——"normal"
ionospheric current system, ----IMF associated current system (direction of current flow for A-days),
——SCV.

MANN (1986) and BUTCHER (1987) suggested that the $Sq(H)$ variation on AQDs may be the low amplitude end of the normal $Sq(H)$ amplitude spectrum and the SCV would then represent a variable current which is present (or strong) on AQDs and absent (or weak) on NQDs. If this is so, then since the SCV covers both hemispheres, there should be some correlation between the Sq variations at stations separated by large distances. SCHLAPP (1968) and GREENER and SCHLAPP (1979) however, found that the correlation between stations of the $Sq(H)$ variation on quiet days fell off quite quickly—the correlation coefficient falling to ~ 0.5 for a station separation of $\sim 15°$ of latitude and $\sim 40°$ of longitude—and suggested that this variability was associated with irregular local wind patterns caused by patchy heating of molecular oxygen by UV radiation at ionospheric heights. HIBBERD (1981) suggested that by subtracting the daily variation of H of one station from another on the same longitude but on the opposite side of the focus the effects of substorms and distant currents may be eliminated and the resultant variation (ΔH) may be considered as proportional to the current in the ionospheric current loop. Using this technique for two pairs of stations, one in each hemisphere, along the same meridian, Hibberd found that for selected single months of data, ΔH in the two hemispheres were highly correlated. BRIGGS (1984) using the same method, but using a full year of data, found a small ($\rho = 0.38$) but significant correlation between ΔH variations for the two pairs of stations, the centres of whose current loops were separated by $\sim 7{,}000$ km. SCHLAPP et al. (1988) selected days of the smallest amplitude in H at Hartland and found that these days were well correlated with days of large D-variation at Hermanus (34°25′S, 19°14′E), a station far removed from Hartland and in the opposite hemisphere. When AQDs and QAQDs (additional AQD type days occurring on quiet days determined using the MAYAUD (1973) *aa* indices) were removed from the sample, the correlation disappeared for D-months and was reduced for E and J months. These results may be explained in terms of the SCV and the magnitude of its effect relative to the normal amplitude of the *Sq*(*H*) variation in a given season. It was concluded that the evidence did not support the presence of the SCV on all quiet days and the day-to-day variability consists of at least two components—one small scale (e.g., local) and one inter-hemispheric scale (e.g., SCV).

The driving forces of the proposed SCV have not yet been identified. However, some recent work by Singh and the author has been investigating the effects of variable diurnal winds on the ionospheric currents by extending the approach taken by SINGH and COLE (1987) to a global situation. Using the wind profiles given by FORBES and GILLETTE (1982) it has been found that an increased wind amplitude in one hemisphere produces an additional (clockwise) current extending over both hemispheres, with a focus near the equator, of the form required on AQDs. At present this work is only qualitative but it is interesting that such an effect may cause an SCV in the right direction since it might be expected that day-to-day variability in the amplitude of the tidal winds in each hemisphere may be independent of each other.

9. Future Work

Although the general features and properties of AQDs are well understood, several problems remain which need to be addressed in order to understand the detailed causes of the observed variations. Some of the problems which need to be considered include:

1. A detailed understanding of the mini-substorm. Although it has characteristics similar to the normal substorms it has a very small amplitude (5–10 nT in mid-latitudes) and its magnitude appears to be correlated with the magnitude of the variation of the *z*-component of the IMF.
2. A study of the relationship of AQDs to the IMF polarity. Although at sunspot minimum the additional field on AQDs is larger on A-days than T-days, there is some evidence to support that this is not always so.
3. The ultimate causes of the additional field. A detailed study of the effect of ionospheric winds is still required.
4. The longitudinal variation of the additional field. Only measurements over a limited longitude range at one latitude and a "spot" measurement at another, have been made. A more detailed analysis over 360° of longitude and preferably in more than one latitude range, is required. The implications of the (limited)

observations of the longitudinal variation of the additional field, have not been addressed.

5. Is the AQD phenomenon observed at low latitudes the same as that observed at mid-latitude? If it is, is it linked via the additional current system or the mini-substorm? If not, the low latitude AQD needs to be looked at in more detail.

Acknowledgement

The author would like to thank Drs. G. M. Brown and D. M. Schlapp for reading the original manuscript and for their helpful suggestions.

REFERENCES

ARORA, B. R. (1972), *On Abnormal Quiet-day Variation in the Low-latitudes*, Indian J. Met. Geophys. *23* (2), 195–198.

BRIGGS, B. H. (1984), *The Variability of Ionospheric Dynamo Currents*, J. Atmos. Terr. Phys. *26*, 419–429.

BROWN, G. M. (1974), *A New Solar-terrestrial Relationship*, Nature *251*, 592–594

BROWN, G. M. (1975), *Sq Variability and Aeronomic Structure*, J. Atmos. Terr. Phys. *37*, 107–117.

BROWN, G. M. (1979), *New Methods for Predicting the Magnitude of Sunspot Maximum*, Solar-Terrestrial Predictions Proc. (Boulder, Colorado), *2*, 264–279.

BROWN, G. M. (1981), *Possible Use of (a) Solar Faculae and (b) the Interplanetary Magnetic Field as Heralds of a Solar Cycle Peak*, Solar Physics *74*, 125–129.

BROWN, G. M. (1986), *The Change in the Sq(H) Amplitude on Abnormal Quiet Days*, Geophys. J. R. Astr. Soc. *86*, 467–473.

BROWN, G. M., and WILLIAMS, W. R. (1969), *Some Properties of the Day-to-day Variability of Sq(H)*, Planet. Space Sci. *17*, 455–470.

BROWN, G. M., and WILLIAMS, D. C. (1971), *Pressure Variations in the Stratosphere and Ionosphere*, J. Atmos. Terr. Phys. *33*, 1321–1328.

BROWN, G. M., and BUTCHER, E. C. (1981), *The Use of Abnormal Quiet Days in Sq(H) for Predicting the Magnitude of Sunspot Maximum at the Time of the Preceding Minimum*, Planet. Space Sci. *29*, 73–77.

BUTCHER, E. C. (1980), *On the Location of the Ionospheric Current System Responsible for the Lunar and Solar Magnetic Variations*, Geophys. J. R. Astr. Soc. *63*, 775–782.

BUTCHER, E. C. (1982a), *An Investigation of the Causes of Abnormal Quiet Days in Sq(H)*, Geophys. J. R. Astr. Soc. *69*, 101–111.

BUTCHER, E. C. (1982b), *On the Latitude of the Sq(H) Focus at Sunspot Minimum*, Geophys. J. R. Astr. Soc. *69*, 113–120.

BUTCHER, E. C. (1987), *Currents Associated with Abnormal Quiet Days in Sq(H)*, Geophys. J. R. Astr. Soc. *88*, 111–123.

BUTCHER, E. C., and BROWN, G. M. (1980), *Abnormal Quiet Days and the Effect of the Interplanetary Field on the Apparent Position of the Sq Focus*, Geophys. J. R. Astr. Soc. *63*, 783–789.

BUTCHER, E. C., and BROWN, G. M. (1981a), *On the Nature of Abnormal Quiet Days in the Sq(H)*, Geophys. J. R. Astr. Soc. *64*, 513–526.

BUTCHER, E. C., and BROWN, G. M. (1981b), *The Variability of Sq(H) on Normal Quiet Days*, Geophys. J. R. Astr. Soc. *64*, 527–537.

CAMPBELL, W. H., and SCHIFFMACHER, E. R. (1985), *Quiet Ionospheric Currents of the Northern Hemisphere Derived from Geomagnetic Field Records*, J. Geophys. Res. *90*, 6475–6487.

CAMPBELL, W. H., and SCHIFFMACHER, E. R. (1987a), *Quiet Ionospheric Currents and Earth Conductivity Profile Computed from Quiet Time Geomagnetic Field Changes in the Region of Australia*, Aust. J. Phys. *40*, 73–87.

CAMPBELL, W. H., and SCHIFFMACHER, E. R. (1987b), *Quiet Ionospheric Currents of the Southern Hemisphere Derived from Geomagnetic Records*, submitted to J. Geophys. Res.

FORBES, J. M., and GILLETTE, D. F. (1982), *A Compendium of Theoretical Atmospheric Total Structures, Part I*, Technical Report AFGL-TR-82-0173 (I).

GREENER, J. G., and SCHLAPP, D. M. (1979), *A Study of the Day-to-day Variability of Sq over Europe*, J. Atmos. Terr. Phys. *41*, 217–223.

HASEGAWA, M. (1960), *On the Position of the Geomagnetic Sq Current System*, J. Geophys. Res. *65*, 1437–1447.

HIBBERD, F. H. (1981), *Day-to-day Variability of Sq Geomagnetic Field Variation*, Aust. J. Phys. *34*, 81–90.

LAST, B. J., EMILIA, A., and OUTHRED, A. K. (1976), *AQD Occurrence at Addis Ababa, Trivandrum and Alibag*, Planet. Space Sci. *24*, 567–572.

MANN, R. J. (1986), Ph.D. Thesis, University of Exeter.

MATSUSHITA, S. (1975), *IMF Polarity Effects on the Sq Current Focus Location*, J. Geophys. Res. *80*, 4751–4754.

MATSUSHITA, S., TARPLEY, J. D., and CAMPBELL, W. H. (1973), *IMF Sector Structure Effects on the Quiet Geomagnetic Field*, Radio Science 8, 963–972.

MAYAUD, P. N. (1973), *A Hundred Year Series of Geomagnetic Data 1868–1967*, IAGA Bulletin No. *33*.

MIZZI, C., and SCHLAPP, D. M. (1971), *On the Cause of a Type of Magnetic Variation on Quiet Days*, Planet. Space Sci. *19*, 273–274.

PENG FENGLIN, and TSCHU, K. K. (1987), *Analyses of Geomagnetic AQD in China Area*, submitted to Acta Geophysica Sinica (in Chinese).

PATIL, A., ARORA, B. R., and RASTOGI, R. G. (1985), *Seasonal Variations in the Intensity of the Sq Current System and its Focus Latitude over the Indian Region*, Indian J. Radio Space Phys. *14*, 131–135.

SASTRI, J. H., and MURTHY, B. S. (1978), *On the Occurrence of AQDs at Kodaikanal*, Indian J. Radio Space Phys. *7*, 62–64.

SCHLAPP, D. M. (1968), *World-wide Morphology of Day-to-day Variability of Sq*, J. Atmos. Terr. Phys. *30*, 1761–1776.

SCHLAPP, D. M., BUTCHER, E. C., and MANN, R. J. (1988), *Days of Small Range in Sq(H) on Quiet Days at Hartland*, Geophys. J. *95*, 633–639.

SINGH, A., and COLE, K. D. (1987), *A Numerical Model for the Ionospheric Dynamo—I. Formulation of a Numerical Technique*, J. Atmos Terr. Phys. *49*, 521–527.

TAKEDA, M. (1982), *Three-dimensional Structure of Ionospheric Currents Produced by Field-aligned Currents*, J. Atmos. Terr. Phys. *44*, 695–701.

TAKEDA, M., and ARAKI, T. (1984), *Time Variation of Instantaneous Equivalent Sq Current System*, J. Atmos. Terr. Phys. *46*, 911–915.

TAKEDA, M., and MAEDA, H. (1980), *Three-dimensional Structure of Ionospheric Currents—I. Currents Caused by Diurnal Tidal Winds*, J. Geophys. Res. *85*, 6895–6899.

THAKUR, N. K., and SONTAKKE, K. G. (1981), *On Abnormal Quiet Day Sq(Z) ranges in the Indian Region*, J. Atmos. Terr. Phys. *43*, 1107–1111.

(Received December 21, 1987, revised/accepted May 16, 1988)

PAGEOPH, Vol. 131, No. 3 (1989)

0033–4553/89/030485–24$1.50 + 0.20/0

The Equatorial Electrojet

C. A. REDDY[1]

Abstract—The typical quiet day variations of the equatorial electrojet (EEJ) current intensity with time of the day, season, sunspot number, and geomagnetic latitude are presented in terms of the corresponding variations of ΔH which is the deviation of the horizontal component (H) of the geomagnetic field from its steady nighttime level. The observed height structure of the current density in the EEJ as measured in rocket flights is presented, along with the theoretically computed structure. Theoretical model results on the polarization electric fields and east-west currents as generated by the local interactions of height-varying winds in the EEJ show large height gradients and reversals for both currents and electric fields; experimental evidence for the reality of such height structures is also shown. The characteristics of the counter-electrojet events are presented and the possible causative mechanisms are discussed critically.

Some typical experimental results are presented on the electric field changes in the EEJ which result from its sensitive response to electrodynamic disturbances in the magnetosphere and the auroral-polar latitude ionosphere during geomagnetic substorms and storms; and their implications are discussed. Possibilities for utilizing the EEJ as a very useful medium for important scientific studies on the larger space domain of ionosphere-magnetosphere system, on plasma waves, and on the earth's conductivity are emphasized.

Key words: Equatorial electrojet, counter-electrojet, electric fields, electric currents, magnetic field variations, wind effects, magnetosphere-ionosphere interactions, substorms, magnetic storm.

Introduction

The magnetometer recording of the daily variation of the earth's magnetic field at Huancayo, Peru, started in 1922 and it soon revealed that the range of daily variation of the horizontal component at this station close to the geomagnetic equator was larger by a factor of about 2.5 compared to that at other equatorial stations beyond several degrees from the geomagnetic equator. This characteristic feature was attributed to the existence of a band of intense electrical current centred at the dip equator and flowing in the 90–130 km height range. This current band was named the "Equatorial Electrojet" by SYDNEY CHAPMAN (1951). The equatorial electrojet (EEJ) turned out to be a multi-faceted phenomenon and it has been a subject for a variety of scientific studies in the last four decades.

---

[1] Space Physics Laboratory, Vikram Sarabhai Space Centre, Trivandrum 695 022, India.

The morphological characteristics of EEJ were studied with the help of ground-based magnetometer recordings; the height structure with rocket-borne magneto-meters; the latitudinal structure and variability with ground-based, rocket-borne and satellite-based magnetometers; the plasma instabilities and the electric fields with VHF and HF radars; and the counter electrojet events, the substorm and storm-related disturbances with ground magnetometers and VHF radars. Theoretical explanations and modeling studies kept up a vigorous pace with the progress of observational knowledge on most aspects of the EEJ. Nevertheless, our understanding of some basic aspects remains incomplete: the nondependence of their phase velocity on the elevation angle for the plasma waves responsible for type I irregularities in the EEJ, the regional or mesoscale character of the counter-electrojet events, the physical process involved in the sensitive response of the EEJ to electrodynamic changes in the magnetosphere and high latitude ionosphere are yet to be understood in quantitative terms. These aspects of the electrojet continue to pose intellectually challenging tasks to researchers on the earth's space environment.

In this article an overview of the equatorial electrojet is presented in such a way that the main features of the EEJ are first summarized, and then the physics of such aspects which are of continuing research interest are discussed in greater detail. It is suggested that the interested readers may read the earlier reviews for greater details of such aspects which are treated briefly here. The most relevant earlier reviews are those by KANE (1976), MAYAUD (1977), FORBES (1981) and REDDY (1981). The more recent paper by STENING (1985) provides useful discussion of some unresolved questions on the structure of the electrojet. In this paper the subject of plasma instabilities and ionization irregularities is not dealt with for two reasons: it is an aspect of EEJ which is not too closely linked with its geomagnetic and ionospheric aspects, and very good papers with brief as well as detailed reviews are available on this aspect (e.g., FEJER and KELLEY, 1980; SUDAN, 1983a,b; FARLEY, 1985; ST.-MAURICE et al., 1986).

Morphology

The main morphological characteristics of the electrojet on magnetically quiet days are summarized in Figure 1 in terms of the variations of (ΔH) or $(\Delta H)_{\text{Max}}$ with time of the day, the season, the sunspot cycle, and the geomagnetic latitude. ΔH is the deviation of the geomagnetic field's horizontal component (H) from the supposedly steady "base level" at night. The base level frequently exhibits deviations of 5–10 gammas (nT) from "steadiness" on a single night and from night to night, even on quiet days with $A_p \leq 6$. This is primarily due to the contribution of the distant magnetospheric currents at night, in addition to the typical 2–5 gamma errors involved in many standard magnetometer recordings and scalings. $(\Delta H)_{\text{Max}}$ is the maximum value of ΔH for a given day and it is denoted sometimes as $r(\Delta H)$, the 'daily range'. The observed ΔH at an electrojet station represents the strength of

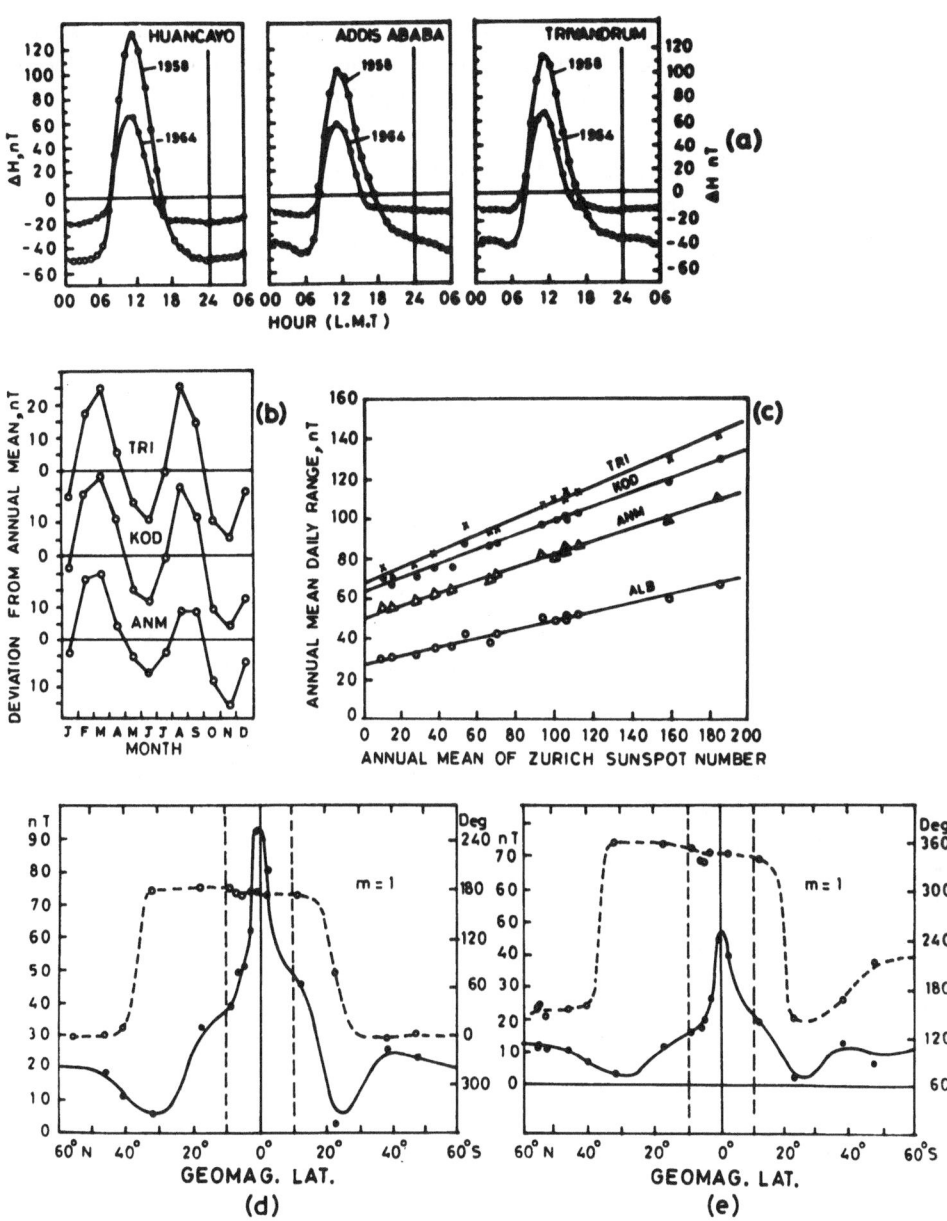

Figure 1

(a) Average daily variations of ΔH at 3 electrojet stations during 1958 (sunspot maximum year) and 1964 (sunspot minimum year), about the daily mean of ΔH as zero level; (b) deviations of monthly mean values of the daily range, $r(\Delta H)$, from its annual mean for the years 1958–1972; (c) variation of annual mean daily range with sunspot number at 4 Indian stations: (All from RASTOGI and IYER, 1976), (d) and (e) show the latitude variations of the amplitude (———) and phase (– – – –) of the 24-hour and 12-hour components of ΔH, respectively (from MATSUSHITA, 1967).

the overlying height-integrated current (J) upto a latitudinal distance of about $\pm 2.5°$ on either side of the recording station plus the contribution of the induced earth current to ΔH which is mostly in the range of 0.4 to 0.6 times the ΔH due to the overhead current.

The following main features of ΔH, representing the EEJ current intensity (J), can be noted from Figure 1:

 (i) The daily range $r(\Delta H)$ is different in different longitude zones, but the daily variation pattern is the same (Fig. 1a);

 (ii) The seasonal variation is such that $r(\Delta H)$ in equinoxes is about 1.7–1.8 times as large as that in solstices (Fig. 1b);

(iii) (ΔH) increases linearly with the annual mean sunspot number to become about twice as large during the sunspot maximum year as that during the sunspot minimum year (Fig. 1c);

(iv) The amplitudes of diurnal and semidiurnal components of the daily variation of ΔH at the magnetic equator are 2.2 and 2.6 times larger, respectively, than those at $10°$ gm latitude (Fig. 1d). The annual mean daily ranges at Trivandrum and Alibag also show an average ratio of 2.2 (Fig. 1c).

Coordinates of the relevent stations are: Huancayo (0.8°S, 75°W) Addis Ababa (5.2°N, 39°E), Trivandrum/Thumba (1.1°S, 77°E), Kodaikanal (0.6°N, 77.5°E), Annamalainagar (1.4°N, 79.7°E) and Alibag (9.2°N, 72.9°E). The geomagnetic dipole latitudes (for 1980) and the geographic longitudes are given.

Equatorial electrojet also exhibits very clear variation with the phase of the moon; and in fact the lunar daily variation of ΔH at the magnetic equator is 4–6 times larger than that at the nonelectrojet equatorial stations (MATSUSHITA and MAEDA, 1965; RAJA RAO, 1961; MATSUSHITA, 1967). Several details of the above morphological features are discussed in the papers referred to in the earlier reviews by KANE (1976), MAYAUD (1977), FORBES (1981), and REDDY (1981). Papers by ONWUMECHILI (1963a,b; 1967), CHAPMAN and RAJA RAO (1965), HUTTON (1967a,b), BURROWS (1970), FAMBITAKOYE and MAYAUD (1976a,b,c), RASTOGI and IYER (1976), STENING (1985) and ANANDA RAO and RAGHAVA RAO (1987) are good sources for information on morphology.

Though the theoretical models of the equatorial electrojet are well developed now (e.g., RICHMOND, 1973a,b; ANANDA RAO and RAGHAVA RAO, 1987), there are a few details which are not yet satisfactorily explained. The theoretical models often predict a larger increase of $r(\Delta H)$ than the observed $r(\Delta H)$ at the equator relative to that at stations away from the equator (e.g. RICHMOND, 1973b). The observed enhancement of $r(\Delta H)$ at the equator is by a factor of 4–6 for the lunar semi-diurnal component in contrast to the factor of about 2.5 for the solar components, and this remains unexplained (MATSUSHITA, 1967). The observed seasonal and sunspot cycle-related changes are caused predominantly by corresponding changes in the strength of the driving electric field (MATSUSHITA, 1977; VIKRAMKUMAR et al., 1984); and the contribution of conductivity changes seems to be rather small, though E-region electron densities are known to change with the solar zenith angle and the sunspot cycle. Thus, some of the morphological features require quantitatively exact theoretical explanations. Significant uncertainties remain about the exact magnitudes of the induced earth currents and their contributions to ground level ΔH. Regional variations in the conductivity of earth crust complicate the problem further (see DUHAU and OSELLA, 1982; DUHAU et al., 1982; STENING, 1985, for details on this aspect).

Fairly extensive measurements at Jicamarca with the coherent VHF backscatter radar showed that the daily variation of the drift velocity of plasma irregularities in the EEJ has an average structure resembling more of a square wave rather than a sinusoid (BALSLEY, 1973). Since the drift velocity is directly proportional to the east-west electric field in the EEJ, the radar observations imply an eastward electric field during 0700–2000 hrs (LT) and a westward field during 2000–0700 hrs (LT) on the average. The morning and evening reversal durations are short (i.e., about one hour) but the times of transition vary with season and sunspot cycle (FEJER et al., 1979a,b). The radar observations at Jicamarca and Thumba also show that the EEJ electric fields have a large day-to-day variability even on quiet days (BALSLEY, 1973; VIKRAMKUMAR et al., 1984; VISWANATHAN et al., 1987) apart from considerable fluctuations on a given day on time scales of half an hour to a few hours.

The morphological characteristics on magnetically disturbed days with different levels of disturbance cannot be discussed on a statistical basis, because the necessary statistical studies are not based on sufficiently extensive data sets. The limited studies indicate that the electric field value in daytime is reduced on the average on disturbed days compared to that on quiet days (REDDY et al., 1981; VISWANATHAN et al., 1987).

Structure of Equatorial Electrojet

The "structure" implies the height and latitude structure of the electric fields and currents in the electrojet. After the early theoretical work of HIRONO (1952), and BAKER and MARTYN (1953), elaborate theoretical models were developed in later years to explain in considerable detail the height structure, the latitude structure, and the longitude-dependent variations of the electrojet (UNTIEDT, 1967; SUGIURA and POROS, 1969; RICHMOND, 1973a,b). The height structure of the current at and near the magnetic equator was measured with rocket-borne magnetometers (MAYNARD and CAHILL, 1965; MAYNARD, 1967; DAVIS et al., 1967; SAMPATH and SASTRY, 1979). The results for the American sector and the Indian sector are shown in Figures 2a and 2b. The vertical structure is the same in both cases, with some minor differences. What is more significant is the 6–7 km difference (in Fig. 2a) between the heights of the computed and observed maxima of the current density (RICHMOND 1973b). The VHF radar measurements at Thumba on the drift velocity

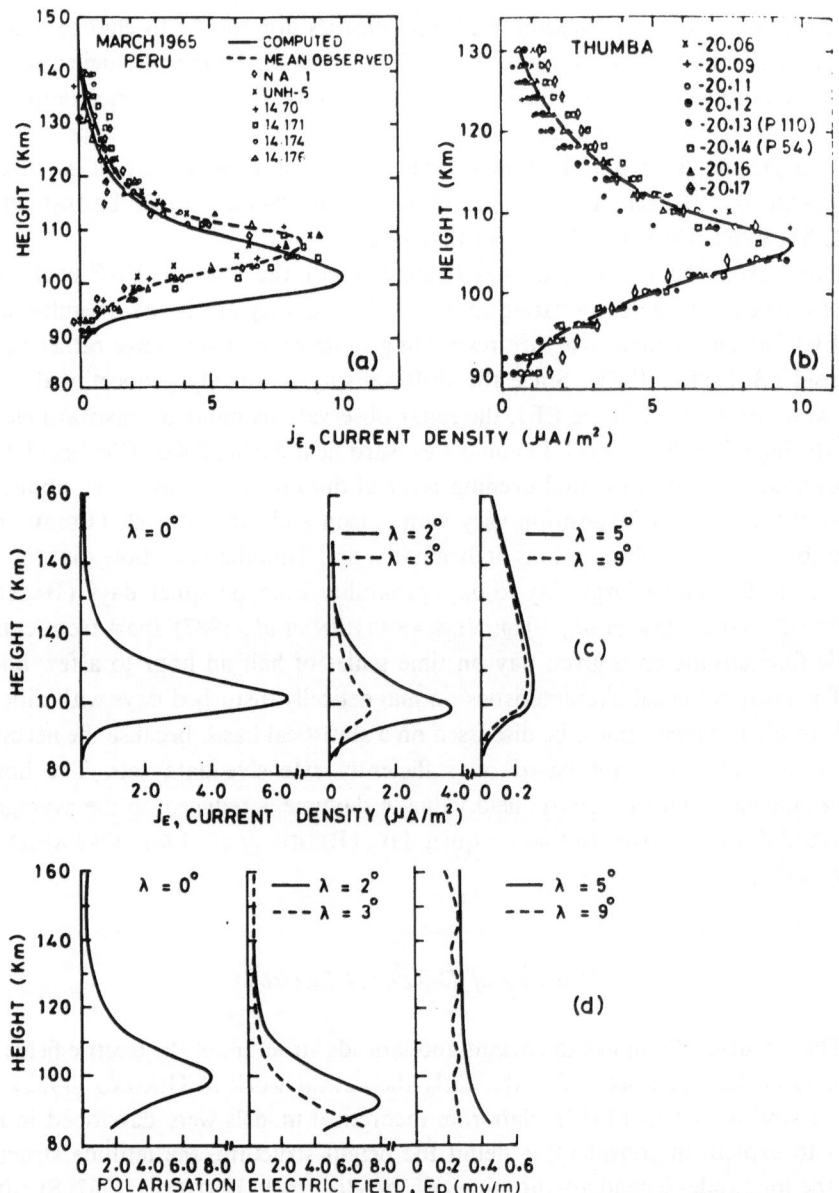

Figure 2

(a) Observed height structure of current density in EEJ near Peru (points and dotted curve) and the computed structure (continuous curve): (from RICHMOND, 1973b); (b) observed height structure of EEJ current density over Thumba (experimental points and their mean shown by continuous curve): (from SAMPATH and SASTRY, 1979); (c) computed height structure of eastward current density and (d) computed polarization electric field at different latitudes for $E_y - 0.30$ mV/m, $B - 0.39$ gauss and noontime conductivities integrated along magnetic flux tubes (from REDDY and DEVASIA, 1981).

profiles also show a similar but a smaller average difference between the theoretically expected and the observed heights of drift velocity maxima (REDDY et al., 1987). This radar result rules out the changes of the electron density (N_e) profiles as the *main* cause of the observed discrepancy, as proposed by STENING (1985), because the changes in N_e cannot cause changes in the mobility of ions or electrons in the E-region where N_e is a very small fraction of the neutral density (n). On the other hand, the changes in neutral density can produce changes in both electron density and ion, electron mobilities (STENING, 1986; REDDY et al., 1987). Thus, REDDY et al. (1987) pointed out that a possible increase of n over the model values by 50% to 100% could account for the observed discrepancy between the observed and computed heights of maximum drift velocity/current density in the electrojet. In a significant study, MAEDA (1981) showed that the neutral wind effects on the EEJ could satisfactorily explain the observed altitude of the current maximum. A third explanation is based on a hypothesis of electron collision frequencies in the EEJ being 4 times larger than the model values (GAGNEPAIN et al., 1977). None of the three explanations can be termed "satisfactory" as yet. An experimental verification of the above explanations is not easy but hopefully it may be expected to materialize soon.

Figure 2c shows the calculated model profiles of current density (j_E) at gm. latitudes of 0°, 2°, 3°, 5° and 9°, while the profiles of the vertical polarization electric field (E_p) at the same latitudes are shown in Figure 2d. The height extent where most of the current flows is the smallest and the peak current density is the largest at the equator; and the broadening of the current flow height range and the reduction of the peak current density with increasing latitude are quite rapid and large beyond 2° (Fig. 2c). The polarization electric field, which is orthogonal to the field lines in the magnetic meridional plane, shows the same characteristic (Fig. 2d). It is to be noted in particular that E_p is about the same as the east-west electric field E_y (which generates E_p) at 9°, whereas it is about 24 times larger than E_y at 0°. The above features of j and E_p have great significance for several scientific studies. For example, this intense and narrow current band has its location so well-defined in latitude and height that it can be used as an excellent source for the earth conductivity studies (e.g., RAJARAM et al., 1979; SINGH and AGARWAL, 1983; CAMPBELL, 1987). Secondly, the fluctuations of the large scale east-west electric field E_y become amplified by a factor of 24 in terms of the corresponding fluctuations in E_p which can be measured with a coherent VHF radar through the corresponding drift velocity fluctuations of the ionization irregularities in the electrojet (e.g., REDDY et al., 1981).

DAVIS et al. (1967) showed that on magnetically quiet days the ground level ΔH can be satisfactorily accounted for by the rocket measured height profiles of the electrojet current within the limits of measurement errors and other uncertainties. SAMPATH and SASTRY (1979) showed the existence of a linear relationship between the ground level ΔH values (in the range of 60–130 nT) and the rocket-measured peak current density values. Using the VHF radar measurements of the height structure of ionization irregularity drift velocities, VIKRAMKUMAR *et al.* (1987) deduced the electrojet current intensities and showed that the current intensities could satisfactorily account for the observed ground-level ΔH values and their time variations. Thus the quantitative relationship between the observed EEJ current and the magnetic field changes caused by it at ground level is firmly established. Discrepancies sometimes arise primarily from uncertainties in the magnitudes of the magnetospheric current contribution and the induced earth current contribution to the observed ΔH. The departures of the height structure of the current from the normal structure shown in Figure 2 are discussed in a later section on wind effects.

No *in situ* measurements of the east-west (E_y) or vertical polarization electric fields (E_p) in the equatorial electrojet are available. But it was demonstrated that the east-west drift speeds of type II irregularities, as measured with the VHF radars, are closely related to those of the east-west electric fields in the electrojet (BALSLEY, 1973; REDDY *et al.*, 1981). More recently, the absolute values of the electrojet electric fields have been deduced successfully from VHF radar measurements with a high time resolution (REDDY *et al.*, 1987; VISWANATHAN *et al.*, 1987). Such high resolution measurements are extremely useful in several scientific studies, including the study of micropulsations.

Neutral Wind Effects

The large scale eastward (westward) electric field driving the electrojet current in daytime (night-time) on quiet days originates in the dynamo action of global scale winds in the 80–150 km height region. Winds well beyond the electrojet latitudes contribute predominantly to the generation of E_y, with winds within electrojet latitudes making negligible contribution, as shown in a study by STENING (1977). However, the local interaction of winds with the magnetoplasma within the electrojet region can generate substantial polarization electric fields (E_p) perpendicular to the geomagnetic field in the magnetic meridional plane if the winds have significant vertical shears (KATO, 1973). Such wind-generated electric fields can in turn modify the vertical and latitude structures of the electrojet current (RICHMOND, 1973a; FAMBITAKOYE *et al.*, 1976; FORBES and LINDZEN, 1976; REDDY and DEVASIA, 1981; ANANDA RAO and RAGHAVA RAO, 1987).

Figure 3 illustrates the typical model results on the wind effects. Figure 3a shows some wind models and Figure 3b shows the height structures of the electrojet current at the magnetic equator primarily due to E_y but modified by the extra electric fields generated by the model winds shown in Figure 3a. The effect of the winds on the current at the equator is seen to be rather negligible. In contrast, at latitudes several degrees away from the equator, the winds are able to alter the current structure greatly and they even reverse its direction in some cases (Fig. 3c).

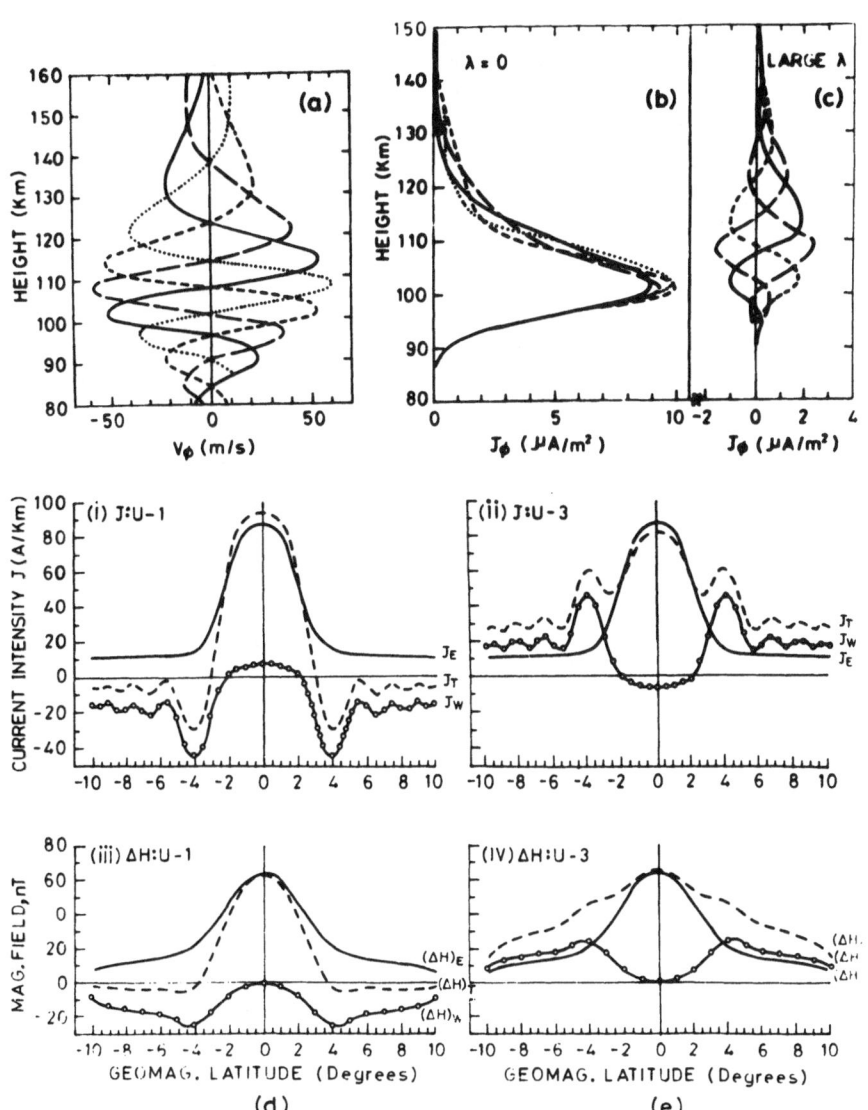

Figure 3

(a) Four models of assumed east-west wind; (b) current density at magentic equator and (c) at several degrees away from it due to an eastward electric field of 0.30 mV/m and $B = 0.293$ gauss plus each of the 4 wind profiles in (a): (from RICHMOND, 1973a). (d) and (e) show latitude structures of height-integrated current. Due to E_y alone (J_E,————), due to wind along (J_W,– – – –) and total current (J_T– – – –) for two wind profiles U-1 and U-2 similar to those in Figure 4a. Corresponding (ΔH) variations at ground level are shown in (iii) and (iv) (from REDDY and DEVASIA, 1981).

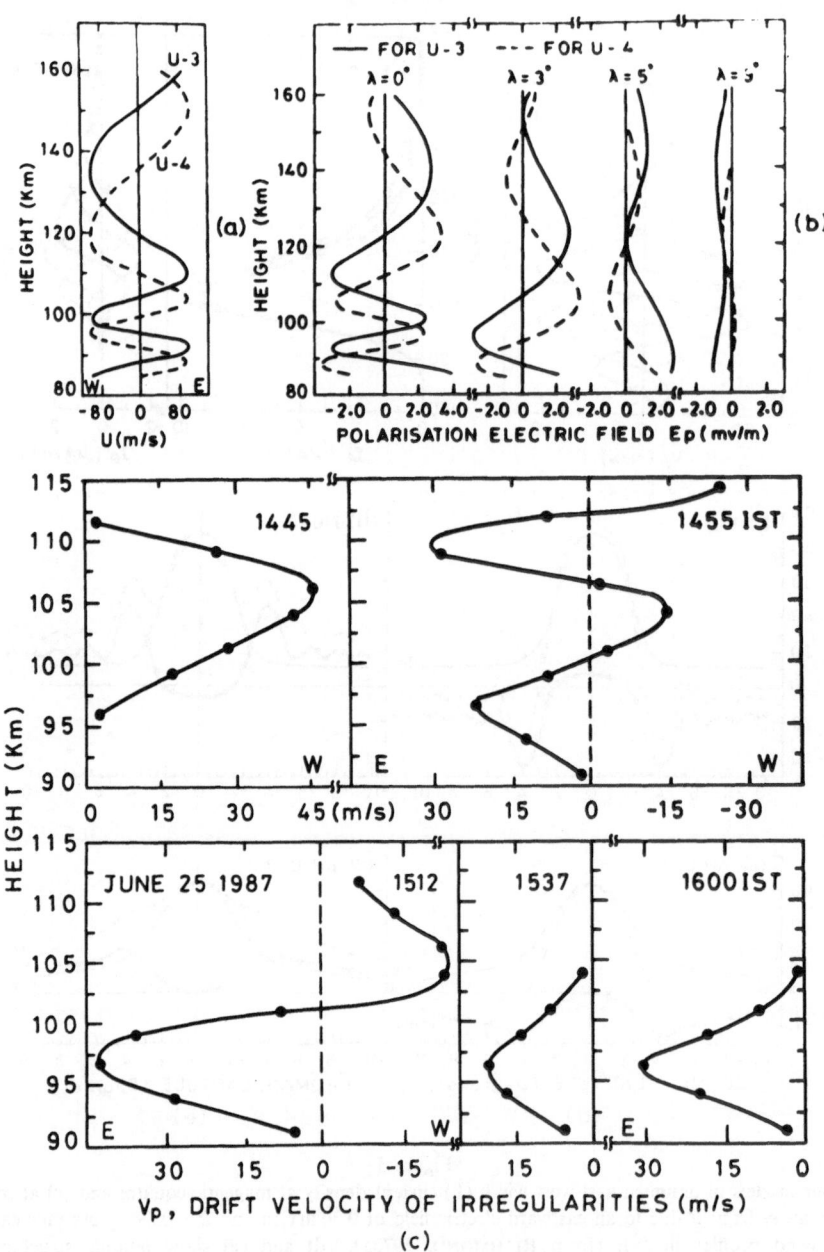

Figure 4
(a) Two assumed models of the east-west wind; (b) the polarization electric fields generated by U-3, U-4
(without E_y) at different gm. latitudes (from REDDY and DEVASIA, 1981); (c) height structures of drift
velocity of ionization irregularities in the EEJ at Thumba as measured with a VHF radar (SOMAYAJULU,
private communication).

Researchers have modeled the wind-induced modifications to the latitude structure of the height-integrated electrojet current intensity (J) and the corresponding surface level ΔH; and Figure 3d shows two typical examples of the effects of two height-varying zonal wind profiles with opposite direction of wind at all altitudes in the 85–160 km range. It is seen that the modification (due to winds) to J or ΔH is insignificant within $\pm 2°$ and it is quite large beyond $\pm 3°$. Thus the height structure and the latitude structure of the electrojet within $\pm 2°$ are very little affected by the local action of vertical wind shears of zonal wind which can, however, produce drastic changes beyond $\pm 3°$ gm. latitude. The drastic change in the $(\Delta H)_T$ variation with latitude from (iii) to (iv) in Figure 3d is to be noted in particular.

Within $\pm 2°$ gm. latitude, the polarization electric field E'_p generated by height varying zonal winds can become a substantial fraction of the corresponding field E_p due to the large scale east-west electric field E_y (REDDY and DEVASIA, 1981). Interestingly, the wind generated E'_p can have strong vertical gradients and reversals within $\pm 3°$ gm. latitude, but the gradients decreases rapidly beyond $\pm 3°$, as shown in Figure 4b. The interrelatedness of the changes in the vertical structure of E'_p with the (phase) changes in the vertical structure of the wind are to be noted from Figures 4a,b. Such reversing electric fields in the E-region can give rise to ionization convergence and the formation of thin ionization layers (known as blanketing sporadic-E layers which were observed at Thumba by REDDY and DEVASIA, 1973). Figure 4c shows a clear example of such height reversals in the drift velocity of ionization irregularities, as observed with a VHF radar at Thumba (unpublished results of VV Somayajulu). The drift velocity reversals occurred during the presence of blanketing sporadic-E layers detected in the simultaneous ionograms at Thumba. The normal westward drift with its typical height structure, under the action of an eastward electric field, is seen at 1445 hours. After 10 minutes, it becomes reversed to eastward drift around 95 km as well as 110 km, and it undergoes further changes during 1455–1600 hrs with no westward drift below 105 km. The rapid, large changes of the drift velocity imply corresponding changes in the polarization electric field E'_p generated by height-varying winds. This experimental result confirms the reality of the theoretically predicted fine structure or distortion of the electrojet by shearing neutral winds associated with gravity waves and tides.

Equatorial Counter-Electrojet

On some days, the post-noon decrease of ΔH at an equatorial electrojet station becomes abnormally rapid and large to such an extent that the ΔH value (with reference to the steady night-time level of ΔH) becomes negative for sometime and then it recovers back to its normal positive value before its decrease towards zero around sunset time (GOUIN, 1962; GOUIN and MAYAUD, 1967). The event lasts about 3 hrs and the maximum negative value of ΔH usually occurs between 1500 and 1600 hours (LT). This phenomenon is termed the "counter electrojet." Later it became known that the counter electrojet also occurs in the morning hours. Figure 5 shows a morning counter electrojet event in terms of latitude structure of ΔH and ΔZ during 0630 to 1230 hours (LT). On this day, instead of the normal increase at EEJ latitudes above the background value, a decrease takes place during 0630 to 0930, after which the normal increase is seen after 1030 hours. ΔZ shows corresponding changes. The most pertinent features to be noted are the reversal of the EEJ current in the entire latitude band of the EEJ and the absence of significant perturbation much beyond the electrojet latitudes. An abnormal decrease and recovery of ΔH, without its value becoming negative, is known as a "partial counter-electrojet." The morphological characteristics of the counter-electrojet (CEJ) are well documented (KRISHNA MURTHY and SEN GUPTA, 1972; ONWUMECHILI and AKASOFU, 1972; KANE 1976; RASTOGI, 1973, 1974a,b; FAMBITAKOYE and MAYAUD, 1976c; MAYAUD, 1977; MARRIOTT et al., 1979; BHARGAVA and SASTRI, 1979; and references therein). The main characteristics can be summarized as follows:

  (i) The occurrence of a full-fledged counter-electrojet (CEJ) is highly unpredictable for any given day;
 (ii) The average frequency and intensity of CEJ events are larger in the afternoon hours than in the morning hours;
(iii) The frequency of occurrence has a primary maximum in the winter season and a secondary maximum in the summer season with near zero values in the equinoxial months;
 (iv) The frequency of occurrence increases substantially with decreasing (annual mean) sunspot number;
  (v) Even during intense CEJ events, corresponding changes are not usually observed at other equatorial or mid-latitudes;
 (vi) Though it occurs usually (in varying intensities) in all longitude sectors, at the appropriate local time, the CEJ is sometimes seen only in one longitude sector (KANE, 1976).

The first and the last features in the above list are the most intriguing. The last feature, however, seems to occur rather rarely.

Before discussing the physics of the phenomenon, it is to be noted that the partial counter-electrojet is a regular phenomenon related to the lunar phase while the full counter-electrojet is an irregularly occurring event unrelated to lunar phase (MAYAUD, 1977). Moreover, the daytime electrojet reversals in association with geomagnetic bay disturbances at high latitudes (e.g., AKASOFU and MENG, 1968) are to be excluded from the definition of counter-electrojet as they constitute a distinctly different phenomenon with an entirely different origin.

Theoretical investigations of the CEJ events have proceeded on two different lines. In one approach, the possible reversal of the vertical polarization electric field

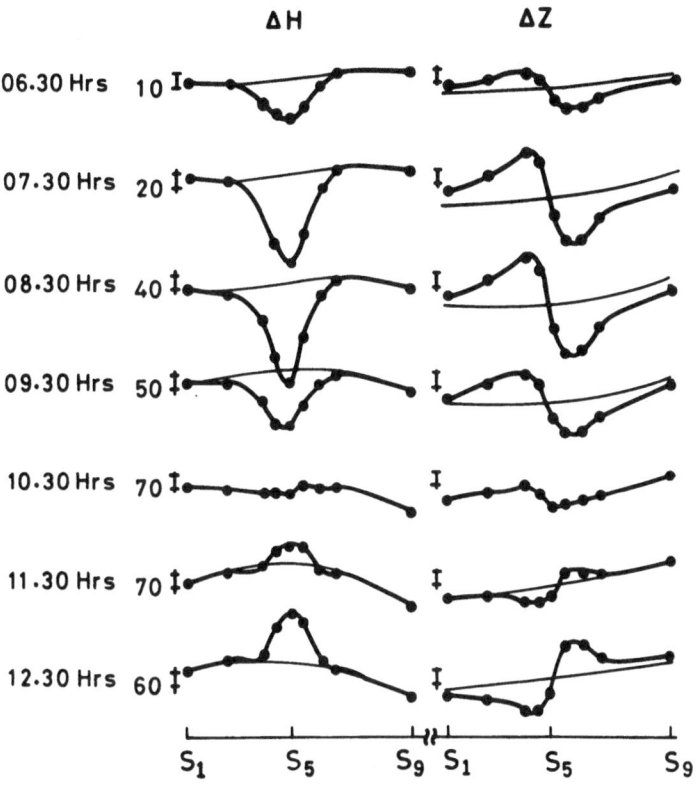

Figure 5

Latitude structure of ΔH and ΔZ during a morning counter-electrojet event in Central Africa. At left is shown the 15°E time. The dots show the measured values at 9 stations S_1 to S_9 covering 3015 km. Dip equator is close to S_5. The thin lines show the latitude variation of the "background" planetary components of ΔH and ΔZ, while the thick lines show the electrojet contribution. S_1 is north of dip equator. Vertical bar to left of each curve represents a scale of 10 nT; the numbers near the bottom of the bar denote the base level of the planetary component for ΔH; and for ΔZ, the base level is zero in all cases. On this day (June 6, 1969), the normal increase of ΔH towards dip equator appears only after 1030 hours. (From FAMBITAKOYE and MAYAUD, 1976c.)

(E_p) in the entire height range of the EEJ due to the local interaction of height-varying winds has been sought, but a negative result was obtained (RICHMOND, 1973a; FAMBITAKOYE et al., 1976; REDDY and DEVASIA, 1981; ANANDA RAO and RAGHAVA RAO, 1987; and references therein). In the other approach the possible reversal of the east-west electric field (E_y) in the EEJ due to an abnormal combination of global scale tidal wind modes generating E_y has been sought with considerable success (FORBES and LINDZEN, 1976; SCHIELDGE et al., 1973; MAR-RIOTT et al., 1979; HANUISE et al., 1983). These theoretical studies lead to a scenario of the counter-electrojet generation as described below.

Globally, the diurnal $S_1(1, -2)$ mode seems to be mainly responsible for the generation of the dynamo region electric fields necessary to drive the normal quiet-day Sq current system and the normal equatorial electrojet in daytime; but the dynamo action of the semi-diurnal tides $S_2(2, 2)$ and $S_2(2, 4)$ also contributes to the electric fields in the dynamo region. All three tidal modes are subject to seasonal changes with regard to their amplitudes and phases, and more importantly, they seem to experience large day-to-day changes, particularly in the December solstice. The model simulations show that the $S_2(2, 2)$ and $S_2(2, 4)$ modes produce electric fields which add up in phase in December solstice and in anti-phase in June solstice, while they produce very small electric fields during equinoxes. The simulation results explain thereby the seasonal characteristics as well as the afternoon plus morning occurrences of the CEJ. A recent analysis by SOMAYAJULU (1988) reveals that the amplitudes and phases of the diurnal as well as the semi-diurnal components of ΔH on the CEJ days are substantially different from those on normal electrojet days only at EEJ latitudes in conformity with the simulation results. The most intriguing feature to be quantitatively explained is the reversal of E_y only at the electrojet latitudes, and over a limited longitude extent sometimes. The abnormal combination(s) of tidal modes and other winds producing the above feature is yet to be identified. But such abnormal combination/behaviour of tidal winds in the dynamo region is quite plausible. Until we gain more observational knowledge of the dynamo region tidal winds, particularly their variabilities, there does not seem to be a necessity to invoke far-fetched or exotic mechanisms to explain the counter-electrojet. All the main features of the CEJ may find a quantitatively satisfactory explanation when a full view is gained of the extreme variabilities of the dynamo region wind field.

The CEJ events have dramatically focussed our attention on the possible large variabilities of tidal winds in the dynamo region, though such variabilities were suspected much earlier as the possible cause of the large shifts in the Sq current foci (HASEGAWA, 1960; MATSUSHITA, 1960). The reported abnormal phase variability of $Sq(H)$ at the equatorial stations (LAST et al., 1976; SASTRI, 1982) and at middle latitudes (BROWN and WILLIAMS, 1969) are also likely to be caused by some abnormal condition of the tidal winds on some days.

Equatorial Electrojet Response to Solar Wind-magnetosphere-ionosphere Interactions

Currently, there is a great scientific interest in understanding the changes in the equatorial electrojet in response to the electrodynamic processes involved in the linked interactions of the solar wind with the magnetosphere and of the magnetosphere with the ionosphere. This is understandable because the dynamo region electric fields are communicated to higher altitudes along the highly conducting geomagnetic field lines. They greatly influence the plasma motions, plasma distributions, temperatures, winds (through plasma motions), and other equatorial ionospheric process up to high altitudes.

Based on ground magnetometer recordings alone, the equatorial dynamo region currents in general and the equatorial electrojet in particular were known, for a long time, to have consistent and near instantaneous response to geomagnetic disturbances at high latitudes (ONWUMECHILI and OGBUEHI, 1962; AKASOFU and MENG 1968; ONWUMECHILI et al., 1973). The close relationship between the high latitude as well as low latitude geomagnetic disturbances and the solar wind-magnetospheric interaction processes was also demonstrated about two decades ago (NISHIDA, 1968a,b, 1971; NISHIDA et al., 1966; NISHIDA and KOKUBUN, 1971). In the last decade the observations with coherent and incoherent scatter radars on the equatorial ionospheric electric fields have confirmed that the disturbances in the dynamo region electric fields at equatorial latitudes originate in corresponding electrodynamic disturbances at high latitudes (MATSUSHITA and BALSLEY, 1972; REDDY et al., 1979, 1980; GONZALES et al., 1979, 1983; FEJER et al., 1979). Specifically, it was pointed out by REDDY et al. (1980) that even minor disturbances at auroral latitudes on comparatively quiet days with $A_p < 6$ produced clear signatures of near simultaneous disturbances in the equatorial electrojet electric fields. Figure 6 shows one example of small fluctuations in ΔH that are highly coherent with each other at widely separated stations at equatorial, middle and high latitudes, and also coherent with the fluctuations of the VHF radar signals backscattered from the ionization irregularities in the EEJ at Thumba. The signal fluctuations represent quite well the electric field fluctuations in the EEJ (REDDY and DEVASIA, 1976).

VHF radar observations at Thumba have also revealed that the magnetospheric and high latitude electrodynamic changes which are concurrent in time with the polar cusp latitude changes during a geomagnetic storm give rise to a large westward electric field in the EEJ varying on a time scale of 9–10 hours (SOMAYA-JULU et al., 1985). Very recently, the analysis of VHF radar data for the March 22, 1979 1020 UT substorm event (selected for CDAW-6) has shown that the magnetospheric convection electric field seems to penetrate to equatorial latitudes before the shielding by ring current particles becomes effective (SOMAYAJULU et al., 1987). Both of the above interesting results are shown in Figure 7. The results in Figures 6 and 7 demonstrate that the equatorial electrojet electric fields respond very sensitively to the electrodynamic changes, on a variety of time scales, in the magnetosphere-ionosphere system arising from solar wind-magnetosphere interactions. Qualitatively this characteristic behaviour is not surprising because the ionospheric dynamo region is a conducting shell (in terms of ionospheric conductivities), and in such a shell any disturbances of electric potential due to changes in the potential, conductivity or current in one part of the system are bound to produce potential distribution changes in all parts of the system.

In an attempt to understand the penetration of the electric fields from high to low geomagnetic latitudes, theoretical models have been developed by different

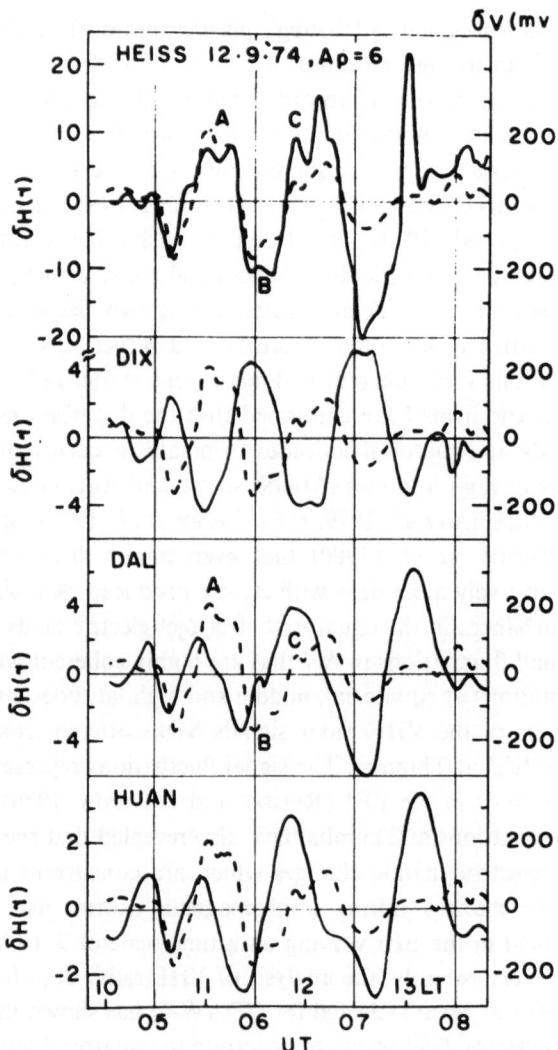

Figure 6

Detrended fluctuations of ΔH at Heiss Island (71°N, 264°E), Dixon (63°N, 81°E), Dallas (43°N, 263°E) and Huancayo (0.8°S, 285°E) are shown by solid lines. The superposed broken line is identical in all cases and it shows the detrended fluctuations of the VHF radar signal strength backscattered from the electrojet over Thumba (1.1°S, 77°E). Geomagnetic latitude and geographic longitude of the stations are given. Signal fluctuations represent electric field fluctuations in EEJ. LT shown is for Thumba (from REDDY et al., 1980).

researcher using different approaches (NOPPER and CAROVILLANO, 1978; NISBET et al., 1978; KIKUCHI et al., 1978; KIKUCHI and ARAKI, 1979; KAMIDE and MAT-SUSHITA, 1979a,b; BLANC and RICHMOND, 1980). More comprehensive models, including the more important physical processes of magnetosphere-ionosphere interactions at high latitudes, have also been developed in the last decade (WOLF,

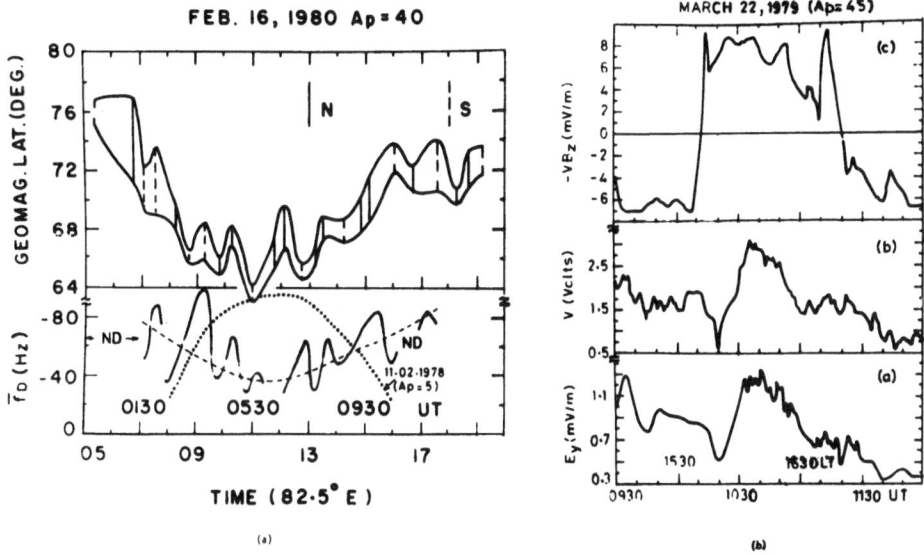

Figure 7

(a) Doppler frequency variations (representing electric field variations in EEJ) of VHF radar backscatter signal at Thumba in phase with polar cusp latitude changes (from SOMAYAJULU et al., 1985). (b) The variations of eastward electric field in the EEJ over Thumba as deduced from VHF radar measurements during 1020 UT substorm event of March 22, 1979 following the southward turning of B_z at 1010 UT (from SOMAYAJULU et al., 1987).

1970, 1974; WOLF et al., 1982; SPIRO et al., 1981; BLANC, 1983; SENIOR and BLANC, 1984, 1987; BLANC and CAUDAL, 1985). While the models are quite successful in reproducing several essential features of electric field characteristics at high latitudes and middle latitudes, the observed characteristics of disturbance electric fields in the equatorial electrojet are not satisfactorily reproduced so far by the models. Figure 8 shows a comparison of the disturbance electric fields at the magnetic equator as predicted by the SENIOR and BLANC (1984) model with the experimental result obtained at Thumba from VHF backscatter radar observations (VISWANATHAN et al., 1987). The large differences between model-predicted and observed values at several local times are obvious. The limitations of the assumptions and approximations in the models in relation to equatorial latitudes are yet to be understood. The observational results are also limited in extent, and statistical significance of the average characteristics of disturbance electric fields in the equatorial electrojet is yet to be assessed. Nevertheless, observations of the type shown in Figures 7a,b, have little ambiguity and successful modeling of such events would result in significant improvement in our understanding of the complex electrodynamics involved in solar wind-magnetosphere-ionosphere interactions.

Since the auroral and polar cap electric fields and currents (in the dynamo region) are known to be closely controlled by solar wind-magnetospheric inter-

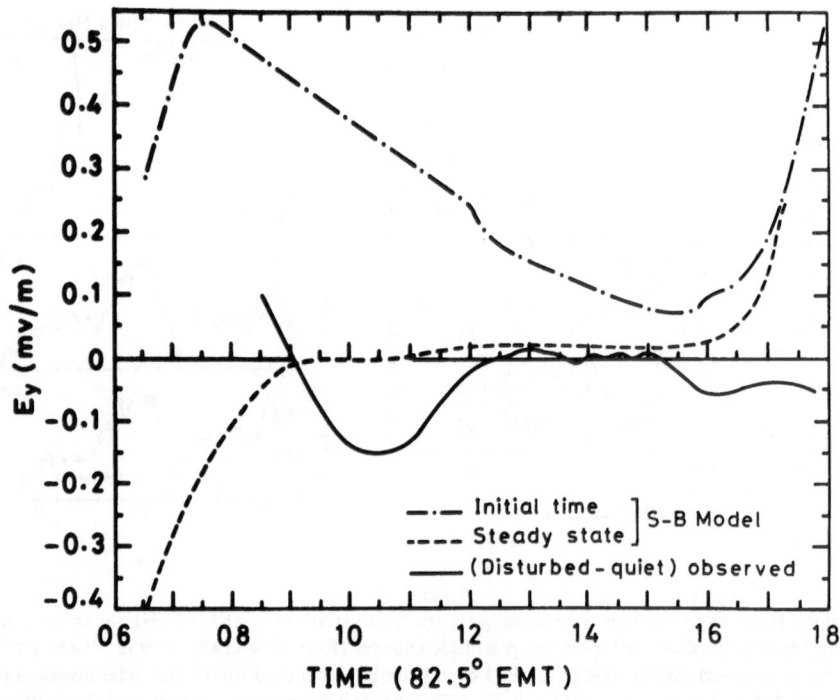

Figure 8
Comparison of the observed disturbance electric field variation in daytime at Thumba, as measured with a VHF radar (derived from VISWANATHAN *et al.*, 1987), with the model values of SENIOR and BLANC (1984).

actions, attempts have been made to assess the direct correlation between the time variations of the interplanetary magnetic field (IMF) as measured by satellites and the radar-measured or the ground level magnetometer-measured changes of electric fields or currents in the equatorial ionosphere (FEJER *et al.*, 1979; PATEL, 1978; GALPERIN *et al.*, 1978; RASTOGI and PATEL, 1975; MATSUSHITA, 1977; KELLEY *et al.*, 1979). The results indicate the presence of good correlation in some cases and also the absence of correlated changes in the IMF and equatorial ionospheric electric fields.

The results of correlation observed in some cases are also found to be difficult to interpret because the IMF effect on the equatorial ionospheric electric fields is not likely to be direct or simple; an intervening chain of physical processes is involved and the knowledge of this chain is as yet largely qualitative as reflected in the arguments of the practitioners of the correlation studies.

A clear semi-quantitative understanding of the inter-relationship between the IMF behaviour and the equatorial ionospheric electric fields has to await a more quantitative understanding of the solar wind-magnetosphere interaction processes. Progress in three types of scientific studies will contribute to the goal of

understanding the interactive processes involving IMF at one end and the equatorial ionospheric changes at the other end: (a) extensive measurements of good
quality on electric fields at a selected number of locations; (b) modeling studies on
magnetosphere-ionosphere interactions; and (c) extensive analyses of magnetic field
data (and related data sets) with a physical insight into the nature of the problem
at hand. The first aspect is already in implementation in terms of radar measurements of ionospheric electric fields at several locations (e.g., GONZALES et al.,
1983). On the second aspect, impressive strides have been made in the last decade
as noted, and much progress can be expected in the coming years. The third aspect
needs more vigorous action, especially with regard to middle and equatorial
latitudes.

Concluding Remarks

Neither the quiet day equatorial electrojet nor the counter-electrojet is an
enigma anymore, but the counter-electrojet still poses some intriguing questions
about its causative physical processes. Most of the main characteristics of the
electrojet and their causative mechanisms are understood in considerable detail. The
absence of quantitative information on the dynamo region winds on a global scale
is conspicuous, and this is the cause of the absence of quantitative accuracies in the
theoretical model results on the observed variabilities of the EEJ which can take
extreme forms like the counter-electrojet. A beginning has been made in understanding the effects of wind variability on global scale currents and electric fields,
and also in understanding the local interactions of the neutral winds with the
electrojet and the modifications resulting therefrom in the structure of its electric
field and current. We have to go a long way still in exploring this aspect more
incisively and understanding the full ramifications of such interactions to the
equatorial ionosphere as a whole.

It is the disturbed day behaviour of the equatorial electrojet for which currently
there is no satisfactory theoretical model. This aspect needs more attention from
theoreticians, experimenters and data analysts, and preferably a close interaction of
all, because the equatorial electrojet disturbances are the extensions of the very
complex electrodynamic changes in the magnetosphere-ionosphere system resulting
from solar wind-magnetosphere interactions. The "amplification" of the global
scale east-west electric field by a factor of about 24 into a vertical polarization field
in the EEJ (causing a high speed Hall drift of the electrons) provides a unique
advantage in monitoring and interpreting some crucial aspects of the very complex
solar wind-magnetosphere-ionosphere interactions.

The equatorial electrojet continues to be a very useful natural plasma laboratory
for the study of plasma instabilities of various types. In spite of the vast progress
made in the last 25 years, several fascinating studies, including a few basic studies,

on the plasma instabilities and plasma wave characteristics are yet to be carried out. Hardly any beginning has been made on active, controlled experiments within the magneto-plasma of the electrojet with its unique geometry and boundary conditions.

Finally, the potential of the equatorial electrojet current as a well-defined source for earth conductivity studies is to be fully exploited. Thus, the equatorial electrojet has acquired new dimensions in terms of the identified new possibilities for fascinating scientific studies on the larger space environment of the earth and the application-oriented studies on the earth's conductivity.

Acknowledgements

The author is thankful to Dr. V. V. Somayajulu for making available the unpublished result in Figure 4c and for useful discussions. The good draughting work for the figures by Mr. K. Thilakan and the able typing by Mr. K. A. Somasundaran have been most helpful.

REFERENCES

ANANDA RAO, B. G., and RAGHAVA RAO, R. (1987), *Structural Changes in the Current Fields of the Equatorial Electrojet Due to Zonal and Meridional Winds*, J. Geophys. Res. *92*, 2514–2526.

AKASOFU, S. I., and MENG, C. I. (1968), *Low Latitude Negative Bays*, J. Atmos. Terr. Phys. *30*, 227–241.

BAKER, W. G., and MARTYN, D. F. (1953), *Electric Currents in the Ionosphere*, Phil. Trans. Roy. Soc. London A*246*, 281–294.

BALSLEY, B. B. (1973), *Electric Fields in the Equatorial Ionosphere: A Review of Techniques and Measurements*, J. Atmos. Terr. Phys. *35*, 1035–1044.

BHARGAVA, B. N., and SASTRI, N. S. (1979), *Some Characteristics of the Occurrence of the Afternoon Counter-electrojet Events in the Indian Region*, J. Geomag. Geoelectr. *31*, 97–101.

BLANC, M. (1983), *Magnetospheric Convection Effects at Midlatitudes, 3. Theoretical Derivation of the Disturbance Convection Pattern Inside the Plasmasphere*, J. Geophys. Res. *88*, 235–251.

BLANC, M., and RICHMOND, A. D. (1980), *The Ionospheric Disturbance Dynamo*, J. Geophys. Res. *85*, 1669–1686.

BLANC, M., and CAUDAL, G. (1985), *The Spatial Distribution of Magnetospheric Convection Electric Fields at Ionospheric Altitudes: A Review*, Ann. Geophys. *3*, 27–42.

BROWN, G. M., and WILLIAMS, W. R. (1969), *Some Properties of the Day-to-day Variability of Sq(H)*, Planet. Space Sci. *17*, 455–466.

BURROWS, K. (1970), *The Day-to day Variability of the Equatorial Electrojet in Peru*, J. Geophys. Res. *75*, 1319–1323.

CAMPBELL, W. H. (1987), *Introduction to Electrical Properties of the Earth's Mantle*, Pure Appl. Geophys. *125*, 193–204.

CHAPMAN, S. (1951), *The Equatorial Electrojet as Detected from the Abnormal Electric Current Distribution above Huancayo, Peru and Elsewhere*, Arch. Meteorol. Geophys. Bioklimatol. A *44*, 368–390.

CHAPMAN, S., and RAJA RAO, K. S. (1965), *The H and Z Variations Along and Near the Equatorial Electrojet in India, Africa and the Pacific*, J. Atmos. Terr. Phys. *27*, 559–581.

DAVIS, T. N., BURROWS, K., and STOLARIK, J. D. (1967), *A Latitude Survey of the Equatorial Electrojet with Rocket-borne Magnetometers*, J. Geophys. Res. *72*, 1845–1861.

DUHAU, S., and OSELLA, A. M. (1982), *A Correlation Between Measured E-region Current and Geomagnetic Daily Variation at Equatorial Latitude*, J. Geomag. Geoelect. *34*, 213–224.

DUHAU, S., ROMANELLI, L., and HIRRSCH, F. A. (1982), *Indication of Anamalous Conductivity at the Nigerian Dip Equator*, Planet. Space Sci. *30*, 97–100.

FAMBITAKOYE, O., and MAYAUD, P. N. (1976a), *Equatorial Electrojet and Regular Daily Variation S_R—I. A Determination of the Equatorial Electrojet Parameters*, J. Atmos. Terr. Phys. *38*, 1–17.

FAMBITAKOYE, O., and MAYAUD, P. N. (1976b), *Equatorial Electrojet and Regular Daily Variation S_R—II. The Centre of the Equatorial Electrojet*, J. Atmos. Terr. Phys. *38*, 19–26.

FAMBITAKOYE, O., and MAYAUD, P. N. (1976c), *Equatorial Electrojet and Regular Daily Variation S_R—IV. Special Features in Particular Days*, J. Atmos. Terr. Phys. *38*, 123–134.

FAMBITAKOYE, O., MAYAUD, P. N., and RICHMOND, A. D. (1976), *Equatorial Electrojet and Regular Daily Variation S_R—III. Comparison of Observations with a Physical Model*, J. Atmos. Terr. Phys. *38*, 113–121.

FARLEY, D. T. (1985), *Theory of Equatorial Electrojet Plasma Waves: New Developments and Current Status*, J. Atmos. Terr. Phys. *47*, 729–737.

FEJER, B. G., FARLEY, D. T., WOODMAN, R. F., and CALDERON, C. (1979a), *Dependence of Equatorial F-region Vertical Drifts on Season and Solar Cycle*, J. Geophys. Res. *84*, 5792–5796.

FEJER, B. G., GONZALES, C. A., FARLEY, D. T., KELLEY, M. C., and WOODMAN, R. F. (1979b), *Equatorial Electric Fields During Magnetically Disturbed Conditions, 1. The Effect of the Interplanetary Magnetic Field*, J. Geophys. Res. *84*, 5797–5802.

FEJER, B. G., and KELLEY, M. C. (1980), *Ionospheric Irregularities*, Reviews of Geophys. and Space Phys. *18*, 401–454.

FORBES, J. M. (1981), *The Equatorial Electrojet*, Rev. Geophys. *19*, 469–504.

FORBES, J. M., and LINDZEN, R. S. (1976), *Atmospheric Solar Tides and their Electrodynamic Effects, II. The Equatorial Electrojet*, J. Atmos. Terr. Phys. *38*, 911–920.

GAGNEPAIN, J., CROCHET, M., and RICHMOND, A. D. (1977), *Comparison of Equatorial Electrojet Models*, J. Atmos. Terr. Phys. *39*, 1119–1124.

GALPERIN, Yu. I., PONOMAREV, V. N., and ZOSIMOVA, A. G. (1978), *Equatorial Ionospheric Anomaly and Interplanetary Magnetic Field*, J. Geophys. Res. *83*, 4265–4272.

GONZALES, C. A., KELLEY, M. C., FEJER, B. J., VICKERY, J. F., and WOODMAN, R. F. (1979), *Equatorial Electric Fields during Disturbed Conditions, 2. Implications of Simultaneous Auroral and Equatorial Measurements*. J. Geophys. Res. *84*, 5803–5812.

GONZALES, C. A., KELLEY, M. C., BEHNKE, R. A., VICKERY, J. F., WAND, R., and HOLT, J. (1983), *On the Latitudinal Variations of the Ionospheric Electric Field during Magnetospheric Disturbances*, J. Geophys. Res. *88*, 9135–9144.

GOUIN, P. (1962), *Reversal of the Magnetic Daily Variation at Addis Ababa*, Nature *193*, 1145–1146.

GOUIN, P., and MAYAUD, P. N. (1967), *A propos de l'existence possible d'un contre electrojet and latitudes magnetiques equatoriales*, Ann. Geophys. *23*, 41–47.

HANUISE, C., MAZAUDIER, C., VILA, P. BLANC, M., and CROCHET, M. (1983), *Global Dynamo Simulation of Ionospheric Currents and their Connection with the Equatorial Electrojet: A Case Study*, J. Geophys. Res. *88*, 253–270.

HASEGAWA, M. (1960), *On the Position of the Focus of the Geomagnetic Sq Current System*, J. Geophys. Res. *65*, 1437–1447.

HIRONO, M. (1952), *A Theory of Diurnal Magnetic Variations in Equatorial Regions and Conductivity of the Ionospheric E Region*, J. Geomag. Geoelec. *4*, 7–21.

HUTTON, R. (1967a), *Sq Currents in the American Equatorial Zone during the I.G.Y. I. Seasonal Effects*, J. Atmos. Terr. Phys. *29*, 1411–1428.

HUTTON, R. (1967b), *Sq Currents in the American Equatorial Zone during the I.G.Y. II Day-to-day Variability*, J. Atmos. Terr. Phys. *29*, 1429–1442.

KAMIDE, Y., and MATSUSHITA, S. (1979a), *Simulation Studies of Ionospheric Electric Fields and Currents in Relation to Field-aligned Currents 1. Quiet Periods*, J. Geophys. Res. *84*, 4083–4098.

KAMIDE, Y., and MATSUSHITA, S. (1979b), *Simulation Studies of Ionospheric Electric Fields and Currents in Relation to Field-aligned Currents, 2. Substorms*, J. Geophys. Res. *84*, 4099–4115.

KANE, R. P. (1976), *Geomagnetic Field Variations*, Space Sci. Rev. *18*, 413–540.

KATO, S. (1973), *Electric Field and Wind Motion at the Magnetic Equator*, J. Geophys. Res. *78*, 757–762.

KELLEY, M. C., FEJER, B. J., and GONZALES, C. A. (1979), *An Explanation for Anomalous Equatorial Ionospheric Electric Fields Associated with a Northward Turning of Interplanetary Magnetic Field*, Geophys. Res. Lett. *6*, 301–304.

KIKUCHI, T., and ARAKI, T. (1979), *Horizontal Transmission of the Polar Electric Field to the Equator*, J. Atmos. Terr. Phys. *41*, 927–936.

KIKUCHI, T., ARAKI, T., MAEDA, H., and MAEKAWA, K. (1978), *Transmission of Polar Electric Field to Equator*, Nature *273*, 650–651.

KRISHNA MURTHY, B. V., and SEN GUPTA, K. (1972), *Disappearance of Equatorial Es Associated with Magnetic Field Depressions*, Planet. Space Sci. *20*, 371–378.

LAST, B. J., EMILIA, D. A., and OUTHRED, A. K. (1976), *AQD Occurrence at Addis Ababa, Trivandrum and Alibag*, Planet. Space Sci. *24*, 567–572.

MAEDA, K. I. (1981), *Internal Structure of the Equatorial Ionospheric Dynamo*, J. Atmos. Terr. Phys. *43*, 393–401.

MARRIOTT, R. T., RICHMOND, A. D., and VENKATESWARAN, S. V. (1979), *On Quiet-time Equatorial Electrojet and Counter-electrojet*, J. Geomagr. Geoelectr. *31*, 311–340.

MATSUSHITA, S. (1960), *Seasonal and Day-to-day Changes of the Central Position of the Sq Overhead Current System*, J. Geophys. Res. *65*, 3835–3839.

MATSUSHITA, S., *Solar quiet and lunar daily variation fields, in phenomena*, In *Phys. Geomagn. Phenomena*, Vol. 1 (Academic Press, New York 1967) pp. 302–420.

MATSUSHITA, S. (1977), *IMF Effects on the Equatorial Geomagnetic Field and Ionosphere—A Review*, J. Atmos. Terr. Phys. *39*, 1207–1215.

MATSUSHITA, S., and BALSLEY, B. B. (1972), *A Question of DP-2*, Planet. Space Sci. *8*, 1259–1267.

MATSUSHITA, S., and MAEDA, H. (1965), *On the Geomagnetic Lunar Daily Variation Field*, J Geophys. Res. *70*, 2559–2578.

MAYNARD, N. C. (1967), *Measurements of Ionospheric Currents off the Coast of Peru*, J. Geophys. Res. *72*, 1863–1875.

MAYNARD, N. C., and CAHILL, L. J. (1965), *Measurements of the Equatorial Electrojet over India*, J. Geophys. Res. *70*, 5923–5936.

MAYAUD, P. N. (1977), *The Equatorial Counter Electrojet—A Review of its Geomagnetic Aspects*, J. Atmos. Terr. Phys *39*, 1055–1070.

NISBET, J. S., MILLER, M. J., and CARPENTER, L. A. (1978), *Currents and Electric Fields in the Ionosphere due to Field-aligned Currents*, J. Geophys. Res. *83*, 2647–2657.

NISHIDA, A. (1968a), *Geomagnetic DP-2 Fluctuations and Associated Magnetospheric Phenomena*, J. Geophys. Res. *73*, 1795–1803.

NISHIDA, A. (1968b), *Coherence of Geomagnetic DP-2 Fluctuations with Interplanetary Magnetic Variation*, J. Geophys. Res. *73*, 5549–5559.

NISHIDA, A. (1971), *DP-2 and Polar Substorm*, Planet. Space Sci. *19*, 205–221.

NISHIDA, A., IWASAKI N., and NAGATA, T. (1966), *The Origin of Fluctuations in the Equatorial Electrojet: A New Type of Geomagnetic Variation*, Ann. Geophys. *22*, 478–484.

NISHIDA, A., and KOKUBUN, S. (1971), *New Polar Magnetic Disturbances: S_q^p, SP, DPC and DP-2*, Rev. Geophys. and Space Phys. *9*, 417–425.

NOPPER, R. W., and CAROVILLANO, R. L. (1978), *Polar-equatorial Coupling during Magnetically Active Periods*, Geophys. Res. Lett. *5*, 699–702.

ONWUMECHILI, A. (1963a), *Separation of Semidiurnal Tidal Effect on Individual Days and Some Equatorial Features of the Geomagnetic Lunar Tide*, J. Geophys. Res. *68*, 2425–2433.

ONWUMECHILI, A. (1963b), *Lunar Effects on the Diurnal Variation of the Geomagnetic Horizontal Field Near the Magnetic Equator*, J. Atmos. Terr. Phys. *25*, 55–70.

ONWUMECHILI, A., *Physics of Geomagnetic Phenomena*, Vol. 1 (ed. Matsushita, S., and Campbell, W. H.) (Academic Press, New York 1967) pp. 425–507.

ONWUMECHILI, A., and AKASOFU, S. I. (1972), *On the Abnormal Depression of Sq(H) under the Equatorial Electrojet in the Afternoon*, J. Geomag. Geoelectr. *24*, 161–173.

ONWUMECHILI, A., KAWASAKI, K., and AKASOFU, S. I. (1973), *Equatorial Electrojet and Polar Magnetic Variations*, Planet. and Space Sci. *21*, 1–16.

ONWUMECHILI, A., and OGBUEHI, P. O. (1962), *Fluctuations in the Geomagnetic Horizontal Field*, J. Atmos. Terr. Phys. *24*, 173–190.

PATEL, V. L. (1978), *Interplanetary Magnetic Field Variations and the Electromagnetic State of the Equatorial Ionosphere*, J. Geophys. Res. *83*, 2137–2144.

RAJARAM, M., SINGH, B. P., NITHYANANDA, N., and AGARWAL, K. (1979), *Effect of the Presence of a Conducting Channel between India and Sri Lanka Island on the Features of the Equatorial Electrojet*, Geophys. J. R. Astron. Soc. *56*, 127–138.

RAJA RAO, K. S. (1961), *On the Seasonal Variation in Lunar and Solar Geomagnetic Tides in the Geomagnetic Equatorial Region*, J. Atmos. Terr. Phys. *20*, 289–294.

RASTOGI, R. G. (1973), *Counter Equatorial Electrojet Currents in Indian Zone*, Planet. Space. Sci. *21*, 1355–1365.

RASTOGI, R. G. (1974a), *Lunar Effects in the Counter Electrojet Near the Magnetic Equator*, J. Atmos. Terr. Phys. *36*, 167.

RASTOGI, R. G. (1974b), *Westward Equatorial Electrojet during Daytime Hours*, J. Geophys. Res. *79*, 1503–1512.

RASTOGI, R. G., and PATEL, V. L. (1975), *Effect of Interplanetary Magnetic Field on the Ionosphere over the Magnetic Equator*, Proc. Ind. Acad. Sci. *74*, 62–67.

RASTOGI, R. G., and IYER, K. N. (1976), *Quiet Day Variation of Geomagnetic H-field at Low Latitudes*, J. Geomag. Geoelectr. *28*, 461–479.

REDDY, C. A. (1981), *The Equatorial Electrojet: A Review of the Ionospheric and Geomagnetic Aspects*, J. Atmos. Terr. Phys. *43*, 557–571.

REDDY, C. A., and DEVASIA, C. V. (1973), *Formation of Blanketing Sporadic E-layers at the Magnetic Equator due to Horizontal Windshears*, Planet. Space Sci. *21*, 811–816.

REDDY, C. A., and DEVASIA, C. V. (1976), *Short Period Fluctuations of the Equatorial Electrojet*, Nature *261*, 396–399.

REDDY, C. A., SOMAYAJULU, V. V., and DEVASIA, C. V. (1979), *Global Scale Electrodynamic Coupling of the Auroral and Equatorial Dynamo Regions*, J. Atmos. Terr. Phys. *41*, 189–201.

REDDY, C. A., DEVASIA, C. V., and SOMAYAJULU, V. V., *Electrodynamic coupling of auroral and equatorial dynamo regions—II. Quiet days*, In *Low Latitude Aeronomical Processes*, COSPAR Symposium Series 8. (ed. Mitra, A. P.) (Pergamon Press 1980) pp. 39–42.

REDDY, C. A., and DEVASIA, C. V. (1981), *Height and Latitude Structure of Electric Fields and Currents due to Local East-west Winds in the Equatorial Electrojet*, J. Geophys. Res. *86*, 5751–5767.

REDDY, C. A., SOMAYAJULU, V. V., and VISWANATHAN, K. S. (1981), *Backscatter Radar Measurements of Storm-time Electric Field Changes in the Equatorial Electrojet*, J. Atmos. Terr. Phys. *43*, 817–827.

REDDY, C. A., VIKRAMKUMAR, B. T., and VISWANATHAN, K. S. (1987), *Electric Fields and Currents in the Equatorial Electrojet Deduced from VHF Radar Observations—I. A Method of Estimating Electric Fields*, J. Atmos. Terr. Phys. *49*, 183–191.

RICHMOND, A. D. (1973a), *Equatorial Electrojet—I. Development of a Model Including Winds and Instabilities*, J. Atmos. Terr. Phys. *35*, 1083–1103.

RICHMOND A. D. (1973b), *Equatorial Electrojet—II. Use of the Model to Study the Equatorial Ionosphere*, J. Atmos. Terr. Phys. *35*, 1105–1118.

SAMPATH, S., and SASTRY, T. S. G. (1979), *Results from in situ Measurements of Ionospheric Currents in the Equatorial Region—I*, J. Geomag. Geoelectr. *31*, 373–379.

SASTRI, J. H. (1982), *Phase Variability of Sq(H) on Normal Quiet Days in the Equatorial Electrojet Region*, Geophys. J. R. Astr. Soc. *71*, 187–197.

SCHIELDGE, J. P., VENKATESWARAN, S. V., and RICHMOND, A. D. (1973), *The Ionospheric Dynamo and Equatorial Magnetic Variations*, J. Atmos. Terr. Phys. *35*, 1045–1061.

SENIOR, C., and BLANC, M. (1984), *On the Control of Magnetospheric Convection by the Spatial Distribution of Ionospheric Conductivities*, J. Geophys. Res. *89*, 261–284.

SENIOR, C., and BLANC, M. (1987), *Convection in the Inner Magnetosphere: Model Prediction and Data*, Ann. Geophys. *5*, 405–420.

SINGH, B. P., and AGARWAL, A. K. (1983), *On Spatial and Temporal Separation of Geomagnetic Anomalies at the South Tip of the Indian Peninsula*, Phys. of the Earth and Plan. Inter. *31*, 59–64.

SOMAYAJULU, V. V. (1988), *Behaviour of Harmonic Components of the Geomagnetic Field During Counter Electrojet Events*, J. Geomag. Geoelectr. *40*, 111–130.

SOMAYAJULU, V. V., REDDY, C. A., and VISWANATHAN, K. S. (1985), *Simultaneous Electric Field Changes in the Equatorial Electrojet in Phase with Polar Cusp Latitude Changes During a Magnetic Storm*, Geophys. Res. Lett. *12*, 473–475.

SOMAYAJULU, V. V., REDDY, C. A., and VISWANATHAN, K. S. (1987), *Penetration of Magnetospheric Convective Electric Field to the Equatorial Ionosphere During the Substorm of March 22, 1979*, Geophys. Res. Lett. *14*, 876–879.

SPIRO, R. W., HAREL, M., WOLF, R. A., and REIFF, P. H. (1981), *Quantitative Simulation of a Magnetospheric Substorm, 3. Plasmaspheric Electric Fields and Evolution of the Plasmapause*, J. Geophys. Res. *86*, 2261–2272.

ST.-MAURICE, HANUISE, J.-P., and KUDEKI, E. (1986), *On the Dependence of the Phase Velocity of Equatorial Irregularities on the Polarization Electric Field and Theoretical Implications*. J. Geophys. Res. *91*, 13493–13505.

STENING, R. J. (1977), *Analysis of Contributions to Ionospheric Dynamo Currents from e.m.f.'s at Different Latitudes*, Planet. Space. Sci. *25*, 587–594.

STENING, R. J. (1985), *Modelling the Equatorial Electrojet*, J. Geophys. Res. *90*, 1705–1719.

STENING, R. J. (1986), *Inter-relations Between Current and Electron Density Profiles in the Equatorial Electrojet and Effects of Neutral Density Changes*, J. Atmos. Terr. Phys. *48*, 163–170.

SUDAN, R. N. (1983a), *Unified Theory of Type I and Type II Irregularities in the Equatorial Electrojet*, J. Geophys. Res. *88*, 4853–4864.

SUDAN, R. N. (1983b), *Nonlinear Theory of Type I Irregularities in the Equatorial Electrojet*, Geophys. Res. Lett. *10*, 983–986.

SUGIURA, M., and POROS, D. J. (1969), *An Improved Model Equatorial Electrojet with a Meridional Current System*, J. Geophys. Res. *74*, 4025–4034.

UNTIEDT. J. (1967), *A Model of the Equatorial Electrojet Involving Meridional Currents*, J. Geophys. Res. *72*, 5799–5810.

VIKRAMKUMAR, B. T., VISWANATHAN, K. S., and RAO, P. B. (1984), *VHF Backscatter Radar Observations of the Equatorial Electrojet Irregularities: Diurnal Seasonal and Solar Cycle Variations*, Annales Geophys. *2*, 495–500.

VIKRAMKUMAR, B. T., RAO, P. B., VISWANATHAN, K. S., and REDDY, C. A. (1987), *Electric Fields and Currents in the Equatorial Electrojet Deduced from VHF Radar Observations—III. Comparison of Observed ΔH Values with those Estimated from Measured Electric Fields*, J. Atmos. Terr. Phys. *49*, 201–207.

VISWANATHAN, K. S., VIKRAMKUMAR, B. T., and REDDY, C. A. (1987), *Electric Fields and Currents in the Equatorial Electrojet Deduced from VHF Radar Observations—II. Characteristics of Electric Fields on Quiet and Disturbed Days*, J. Atmos. Terr. Phys. *49*, 193–200.

WOLF, R. A. (1970), *Effects of Ionospheric Conductivity on Convective Flow of Plasma in the Magnetosphere*, J. Geophys. Res. *75*, 4677–4698.

WOLF, R. A., *Calculation of magnetospheric electric fields*, In *Magnetospheric Physics* (ed. McCormack, B. M.) (D. Riedel, Hingham, Mass. 1974) pp. 167–192.

WOLF, R. A., HAREL, M., SPIRO, R. W., VOIGT, G. H., REIFF, P. H., and CHEN, C. K. (1982), *Computer Simulation of Inner Magnetospheric Dynamics for the Magnetic Storm of July 29, 1977*, J. Geophys. Res. *87*, 5949–5962.

(Received January 25, 1988, revised/accepted May 4, 1988)

PAGEOPH, Vol. 131, No. 3 (1989)

0033–4553/89/030509–06$1.50 + 0.20/0

Subsolar Elevation of the Equatorial Electrojet

C. A. ONWUMECHILI[1] and P. C. OZOEMENA[1]

Abstract—The subsolar elevation of the equatorial electrojet has been produced from satellite solstitial data available from 09 to 15 hr LT using a new approach with the general style of the overhead equivalent current system. It shows the bunching of the current around the dip equator; the return currents of the equatorial electrojet close to the flanks of the dip equator; the fast growth of the electrojet to its diurnal peak followed by a slow decay; and the contraction of its latitudinal extent around the meridian of its highest intensity. Comparison with the results of other workers using ground data suggests that the elevation from satellite data agrees better with that from ground data when the worldwide *Sq* is removed from the ground data.

Key words: Equatorial electrojet, ionospheric electrical current.

Introduction

Even before the equatorial electrojet was properly articulated and named, MCNISH (1937) produced an overhead equivalent current system for what was then known as the Huancayo anomaly, using five American observatories then available. The overhead equivalent current, expected to flow on a spherical surface, is usually projected on a horizontal plane tangential at the equator, thus producing a view that may be technically called an elevation. It is also noted that the overhead equivalent current depicts the height integrated current flowing on a sheet, at an altitude of about 106 km.

YAREMENKO (1978) has produced an overhead equivalent current using 33 stations from the equatorial zone and called it the overhead equivalent current of the equatorial electrojet. Both MCNISH (1937) and YAREMENKO (1978) used spherical harmonic analysis but VESTINE (1941) and ONWUMECHILI (1967) have pointed out the difficulties of using spherical harmonic analysis to represent any localized features like the equatorial electrojet.

Furthermore, the magnetic field measured at any given point represents the integrated magnetic fields permeating the point but arising from currents flowing in a very large area. Therefore the magnetic field at a station in the equatorial zone

---

[1] Anambra State University of Technology, P.M.B. 01660, Enugu, Nigeria.

includes, not only the field of the equatorial electrojet flowing near it but also the fields of currents flowing distantly from the equator. Thus SUZUKI (1973) removed the magnetic fields arising from distant currents, before producing an elevation of the equatorial electrojet and its return currents by imposing the constraint $\nabla \cdot J = 0$. On the other hand, YAREMENKO (1978), who used the compound magnetic field in the equatorial zone, had to truncate the streamlines in order to cut off the currents flowing far away, whose magnetic fields he had already included.

We have analysed the magnetic field of the equatorial electrojet measured by POGO satellites in December 1968 and June 1969 (OZOEMENA and ONWUMECHILI, 1987) and have used the resulting parameters to construct the subsolar elevation of the equatorial electrojet.

Treatment of Data

The analysis is based on the model of the equatorial electrojet as a two-dimensional continuous distribution of line currents whose current density j is given by

$$j = j_0 \frac{a^2(a^2 + \alpha x^2)}{(a^2 + x^2)^2} \cdot \frac{b^2(b^2 + \beta z^2)}{(b^2 + z^2)^2} \tag{1}$$

where j_0 is the peak density on the current axis $x = 0$, $z = 0$; a and b are scale lengths along x and z measured from the axis northwards and vertically downwards respectively; and α and β are distribution parameters along x and z, respectively (ONWUNECHILI, 1965, 1966, 1967). For the purpose of overhead equivalent current or elevation or indeed analysis of meridional profiles, the density is integrated through the thickness (altitude) of the current to obtain the intensity on a horizontal plane,

$$J(x) = J_0(b) \frac{a^2(a^2 + \alpha x^2)}{(a^2 + x^2)^2} \tag{2}$$

where $J_0(b)$ is the height integrated density depending on the vertical scale length or apparent thickness b. The primary analysis of the magnetic field of the electrojet measured on the satellite along a given meridian across the dip equator yields the values of the parameters α, a and $J_0(b)$ (or for convenience J_0) along the meridian (ONWUMECHILI and AGU, 1981). For this work we have used the combined December and June solstitial data because they extend from 09 to 15 hr LT excelling the equinox data which were restricted from 11 to 15 hr LT. The primary analysis of OZOEMENA and ONWUMECHILI (1987) thus provided the values of the parameters α, a and J_0 for 09, 10, 11, 12, 13, 14 and 15 hr LT.

If along a given local time meridian ΔI amperes flows between the points x_1 and x_2 then

$$\Delta I = \int_{x_1}^{x_2} J(x) \, dx = \tfrac{1}{2} a J_0 [S(a, \alpha, x_2) - S(a, \alpha, x_1)].$$

In practice, it is more convenient to measure x in terms of the scale length a using $E = x/a$.

Then

$$2\Delta I = aJ_0[S(\alpha, E_2) - S(\alpha, E_1)],\tag{3}$$

where

$$S(\alpha, E) = \frac{(1-\alpha)E}{(1+E^2)} + (1+\alpha)\tan^{-1} E.\tag{4}$$

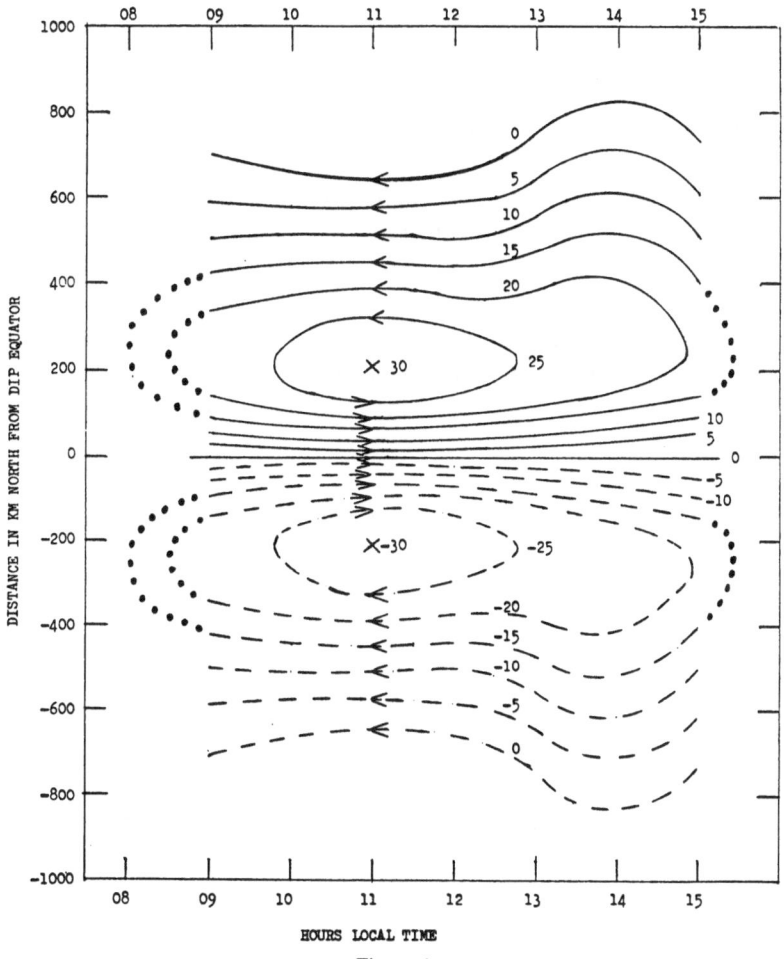

Figure 1

Subsolar elevation of equatorial electrojet current in the style of overhead equivalent current for the combined solstices of December 1968 and June 1969, derived from POGO satellite data. 5000 amperes flow between adjacent lines. Continuous lines indicate northern circuit, broken lines indicate southern circuit, and dotted lines show extrapolation beyond available data.

Thus

$$S(\alpha, E_2) = S(\alpha, E_1) + \frac{2\Delta I}{aJ_0} = S(\alpha, E_1) + C(t). \tag{5}$$

$C(t)$ depends on time t because a and J_0 are different for each local time meridian (0900 to 1500).

Fixing the current interval ΔI, in our case, 5000 amperes, and starting from $E_1 = 0 = x_1$ at the current axis (dip equator) for a selected local time meridian, E_2 and thus x_2 is found from (5). Using this as the new initial position the next point is easily determined from (5), noting that α is constant for the selected meridian. Even without a computer, Eq. (5) is easily solved such as by tabulating $S(\alpha, E)$ from $E = 0$ to 3 at intervals of 0.02 and then interpolating. Indeed, since α is independent of local time (ONWUMECHILI and OZOEMENA, 1985), an average value of $\alpha = -1.863$ (or approximately -2) may be used and one tabulation suffices for all the local time meridians. The determination of successive values of x_2 continues as closely possible to the focal distance w along the selected meridian given by $w^2 = -a^2/\alpha$. Thereafter imposing the condition that all the forward currents across the meridional plane must also cross the same meridional plane on their return, the latitudinal extent of the electrojet, L_1 km, at $S(\alpha, E_L) = 0$, along the meridian is determined (ONWUMECHILI and OZOEMENA, 1985). Now using L_1 as x_1, the value of x_2 is found from (5) such that ΔI flows between x_2 and L_1. Subsequently using x_2 as the initial point, the process is continued down as closest possible to the focal point along the meridian. The result of the process is plotted in Figure 1.

Results and Discussion

The result in Figure 1 is interesting. It has, however, been impaired by the absence of data from the formation of the electrojet in the morning until 09 hr LT and from 15 hr LT to the final demise of the electrojet at night. The process is, however, capable of depicting the electrojet as it is known: a current system that exists in the daytime. Among other features of the results are: the bunching of the current around the axis or dip equator, and the contraction of its extent towards the noontime meridian when the current is most intense. The asymmetry about the meridian of highest intensity (here 11 hr LT) especially as seen from the complete loop, demonstrates the well-known feature from ground data that the electrojet grows fast to its diurnal peak near noon and then declines slowly towards evening. It is of particular interest that the equatorial electrojet returns quite close to the dip equator long before the Sq current focus at about $35°$ latitude from the dip equator.

Our elevation of the electrojet given in Figure 1 should be compared with that of SUZUKI (1973) given in Figure 2. Although SUZUJI's (1973) elevation is in terms of current intensity and ours is given in the style of overhead equivalent current, the

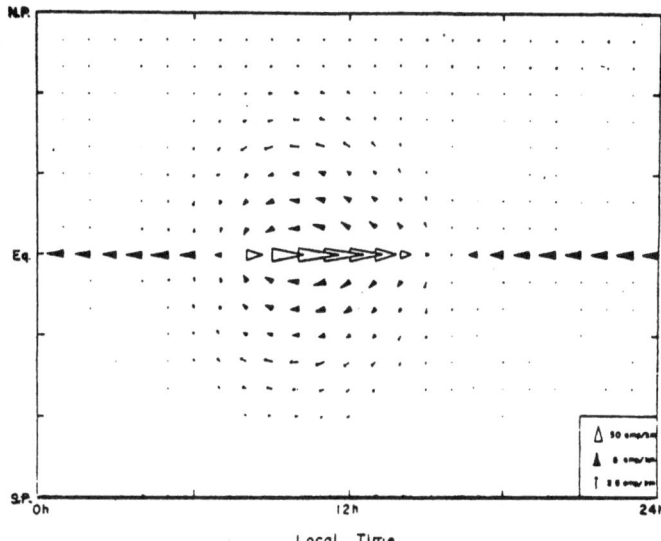

Figure 2
The distribution of the intensities of return currents of the equatorial electrojet. (After SUZUKI, 1973.)

features mentioned in the preceding paragraph are generally similar to the features of Figure 2, with the exception of the contraction of the extent towards the meridian of highest intensity. Indeed, SUZUKI's (1973) elevation might have resembled ours even more closely with respect to the prevalently daytime feature of the electrojet if he had used the nighttime level, instead of the daily mean as the datum level, as is sometimes done in spherical harmonic models.

Conclusions

The subsolar elevation of the equatorial electrojet has been constructed by a new approach with the general style of the overhead equivalent current, and the approach is capable of preserving daytime features of the electrojet while avoiding the smearing of its current into the night and places distant from the dip equator where the existence of the electrojet is in doubt.

The major features of the elevation are: the bunching of the forward current around the dip equator; the return currents of the electrojet close to the flanks of the dip equator and long before the Sq current focus at about 35° dip latitude; the fast growth of the electrojet to its diurnal peak around noon followed by a slow decay towards night; and the contraction of its latitudinal extent towards the meridian of highest intensity.

The subsolar elevation of the equatorial electrojet produced by SUZUKI (1973) from ground data, after removing the worldwide Sq, agrees with most features of our elevation produced from satellite data, unlike the elevation produced by YAREMENKO (1978) with the spherical harmonic model from ground data without the removal of the worldwide Sq. It should be stressed that we analysed the satellite data as tabulated by CAIN and SWEENEY (1972). The comparison seems to show that the elevation of the equatorial electrojet produced from satellite data agrees better with the elevation produced from ground data when the worldwide Sq is removed from the ground data. This may be considered significant when it is recalled that several comparisons of satellite data with ground data by different investigators proved that the satellite field of the electrojet correlated with only the portion of the ground data produced by removing the worldwide Sq (see their review by ONWUMECHILI, 1985).

REFERENCES

CAIN, J. C., and SWEENEY, R. E. (1972), *POGO Observations of the Equatorial Electrojet*, Goddard Space Flight Center Publication, X–645–72–299, pp. 1–54.

MCNISH, A. G. (1937), *Progress of Research in Magnetic Diurnal Variations at the Department of Terrestrial Magnetism*, Carnegie Institution of Washington, in International Union of Geodesy and Geophysics, Association of Terrestrial Magnetism and Atmospheric Electricity Bulletin No. 10, Transactions of Edinburgh Meeting, 1936; pp. 271–280, Copenhagen.

ONWUMECHILI, C. A., *A Three-dimensional Model of Density Distribution of Ionospheric Current Causing Part of Quiet Day Geomagnetic Variations*, Proceedings of the Second International Symposium on Equatorial Aeronomy (ed. de Mendonca, F.) (Brazilian Space Commission, Sao Paulo 1965) pp. 384–386.

ONWUMECHILI, C. A. (1966), *A New Model of the Equatorial Electrojet*, Nigerian J. Sci. *1*, 11–19.

ONWUMECHILI, C. A. (1967), *Geomagnetic variations in the equatorial zone*, In *Physics of Geomagnetic Phenomena* (eds. Matsushita, S., and Campbell, W. H.) Chap. II-2 (Academic Press, New York 1967) pp. 425–507.

ONWUMECHILI, C. A. (1985), *Satellite Measurements of the Equatorial Electrojet*, J. Geomag. Geoelectr. *37*, 11–36.

ONWUMECHILI, C. A., and AGU, C. E. (1981), *Longitudinal Variation of Equatorial Electrojet Parameters Derived from POGO Satellite Observations*, Planet. Space. Sci. *29*, 627–634.

ONWUMECHILI, C. A., and OZOEMENA, P. C. (1985), *Latitudinal Extent of the Equatorial Electrojet*, J. Geomag. Geoelectr. *37*, 193–204.

OZOEMENA, P. C., and ONWUMECHILI, C. A. (1987), *Global Variations of the POGO Electrojet Parameters*, J. Geomag. Geoelectr. *39*, 625–636.

SUZUKI, A. (1973), *Returning Flow of the Equatorial Electrojet Currents*, J. Geomag. Geoelectr. *25*, 249–258.

VESTINE, E. H. (1941), *On the Analysis of Surface Magnetic Fields by Integrals. Part I*, Terrest. Magn. Atmosph. Elec. *46*, 27–41.

YAREMENKO, L. N. (1987), *Equivalent Current Systems of the Equatorial Electrojet*, Geomagnetism and Aeronomy *18*, 770–771.

(Received December 16, 1978, revised/accepted April 26, 1988)

PAGEOPH, Vol. 131, No. 3 (1989)

0033–4553/89/030515–11$1.50 + 0.20/0

On Sq and L Current Systems in the Ionosphere

WEN-YAO XU[1]

Abstract—Ionospheric dynamo current system representing geomagnetic lunar variation on disturbed days is calculated. A characteristic current system L^p is formed at high latitudes owing to conductivity enhancement in the auroral belt, which also influences the L currents at mid and low latitudes: the foci of mid-low latitude current vortices shift $4°$ toward high latitude. IMF B_z and B_y effects on mid-low latitude Sq are studied. Rapid variation of magnetospheric convection electric field caused by B_z variation will penetrate into the inner magnetosphere and then map onto the ionosphere along geomagnetic field lines, producing a characteristic electric field and current system, which will move the Sq foci and produce a magnetic disturbance of a few nT on ground surface. IMF sector effect on mid-low latitude magnetic filed is examined by using data from Beijing and Gangzhou in China. 3–5 nT effect is recognized.

Key words: Ionospheric dynamo, L current system, IMF B_z effect, IMF sector effect, mid and low latitudes.

1. Introduction

The ionospheric current systems representing geomagnetic Sq and L variations have been studied extensively. It is believed that Sq at mid-low latitudes and L variations in whole globe are caused by tidal wind dynamo action in the ionosphere, while Sq in the polar region is attributed to field-aligned currents. There are three factors affecting the dynamo process: ionospheric wind (lunar tidal wind for L variation and solar tidal wind for Sq variation), ionospheric conductivity, and geomagnetic field configuration, among which the geomagnetic field is the most stable factor, but ionospheric wind and conductivity vary with both season and geomagnetic activity. The seasonal variations of the wind and conductivity depend upon the solar zenith angle, while their variability related with geomagnetic activity depends upon solar wind state, in particular, interplanetary magnetic field (IMF). Furthermore, solar wind state controls field-aligned currents. Consequently, both Sq and L variations would exhibit IMF-dependence.

---

[1] Institute of Geophysics, Academia Sinica, Beijing, China.

In this paper, the dynamo current systems for L variations on disturbed days are calculated for an ionospheric conductivity model with an auroral enhancement, and compared with quiet day current system (Section 2). In Section 3, IMF B_z effect on mid-low latitude Sq is studied. In Section 4, IMF B_y effect (or sector effect) on mid-low latitude Sq is discussed by using geomagnetic data from two observatories in China.

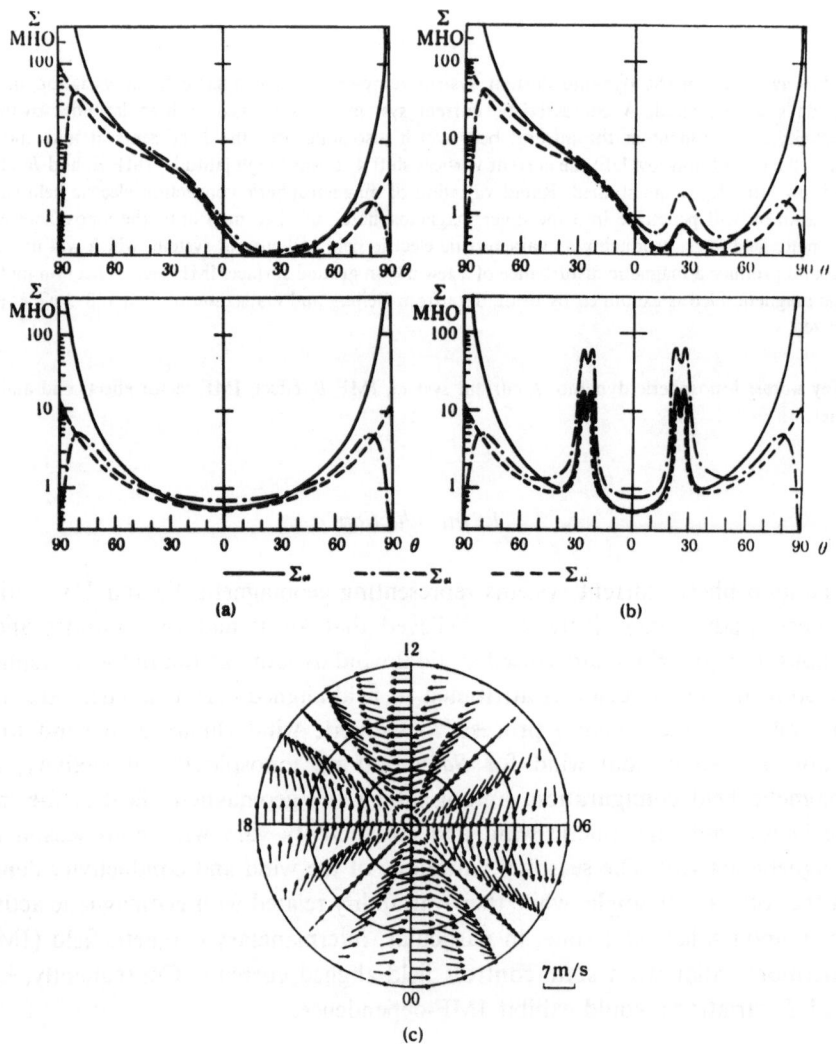

Figure 1
The ionospheric conductivity models for quiet days (a) and disturbed days (b), and wind speed vectors for lunar tide mode (2, 2).

2. L Current System for Disturbed Days

Geomagnetic *L* variation and its equivalent ionospheric current system have been studied by many scientists (e.g., CHAPMAN and BARTELS, 1940; MATSUSHITA, 1966; MATSUSHITA and XU, 1982). Ionospheric dynamo current responsible for *L* variation has been calculated for lunar tide and quiet ionospheric conductivity model (MATSUSHITA, 1969; TARPLEY, 1970; EVANS, 1978; FORBES, 1982). Since *L* variation is very small, studies of *L* and its dynamo process have been commonly confined to quiet conditions. On disturbed days, *L* variation is hard to recognize from noise background and studies of its characteristics are insufficient. In this case, a theoretical calculation of disturbed *L* current system and its geomagnetic

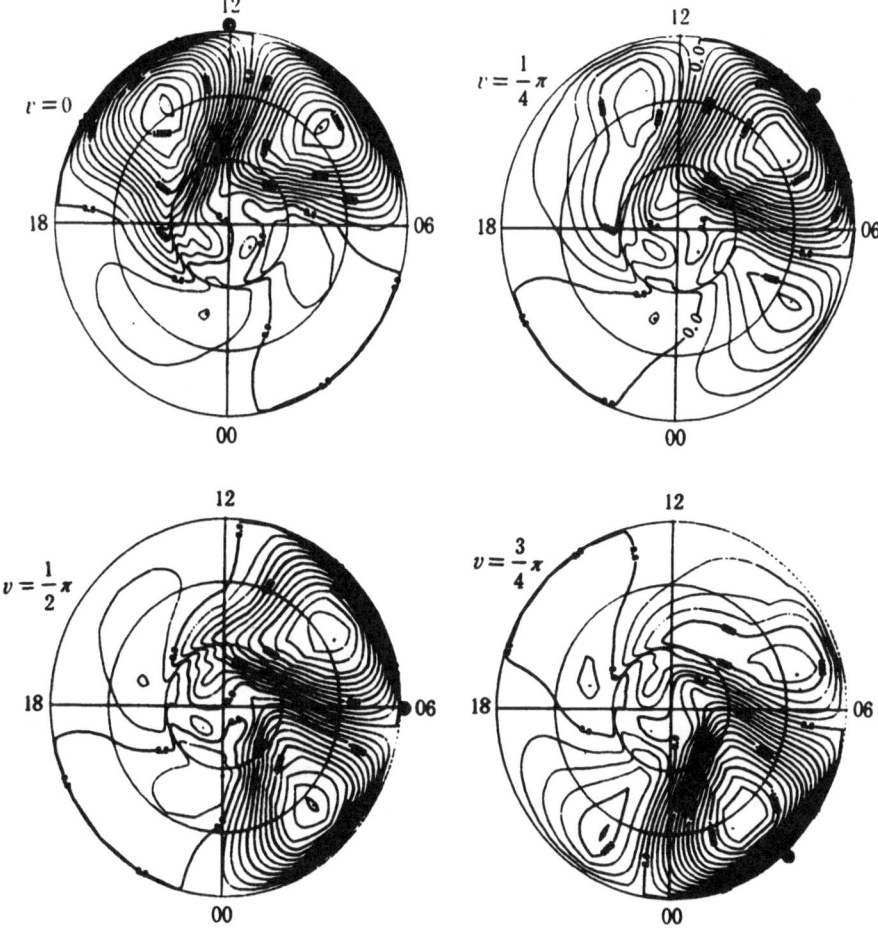

Figure 2
Dynamo current systems of disturbed *L* variation for four lunar phases.

signature is a useful reference for studying geomagnetic variation and related phenomena on the disturbed days.

As mentioned above, among three factors affecting L dynamo process, the ionospheric conductivity is highly variable on disturbed days. An auroral enhancement of the conductivity is a major feature. In this study the ionospheric conductivity model for disturbed days consists of a background quiet conductivity (Figure 1a) and auroral enhancement (Figure 1b). In Figure 1c, the wind speed vectors are depicted for the lunar (2,2) mode which will be used in this calculation.

On the basis of wind dynamo theory, L current systems for the disturbed conductivity model in Figure 1b and the wind system in Figure 1c are calculated for 8 lunar phases. In Figure 2 the current systems for lunar phases $v = 0°, 45°, 90°, 135°$ are shown, the small full circles in the diagram indicate the position of the sun. Other four current systems for $v = 180°, 225°, 270°, 315°$ can be easily obtained by rotating the corresponding systems by $180°$ in a clockwise direction. For comparison, L current systems on quiet days are also calculated (Figure 3) for the quiet conductivity model in Figure 1a and the same wind system in Figure 1c. It is noted in Figures 2 and 3 that quiet L currents are mainly confined in the dayside sector between $0°$ and $60°$ latitudes due to high conductivity there. On disturbed days, however, the dynamo action at high latitudes becomes relatively important, owing to the auroral conductivity enhancement, and a characteristic current system L^p is formed at the polar region. At the same time, L current system in the whole globe is distorted. By comparison with the quiet L current system, the characteristics of disturbed L currents are as follows:

(1) There is a complicated current pattern L^p in the polar region, including the auroral belt.
(2) The total current at mid and low latitudes is enhanced by 20–30% and the total current in the equatorial belt is increased by 20%.
(3) The foci of mid-low latitude current vortices shift $4°$ toward high latitude.

3. IMF B_z Effect on Sq Variation at Mid and Low Latitudes

In the magnetosphere there exists a large-scale convection electric field from dawn to dusk (E_m). In steady state this field cannot penetrate into the inner magnetosphere due to the Alfven layer shielding. When the convection electric field varies rapidly with a time scale less than the characteristic time required for building up the Alfven layer shielding (3 minutes for nighttime, and 5 hours for daytime, JAGGI and WOLF, 1973), the variation of the convection electric field (ΔE_m) will penetrate into the inner magnetosphere, and then map onto the mid-low ionosphere along geomagnetic field lines, producing a characteristic ionosphere electric field and current system, which will be superimposed on normal Sq current system and

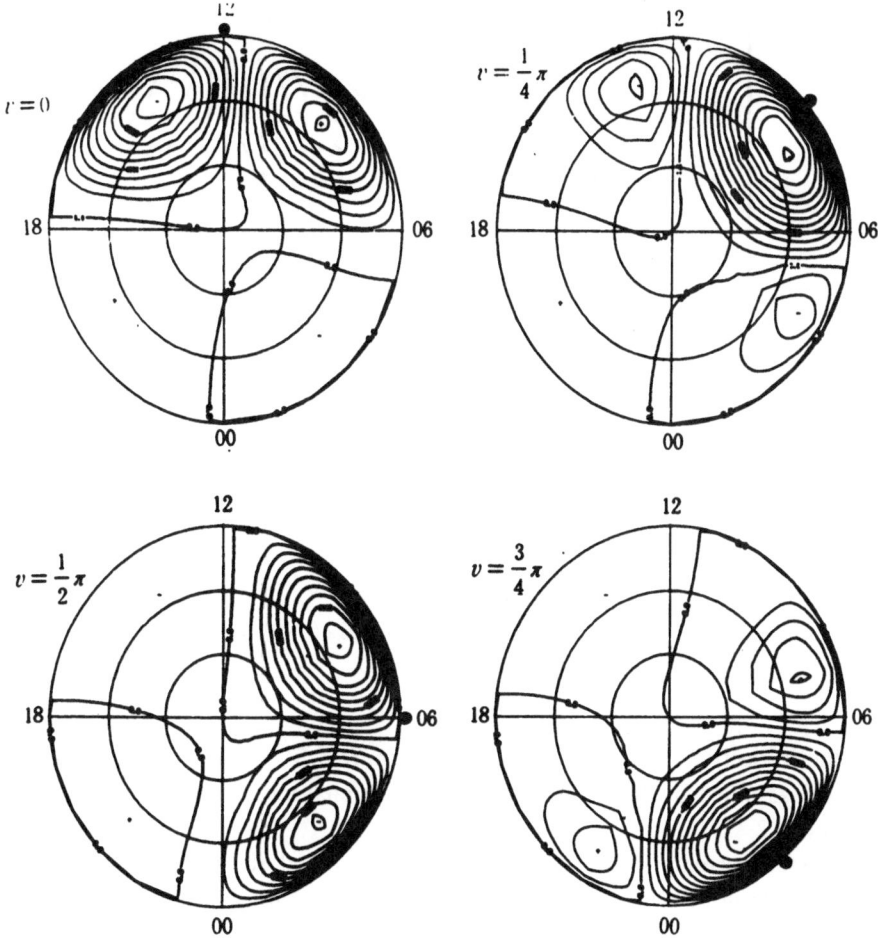

Figure 3
Dynamo current systems of quiet *L* variation for four lunar phases.

distort it. Along with the gradual buildup of the Alfven layer shielding, the electric field penetrating into the inner magnetosphere will decay.

Rapid variations of the large-scale convection electric field in the magnetosphere are closely correlated with variations of the interplanetary electric field (IEF), in particular, its eastward component (E_y). WU *et al.* (1981) compared variations of E_y and E_m, and obtained an empirical relation:

$$\Delta E_m = 0.13\Delta E_y = -0.13v \cdot \Delta B_z$$

where v is the solar wind velocity, ΔB_z is the variation of IMF B_z.

Figure 4 presents an ionospheric electric field associated with $B_z = -5\,\text{nT}$ (for instance, IMF B_z changes from northward to southward). Corresponding

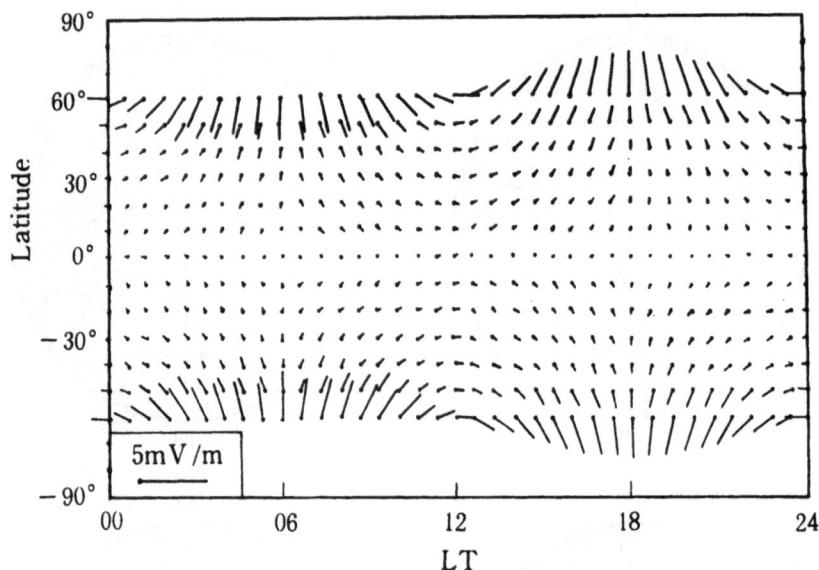

Figure 4
Ionospheric electric field caused by variation of IMF $B_z = -5\,nT$.

ionospheric electric currents are depicted in Figure 5 for an ionospheric conductivity model assumed by MATSUSHITA (1969). The characteristics of the currents are as follows:

(1) Pedersen currents flow from high latitudes toward the equator in the morningside, then turn to eastward near noon, and flow toward high latitudes again during afternoon. On average the current density is 4 mA/m at mid latitudes, the equatorial current density reaches 50 mA/m.

(2) Hall currents flow toward the equator near noon, and turn westward in the morningside and eastward during afternoon, respectively. On average the current density is 8 mA/m at mid latitudes.

(3) Total current pattern is similar to Hall currents. The equatorward currents reach maximum around 11 hr local time. Average current density at mid latitudes is 10 mA/m.

Magnetic disturbances arising from this current system on the ground surface can be estimated. Near noon, Pedersen currents will cause a shift of Sq foci toward high latitudes, while Hall currents will push the foci eastward. Numerical calculation for $B_z = -5\,nT$ shows a magnetic disturbance of 10 nT on the ground surface and a Sq focus shift of a few degrees in both latitude and longitude. If $B_z > 0$ the directions of the currents and geomagnetic disturbances are opposite.

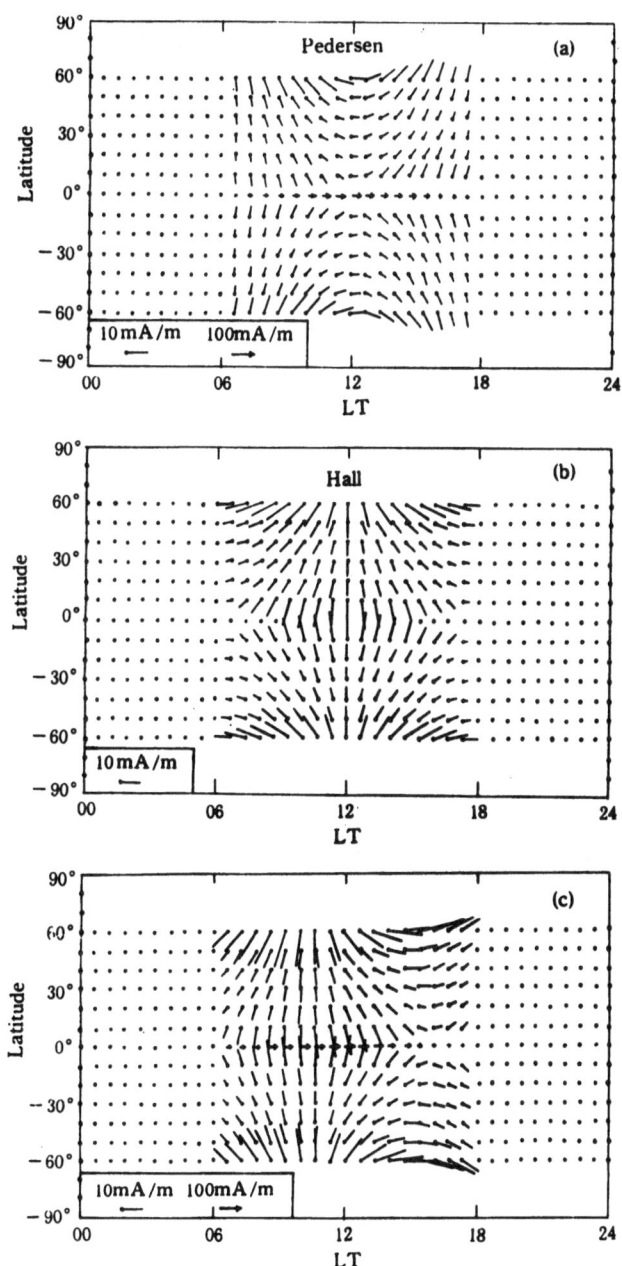

Figure 5
Ionospheric current vector distributions arising from the electric field in Figure 4.

4. IMF Sector Effects on Geomagnetic Field at Mid and Low Latitudes

Effects of interplanetary magnetic field (IMF) sector structures, such as toward or away from the sun, on the polar cap magnetic field, have been fairly well established owing to studies by several scientists during past decades (e.g., FRIIS-CHRISTENSEN, 1984). Namely in the Northern Hemisphere, the toward (or away) IMF sector structure correlates with positive (or negative) deviations of the geomagnetic downward component Z at stations near 90° invariant latitude and with negative (or positive) deviations of the horizontal component H at stations near 80° invariant latitude. In the Southern Hemisphere, the same story holds for Z, but the sign reverses for H component.

As for mid and low latitudes, the IMF sector effects are very small and hard to recognize. In 1973, MATSUSHITA et al. studied the IMF sector effects in the whole globe and found that the focus of the Sq current system will shift toward a higher (or lower) latitude when the IMF sector is away from (or toward) the sun. This result gives an average IMF sector effect for a certain period.

In order to study characteristics of the IMF sector effects for an individual day, the method of natural orthogonal components (MNOC) is used in this section to separate the contributions to total geomagnetic daily variation from different processes, such as dynamo process in the ionosphere, ring current, sector effect, and others (KENDALL and STUART, 1976).

First of all, the average behaviour of the IMF sector effects on geomagnetic horizontal component H at two observatories in China (Beijing and Guongzhou) is studied by the use of conventional epoch superposition method for three seasons of 1973. Only the results for Beijing are presented here. In Figure 6a three curves for each of the three seasons represent the average daily variations for away-sector days (A-days) by solid lines with points, toward-sector days (T-days) by dashed lines with open points, and all days (including both A- and T-days) by solid lines. After removing the all-day average daily variation, the sector effects are shown more clearly (Figure 6b). A small but definite sector effect can be seen in this diagram with the following characteristics:

(1) There exists a daily variation of the sector effect with daytime predominance.
(2) There is a seasonal variation with a maximum in summertime (about 5 nT) and a minimum in winter (about 3 nT).

In order to study the IMF sector effects for an individual day, the MNOC is used to same data set. The eigenvalues are calculated for Beijing Observatory, as shown in Figure 7. It is noted in the figure that only the first few eigenvalues are meaningful. In Figure 8 the first 6 eigenvectors $Z1$, $Z2$, $Z3$, $Z4$, $Z5$, and $Z6$ are shown in a sequence of their 24 components for three seasons. As mentioned above, each of these eigenvectors may be attributed to a certain physical process. It is very interesting to note that the curves for the first eigenvector are similar to those in Figure 6b. It is plausible to suppose that the first eigenvector describes IMF

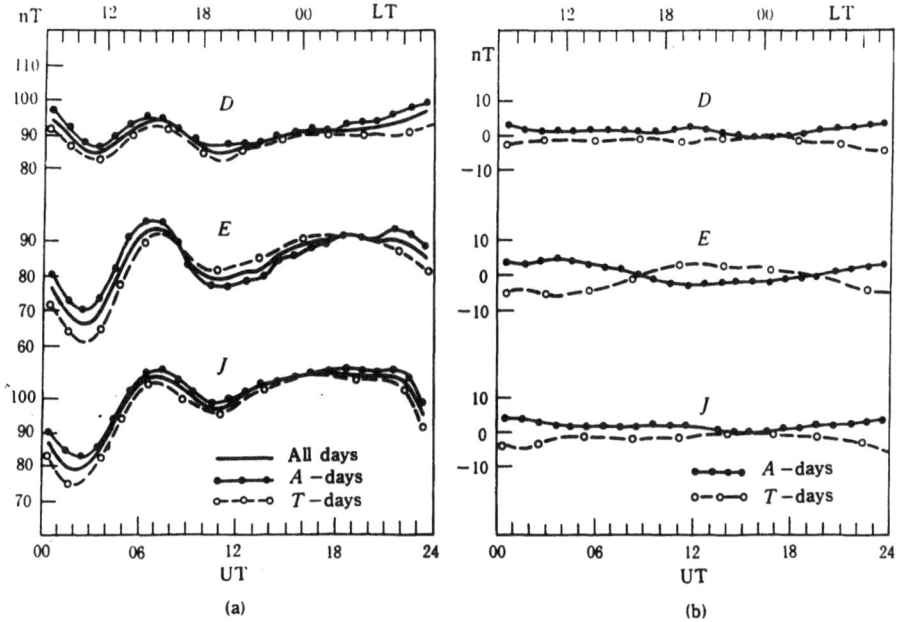

Figure 6
The average IMF sector effects at Beijing Observatory for 1973.

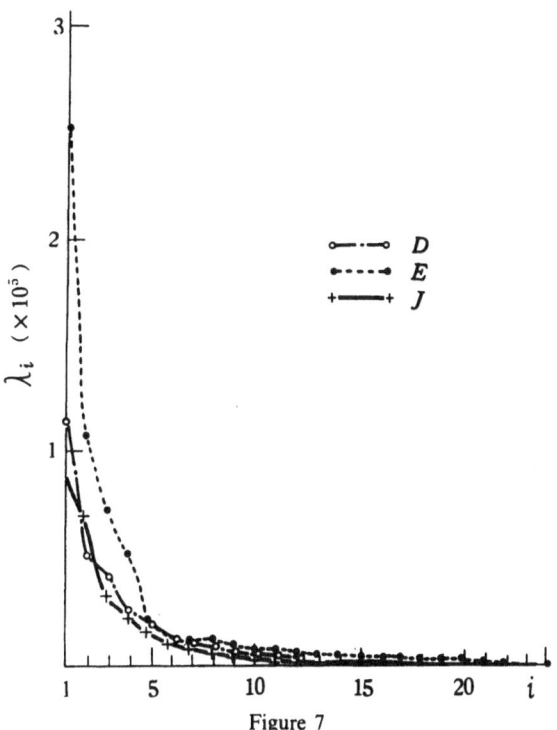

Figure 7
Eigenvalues of IMF sector effects at Beijing.

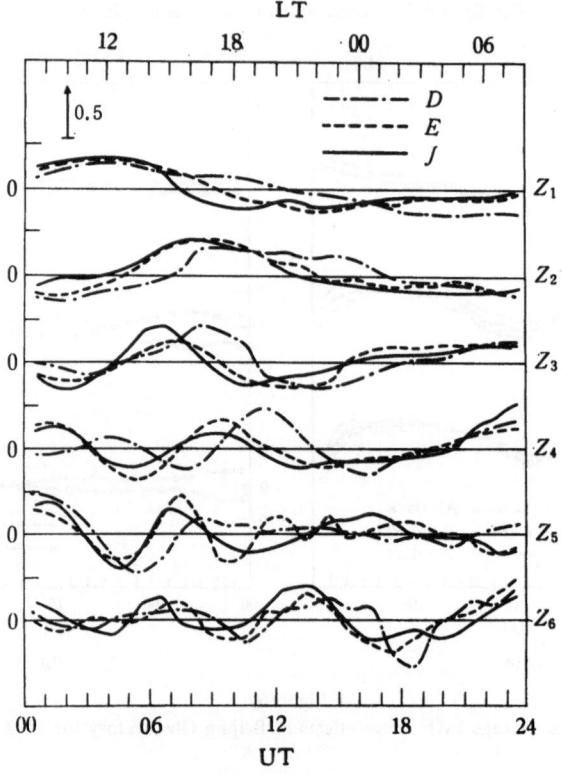

Figure 8
The first six eigenvectors for three seasons.

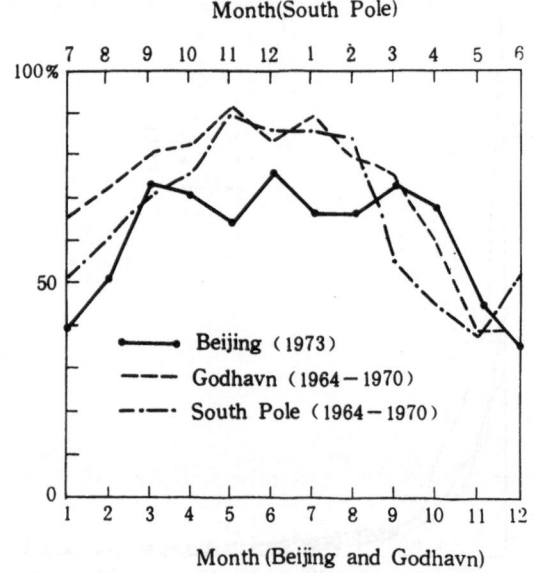

Figure 9
Percentages of agreement between estimated IMF sector polarities and satellite observations at three stations.

sector effects. Furthermore, the amplitudes $Y1(i)$ corresponding to the first eigen-vector $Z1$ can be used to estimate the sector polarity. According to Figure 6b, $Y1(i) > 0$ (or <0) corresponds to away (or toward) sector. This criterion is used to determine the sector polarity for each of the days in 1973. The estimated sector polarities are then compared with satellite observations, and give an agreement of 70% for summer and equinoctial months, as shown in Figure 9, in which the same curves for Godhvan and South Pole are also given for comparison (MATSUSHITA and XU, 1981). Three curves show the same seasonal variation.

Acknowledgment

This work was supported by the National Science Foundation of China (NSFC).

REFERENCES

CHAPMAN, S., and BARTELS, J., *Geomagnetism* (Clarendon Press, Oxford 1940).

EVANS, J. V. (1978), *A Note on Lunar Tides in the Ionosphere*, J. Geophys. Res. *83*, 1647–1652.

FORBES, J. M. (1982), *Atmospheric Tides, 2. The Solar and Lunar Semidiurnal Components*, J. Geophys. Res. *87*, 5241–5252.

FRIIS-CHRISTENSEN, E., *Polar cap current system*, In *Magnetospheric Currents* (ed. Potemra, T. A.) (AGU, Washington, DC 1984) pp. 86–95.

JAGGI, R. K., and WOLF, R. A. (1973), *Self-consistent Calculation of the Motion of a Sheet of Ions in the Magnetosphere*, J. Geophys. Res. *78*, 2852–2866.

KENDALL, M. G., and STUART, A., *The Advanced Theory of Statistics*, v. 3 (Charles Griffin, High Wycomb 1976).

MATSUSHITA, S. (1966), *Lunar Geomagnetic Variations*, J. Geomag. Geoelectr. *18*, 163–191.

MATSUSHITA, S. (1969), *Dynamo Currents, Winds and Electric Fields*, Radio Sci. *4*, 771–780.

MATSUSHITA, S., and XU, WEN-YAO (1981), *IMF Behavior Estimated from Geomagnetic Data at South Pole*, J. Geophys. Res. *86*, 3628–3634.

MATSUSHITA, S., and XU, WEN-YAO (1982), *Sq and L Currents in the Ionosphere*, Ann. Geophys. *38*, 295–305.

MATSUSHITA, S., TARPLEY, J. D., and CAMPBELL, W. H. (1973), *IMF Sector Structure Effect on the Quiet Geomagnetic Field*, Radio Sci. *8*, 1075–1090.

TARPLEY, J. D. (1970), *The Ionospheric Wind Dynamo-1. Lunar Tide*, Planet. Space Sci. *18*, 1075–1090.

WU, L., GENDRIN, R., HIGEL, B., and BERCHEM, J. (1981), *Relationship Between the Solar Wind Electric Field and the Magnetospheric Convection Electric Field*, Geophys. Res. Lett. *88*, 1099–1102.

(Received/accepted January 12, 1988)

PAGEOPH, Vol. 131, No. 3 (1989)

0033–4553/89/030527–05$1.50 + 0.20/0

On the Evolution of Methods of Determining L

KANG-KUN TSCHU[1]

Abstract—A brief description of existing methods of determining L of geophysical elements is given in this paper. Their evolution or various modifications are briefly mentioned. It is hoped that scientific studies of lunar perturbations in geophysics and aeronomy should be encouraged and strengthened during future years.

Key words: Lunar geomagnetic variations, L.

Lunar variations of tides have been detected in a number of geophysical elements over a long time. Although the most obvious example, the tides in the sea, is very evident, the lunar tides in other quantities are generally small but they are nevertheless of theoretical interest since they may be presumed to be caused by the varying gravitational attraction of the Moon which is well-known.

Of the tidal spectra of the gravitational tide-producing potential, the first seven components are most important: they are semidiurnal and diurnal, denoted by conventional symbols $M2$ (period 12.4 solar hours), $S2$ (12.0), $N2$ (12.7), $K2$ (11.97), $K1$ (23.9), $O1$ (25.8) and $P1$ (24.1).

According to the nomenclature adopted by CHAPMAN and MALIN (1969), $M2$, $O1$, and $N2$ are called lunar tidal (potential) harmonics, $S2$ the solar tidal (potential) harmonic, and $K1$ and $K2$ sidereal-time tidal (potential) harmonics.

Since IGY and IMS the geophysical data, both on global surface and in near-earth space, have been greatly accumulated. In view of this vast amount of scientific data and the importance of the problem involved, we propose to discuss scientific studies of lunar perturbations of geophysical elements and their implications during future years.

The methods used for the determination of L may be divided broadly into two categories: a) grouping harmonic analysis methods, and b) spectral analysis and other methods. The former have been widely used for most atmospheric and geomagnetic tides, the latter were used by some workers in recent years. Grouping

[1] Institute of Geophysics, Academia Sinica, P.O. Box 928, Beijing, China.

and harmonic analysis methods which involve grouping the data by lunar time and solar time can be further classified under two main headings. They are (i) the fixed lunar age method—described in CHAPMAN and BARRELS (1940) as Brown's methods; (ii) the fixed solar hour method—described by Van der Stoke's method in Chapman and Bartels.

In the case of the fixed-age method three main stages in its evolution may be conveniently indicated. They are represented, in the order of the evolution, by the lunar-sheets method, the hour-to-hour differences method, and the Chapman-Miller method. It can be noted that each main stage or its representation method may still be subclassified, according to the subsidiary modifications actually made.

The lunar-sheets method is an old but direct (retabulating) method. It is so termed because the original data, or numbers based on them, are written according to lunar time on "lunar sheets." In connection with the lunar reductions for Greenwich geomagnetic declination and others (CHAPMAN, 1925), the method was fully developed and put in a convenient form. This consists of three main things, namely: (1) removal of the solar diurnal variation, (2) retabulating according to lunar time, and (3) harmonic analysis of the lunar hourly inequalities thus formed. The lunar day is taken to be 25 solar hours of length, but 26 hourly entries are usually made, in order that any noncyclic or progressive variation may subsequently be removed from the lunar hourly inequalities from the various groups of data, before harmonic analyses are applied.

The hour-to-hour differences method was used in succession to the lunar-sheets method. This method, briefly the Δ method, was described in some detail by CHAPMAN (1930) in the Adolf Schmidt Festschrift of the Zeitschrift für Geophysik. The important feature of this method is of course that instead of using the hourly values themselves, the hour-to-hour differences (Δ's) are tabulated on the computation sheets. As regards the solar correction, convenient steps are also taken so that it may be applied all at once instead of to individual days, with a great saving of time. This method has two further advantages: (1) the hour-to-hour differences are usually smaller, and are very conveniently checked; (2) the noncyclic parts of the variation add a constant to each difference, the analysis being thus unaffected. However, the disadvantages of hour-to-hour differences are: (1) Hourly values generally have to be differenced; (2) for use with the Hollerith Machines all differences must be made positive by addition of a constant.

Later, CHAPMAN (1930, 1932) modified the Δ's method further, resulting in so-called "625 sums" method, i.e., 25 fixed-age solar-time sequences of 25 hourly sums were formed and used, and array of 625 sums resulted. The Chapman-Miller method evolved partly through a simplified application of some principles used in the Δ's or 625 sums method. This method is of analytical and wholesale character, very suitable and convenient for determining more rapidly and accurately the lunar daily variations in geophysical elements. It is applicable to hourly, bi-hourly, or tri-hourly data, but in practice the bi-hourly one is preferable. Also the transits of the mean moon are adopted; this is simpler than extracting the times of transits from a Nautical Almanac. Besides, this method also has the same merits of previous methods, e.g., checking facility and others.

The essence of the Chapman-Miller method may be outlined here. Data for days of the same lunar age are grouped together and the resultant daily variations are Fourier analyzed, giving a set of harmonics for each lunar age which represent the joint solar and lunar effects.

Usually just the first four harmonics are taken as being sufficient to synthesize a closed approximation to the actual variations. Harmonic analysis of these coefficients in terms of lunar age then leads, by way of a set of linear equations, to the harmonic coefficients of luni-solar variations.

Several papers on the subject have been published; the earliest one (CHAPMAN and MILLER, 1940) gives the mathematical theory and outlines the essentials of the method. Two others (TSCHU, 1949; MALIN and CHAPMAN, 1970) supply a straight-forward account of the practical details and include the determination of probable errors. With the advent and wide use of electronic computers various refinements and further developments of this method have been adopted and published (WINCH, 1970; SCHLAPP and WEEKS, 1973). In addition, a slightly modified, fast lunar analysis method, designated as "FAR" technique for Fourier Analysis of Residuals, for a high resolution, short data series was introduced by MATSUSHITA and CAMPBELL (1972), mainly due to that in recent years magnetic tapes of scaled field values for every 2.5 min have become available for a number of world stations.

In determining and discussing L for Huancayo H (geomagnetic horizontal intensity), BARTELS and JOHNSON (1940) adopted the "fixed hour" method, so-called in contrast to the "fixed-age" method. These two different ways of computing L have long been used in their various forms by various investigators and both are valuable in advancing our knowledge of L. It can also be noted (CHAPMAN, 1942) that although the difference in spirit and purpose between the two methods affects most of the details of the work, the final results can be put in the same form, whichever method is adopted.

In recent years there were the techniques of time series analysis such as Fourier and other least-squares analysis, power spectrum analysis, etc. For instance, BLACK (1970) carried out discrete Fourier transformation on geomagnetic data using the Cooley-Tukey algorithm which enabled him to search for undetected spectral lines. GUPTA and CHAPMAN (1969) made a power spectrum analysis of a lunar daily harmonic geomagnetic variation for 54 stations. Use of maximum entropy rather than the conventional method of power spectrum analysis can improve component resolution (e.g., CURRIE, 1975). Another technique which is occasionally used is that of least-squares analysis, such as initially adopted by LARSEN (1968) and later developed by MALIN and SCHLAPP (1980) for lunar analysis. This method can cope with unequal sampling intervals so that missing data are not a problem. Additionally, the analysis can be accurately tuned to the required frequencies. Since the data are dealt with sequentially, the demands on core memory are not excessive, and are independent of the length of the data series. Great caution must be exercised, however, when applying the method to short runs of data, because spurious results can easily be produced by the presence of a small amount of noise that results from the similarity of the solar and lunar semidiurnal periods.

In concluding this brief paper I would like to mention seveal recent papers. For research directions in recent studies of the lunar geomagnetic effects, CAMPBELL (1983) emphasized the following: the annual variability, the relationship to solar-terrestrial activity, the contribution of ocean tidal motion to magnetic observatory data, and the source processes of the principal lunar current in the ionosphere. Australian authors R. J. STENING and D. E. WINCH (1987) deal with nighttime geomagnetic variations at low latitudes. Using hourly values from 1964–65 of H, D, Z from a large number of stations, their variation during nighttime hours is examined from both their monthly means and from a previously used harmonic analysis method. Consistent changes during the night are often found. In a paper by R. J. STENING et al. (1986) a technique is applied to determine lunar tides at each of ten heights in the virtual height range 82–109 km, using two years of wind data at a time from the partial reflection drifts experiment at a Canadian station Saskatoon. This appears to be the first detailed determination of this kind in the neutral atmosphere at these heights. In addition, both STENING (1986) and BHUYAN et al. (1987) deal with the lunar effect in the ionosphere; the latter, using the Chapman-Miller method, discuss lunar and solar daily variations of equivalent slab thickness at Delhi.

Acknowledgement

The author wishes to thank Dr. W. H. Campbell for valuable remarks for this edited version of the paper.

REFERENCES

BARTELS, J., and JOHNSTON, H. F. (1940), *Geomagnetic Tides in Horizontal Intensity at Huancayo; Part II*, Terr. Magn. Atmos. Elect. *45*, 269–308 and 485–512.

BHUYAN, P. K., and TYAGI, T. R. (1987), *Lunar and Solar Daily Variations of Equivalent Slab Thickness at Delhi*, Geophys. J. R. Astr. Soc. *88*, 487–493.

BLACK, D. I. (1970), *Lunar and Solar Magnetic Variations at Abinger: Their Detection and Estimation by Spectral Analysis via Fourier Transforms*, Phil. Trans. Roy. Soc. *A268*, 233–263.

CAMPBELL, W. H. (1983), *Research Directions in Recent Studies of the Lunar Geomagnetic Effects*, Pub. en volumen conmemorativo 75 Aniversario del Observatorio del Ebro, 147–154.

CHAPMAN, S. (1925), *The Lunar Diurnal Magnetic Variation at Greenwich and other Observatories*, Phil. Trans. Roy. Soc. *A225*, 49–91.

CHAPMAN, S. (1930), *On the Determination of the Lunar Atmospheric Tide*, Zeits. Geophys. *6*, 396–420.

CHAPMAN, S. (1932), *The Lunar Diurnal Variation of Atmospheric Temperature at Batavia, 1866–1928*, Proc. Roy. Soc. *A137*, 1–24.

CHAPMAN, S. (1942), *Notes on the Lunar Geomagnetic Tide: I. Its Mathematical and Graphical Representations, and their Significance*, Terr. Magn. Atmos. Elect. *47*, 279–294.

CHAPMAN, S., and BARTELS, J., *Geomagnetism* (Clarendon Press, Oxford 1940) 2 vols.

CHAPMAN, S., and MALIN, S. R. C. (1969), IAGA Bull. *27*, 116–118.

CHAPMAN, S., and MILLER, J. C. P. (1940), *The Statistical Determination of Lunar Daily Variations in Geomagnetic and Meteorological Elements*, Monthly Not. R. Astr. Soc. Geophys. Suppl. *4*, 649–669.

CURRIE, R. G. (1975), *Lunar Terms in the Geomagnetic Spectrum at Hermanus*, J. Atmosph. Terr. Phys. *37*, 439–446.

GUPTA, J. C., and CHAPMAN, S. (1969), *Lunar Daily Harmonic Geomagnetic Variation as Indicated by Spectral Analysis*, J. Atmosph. Terr. Phys. *31*, 233–252.

LARSEN, J. C. (1968), *Electric and Magnetic Fields Induced by Deep Sea Tides*, Geophys. J. R. Astr. Soc. *16*, 47–70.

MALIN, S. R. C., and CHAPMAN, S. (1970), *The Determination of Lunar Daily Geophysical Variations by the Chapman-Miller Method*, Geophys. J. R. Astr. Soc. *19*, 15–35.

MALIN, S. R. C., and SCHLAPP, D. M. (1980), *Geomagnetic Lunar Analysis by Least-squares*, Geophys. J. R. Astr. Soc. *60*, 409–418.

MATSUSHITA, S., and CAMPBELL, W. H. (1972), *Lunar Semidiurnal Variations of the Geomagnetic Field Determined from the 2.5-min Data Scalings*, J. Atmosph. Terr. Phys. *34*, 1187–1200.

SCHLAPP, D. M., and WEEKS, K. (1973), *The Determination of Lunar Tides: I. Methods of Analysis*, J. Atmosph. Terr. Phys. *35*, 1811–1831.

STENING, R. J., and WINCH, D. E. (1987), *The Lunar Geomagnetic Tide at Night*, Geophys. J. R. Astr. Soc. *88*, 461–476.

STENING, R. J., MEEK, C. E., and MANSON, A. H. (1986), *Middle Atmosphere Lunar Tides at Saskatoon (52°N, 107°W)*, 7th National Congress, Austr. Inst. of Physics, p. 86. (cf., Meek, C. E. and Manson, A. H., 1987 Planet. Space Sci. *35*, 445–449).

STENING, R. J. (1986), *Lunar Effects in the F Region of the Ionosphere*, J. Geophys. Res. *91*, 4581–4584.

TSCHU, K. K. (1949), *On the Practical Determination of Lunar and Luni-solar Daily Variations in Certain Geophysical Data*, Austr. J. Sci. Res. *A2*, 1–24.

WINCH, D. E. (1970), *Geomagnetic Lunar Partial Tides*, J. Geomagn. Geoelec. *22*, 291–318.

(Received December 3, 1987, revised/accepted April 27, 1988)

Grünbaum, F. (1981), Notes on the Inverse Conductance Problem. In J. R. Math, ... and Graphics, Semiconductor... and Inverse Scattering, Inst. Math., Article, Birkhäuser, pp. 279-291.

Lifshits, A., and Glazman, I., ... of Linear Operators, Oxford Univ. Press, Oxford (1965) ...

...

Pöschel, J., and Trubowitz, E. (1986), Inverse Spectral Theory, Academic Press, Boston.

Sabatier, P. C. (1980), Spectral and Scattering Inverse Problems. Resonance Phenomena, ...

...

Marchenko, V. A. (1986), Sturm-Liouville Operators and Applications, Birkhäuser, Basel.

...

(Received December 3, 1987; Accepted April 27, 1988)

PAGEOPH, Vol. 131, No. 3 (1989)

0033–4553/89/030533–17$1.50 + 0.20/0

Lunar Magnetic Variations

Denis E. Winch[1]

Abstract—Displayed daygraphs of magnetic observatory hourly mean values and of lunar magnetic variations reconstructed from spherical harmonic coefficients are used to illustrate the difficulties that arise in separating lunar magnetic effects from those associated with the 27 day recurrence tendency in magnetic activity.

Key words: Daily magnetic variations, ocean dynamo.

Introduction

Since the formation of the Göttingen Magnetic Union some 150 years ago magnetic observatories throughout the world have recorded values of the earth's magnetic field and its variations throughout the day. Hourly mean values were the preferred form in which the data were made available and analyses of hourly mean values have tended to proceed along very traditional lines. For example, vectograms are used to represent daily variations in two magnetic elements simultaneously; daygraphs are formed from monthly mean values over disturbed, quiet or all days and are used to bring out various features of the magnetic variations such as change with season, or the disturbance daily variation, or noncyclic variation.

With the advent of automatic digital recording by observatories and an ever-improving distribution of such observatories together with the availability of computerised graph plotting facilities, it seemed worthwhile to present a large amount of magnetic data on a single graph and observe what might be inferred from the appearance of the data. Such graphs give an interesting overview of seasonal changes and as will be seen, also show the influence of periodicities associated with both the 27 day recurrence tendency of magnetic activity and the lunar magnetic variations dependent on the synodic month of 29.53 days.

Observatory data chosen for the present experiment are from the magnetic observatories at Toolangi, Australia (TOO), Port Moresby, Papua New Guinea (PMG) and the Ethiopian observatory at Addis Ababa (AAE).

---

[1] Department of Applied Mathematics, University of Sydney, Sydney, N.S.W., 2006, Australia.

The Solar Daily Variation

The procedure used was to give each successive daygraph of twenty-four hourly values a small displacement down the page so that the form of the daily variation, as it changes throughout the year, takes on the appearance of a curved surface. The results are given for elements *D*, *H*, *Z*, for all three observatories, AAE, PMG, TOO, in the diagrams on the left of Figures 1, 2, 3. The actual value used for the displacement of each day is given in the caption for each figure and can be determined by the ratio between the day number which appears on the right-hand side and the corresponding units given on the left of each graph. Due to the linear trend of secular variation and to magnetic disturbance, particularly on the first few days of 1964, the daygraphs do not necessarily correspond exactly to the day number. Magnetic disturbance appears as "noise" in each figure, with the smoother quiet day variations immediately evident. Experimental graphs showing the magnetic variations as a surface, using the "hidden line removal" feature available in the plotting package, were not successful because the disturbed day peaks tended to conceal the quiet day variations in the "troughs" whilst showing the disturbance "peaks" as the dominant feature.

The graphs on the left of Figures 1, 2, 3, are plots of "raw" observatory data and at first glance, do not appear to exhibit anything more than a regular change throughout the year, as well, of course, as the expected effect of magnetic disturbance. However, if the graphs are inspected at eye level, along the curves, it is apparent that the curves are not uniformly displaced down the page but are grouped more closely together at some epochs and are spaced further apart at others. In order to enhance this effect, figures have been included to the right of the raw data plots in Figures 1, 2, 3, in which the value at any given hour has been derived as an average of its five neighbouring values at the same hour. The effect is more pronounced in graphs of horizontal intensity than in declination or vertical intensity. Graphs of daily mean values for each day have been plotted separately in Figures 1b, 2b, 3b and those days on which the daygraphs of horizontal intensity appear bunched together are those on which the daily mean value is increasing and such days are of course influenced by both magnetic disturbance and, to a lesser extent, discontinuities associated with baseline control difficulties. The hourly mean values were tabulated for 1964 with a common tabular base value for the entire year, and the tabular base value (which is not required for the present purpose) is to be added to the computed daily mean value. The regularity of the darker bands in the smoothed graphs on the right in Figures 1, 2, 3, for the Declination element seems to indicate a component associated with an origin which is rather more cyclic than an effect associated only with the 27 day recurrence tendency in magnetic activity. The lunar magnetic variation has an amplitude which would give rise to this effect and would be swamped by the 27 recurrence tendency effect in the Horizontal Intensity element.

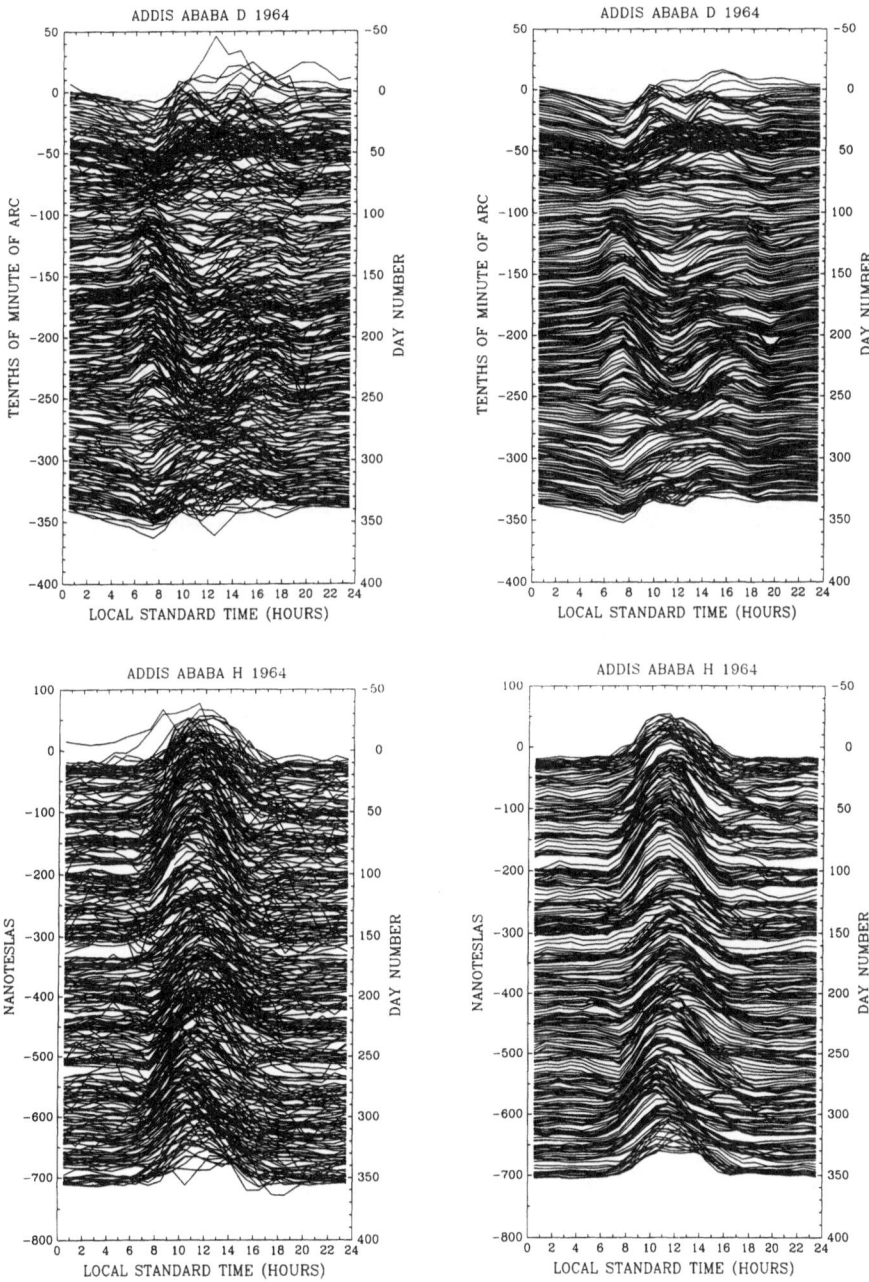

Figure 1a

Displaced daygraphs for Declination and Horizontal Intensity at Addis Ababa, Ethiopia (AAE). Graphs on the left are observatory data, graphs on the right have been obtained by averaging five neighbouring values at each fixed hour. Daily displacements for *D* and *H* are a tenth of minute of arc and 2 nT per day, respectively.

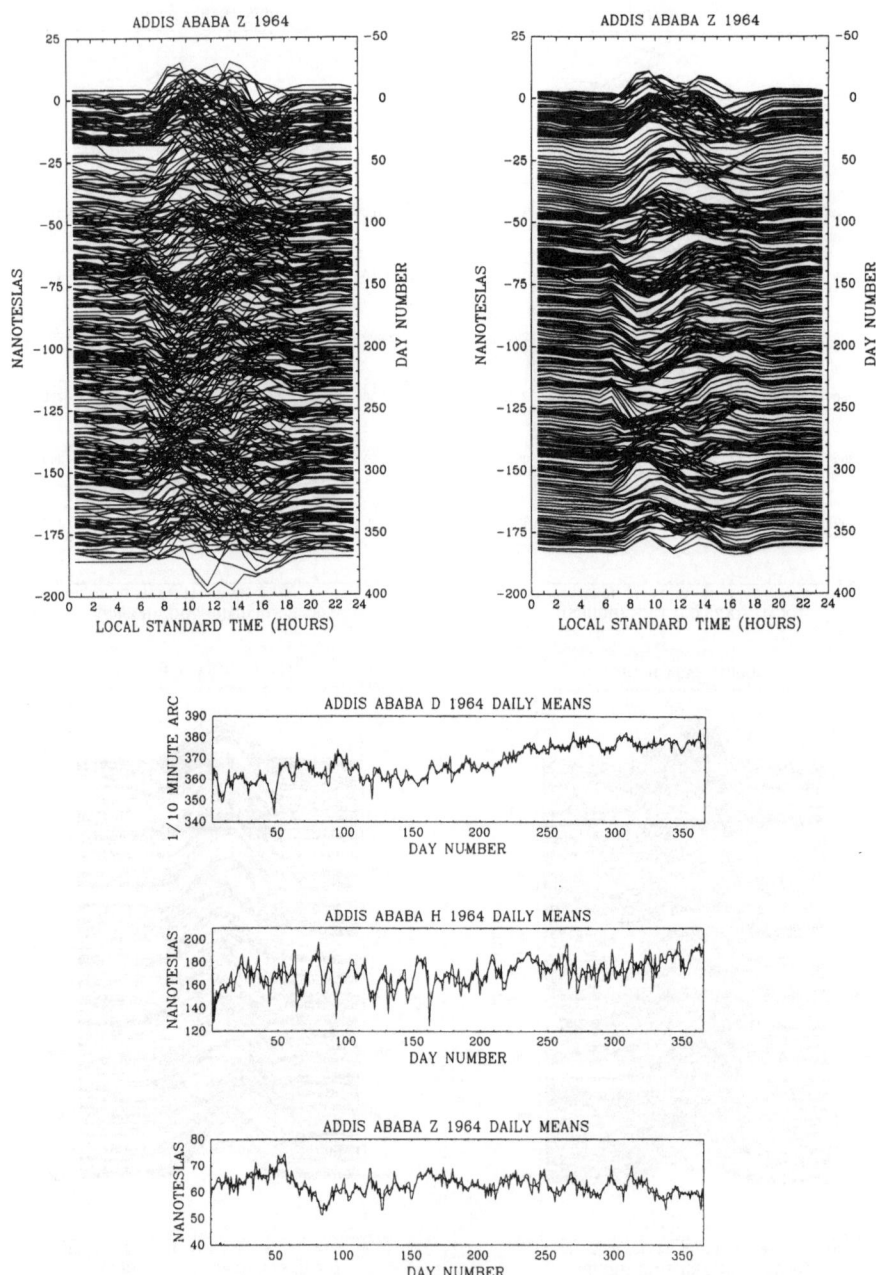

Figure 1b

Displaced daygraphs for Vertical Intensity at Addis Ababa, Ethiopia (AAE); daily displacement is
0.5 nT per day. Daily means for Declination, Horizontal Intensity and Vertical Intensity at AAE, relative
to the yearbook tabular base value are also given; running means of five values are indicated by the
smooth line. Banding of displaced daygraphs occurs when the daily mean values are increasing.

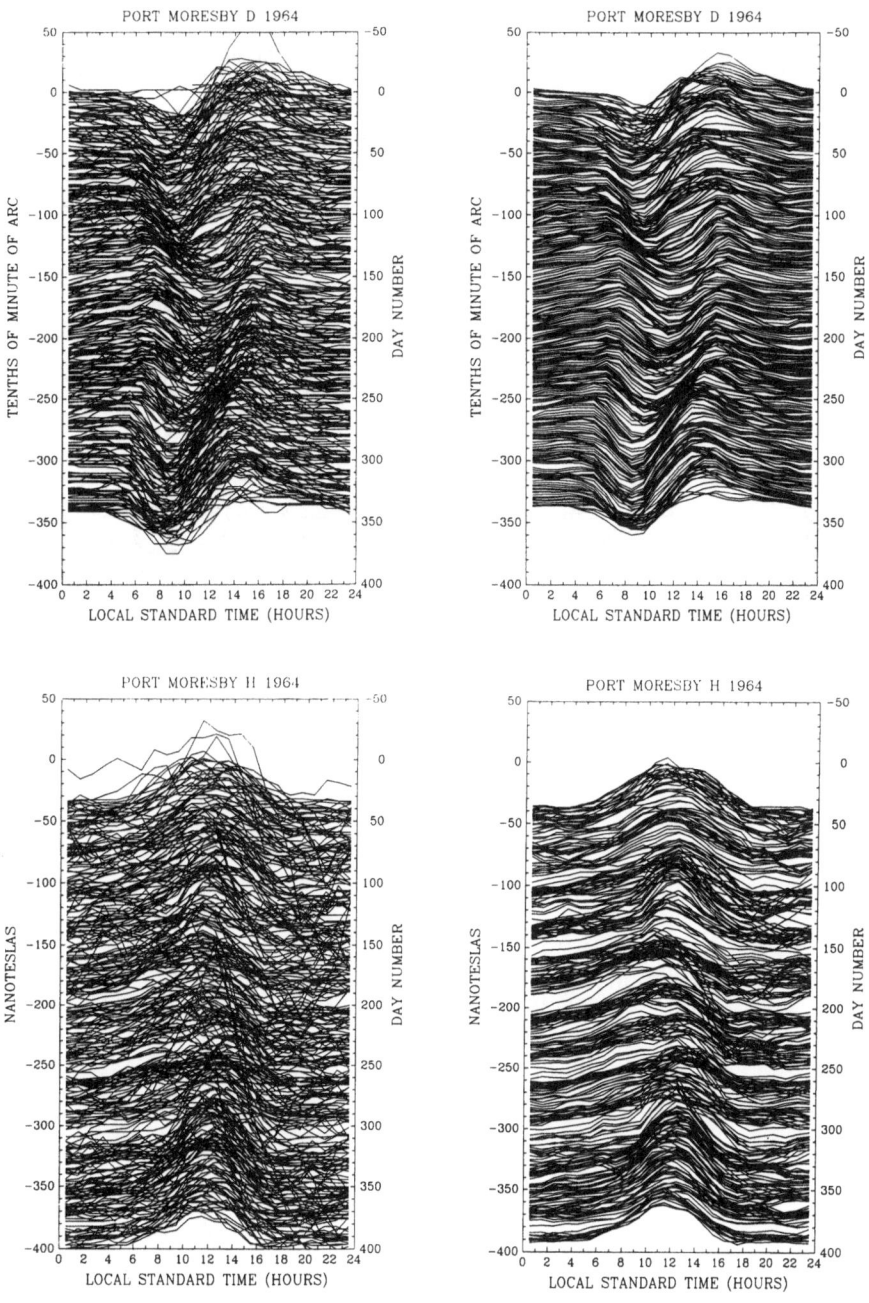

Figure 2a

Displaced daygraphs for Declination and Horizontal Intensity at Post Moresby, Papua New Guinea (PMG). Graphs on the left are observatory data. Graphs on the right have been obtained by averaging five neighbouring values at each fixed hour. Daily displacements for *D* and *H* are a tenth of a minute of arc and 1 nT per day, respectively.

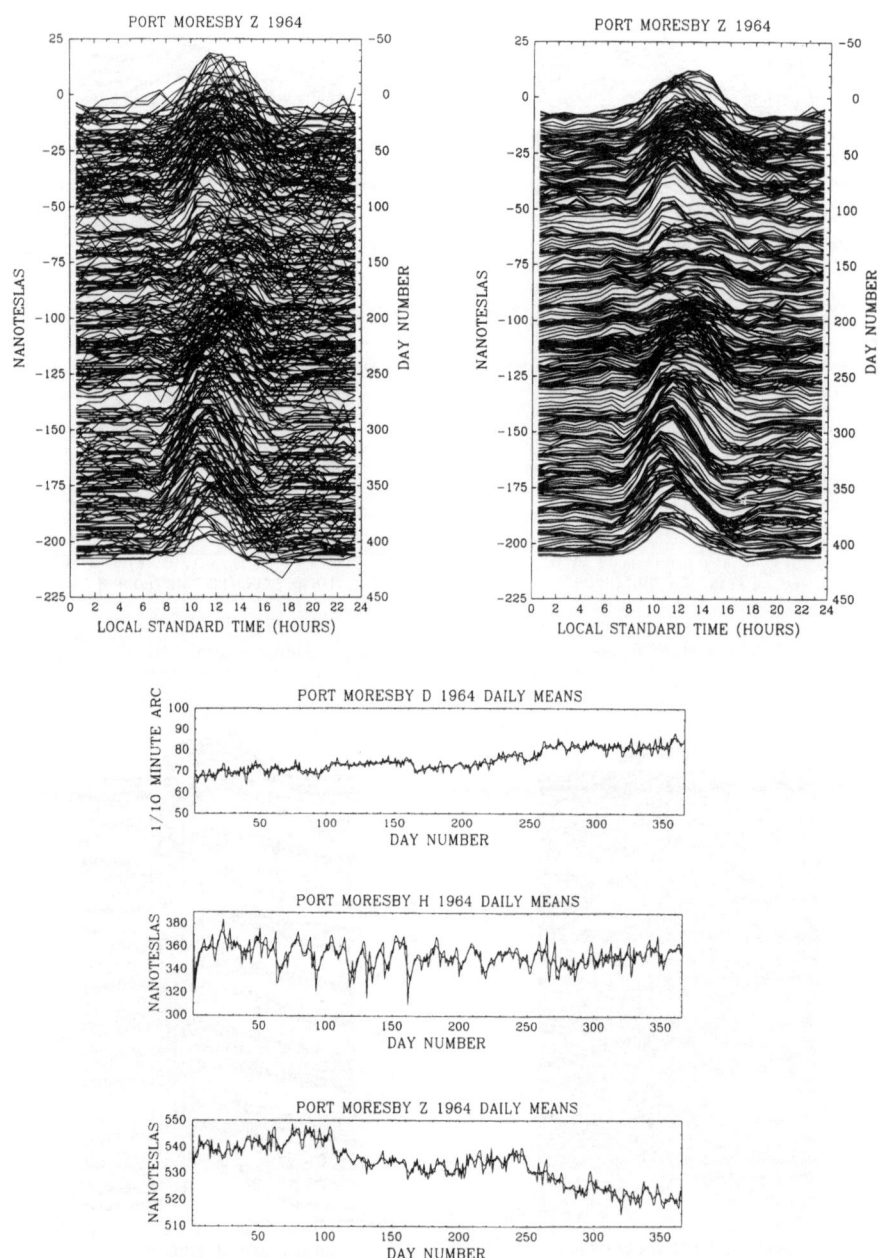

Figure 2b
Displaced daygraph for Vertical Intensity at Port Moresby, Papua New Guinea (PMG); daily displacement is 0.5 nT per day. Daily means for Declination, Horizontal Intensity and Vertical Intensity at PMG
are also given; running means of five values are indicated by the smooth line. Banding of daygraphs
occurs when the daily mean values are increasing.

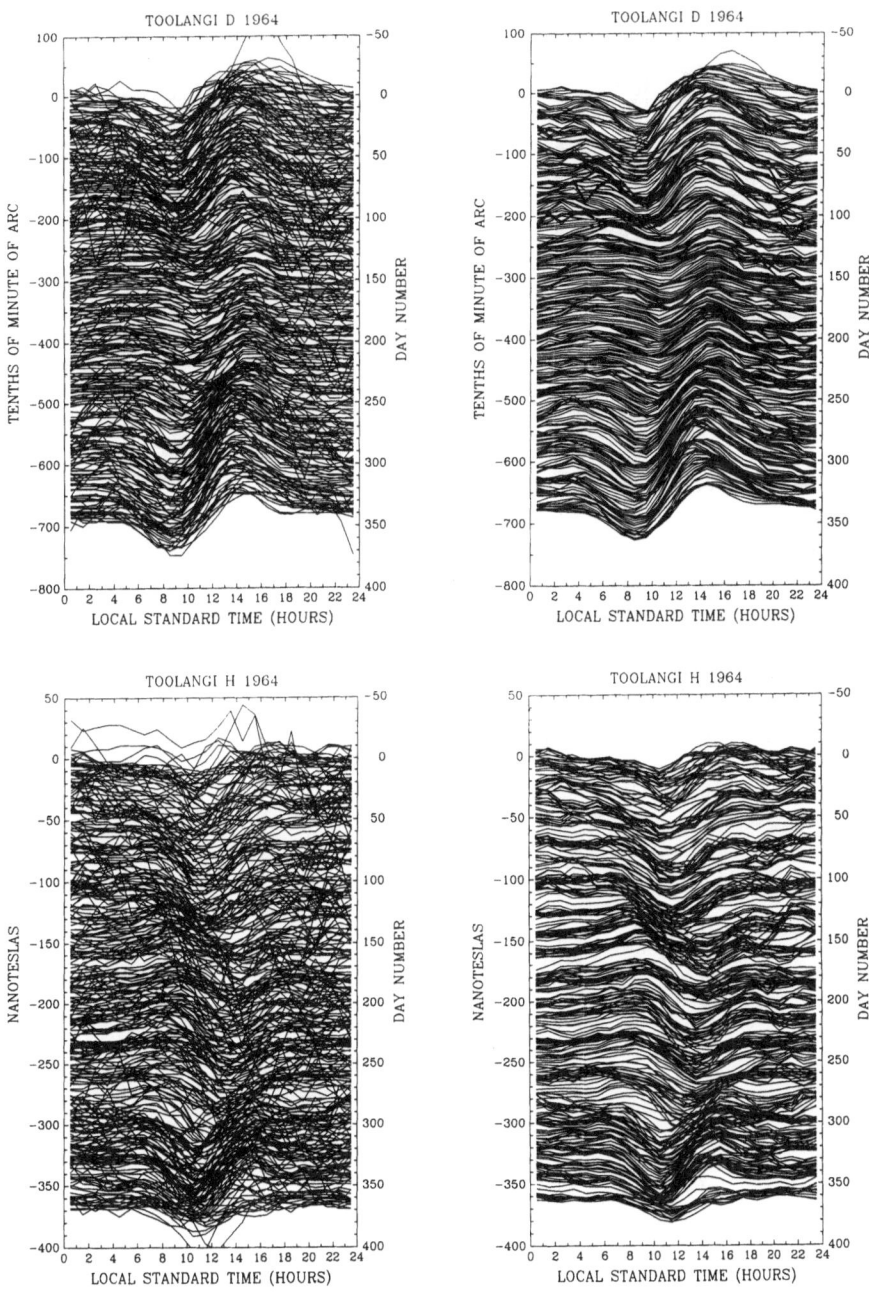

Figure 3a

Displaced daygraphs for Declination and Horizontal Intensity at Toolangi, Australia (TOO). Graphs on the left are observatory data. Graphs on the right have been obtained by averaging five neighbouring values at each fixed hour. Daily displacements for *D* and *H* are 2 tenths of minute of arc and 1 nT per day, respectively.

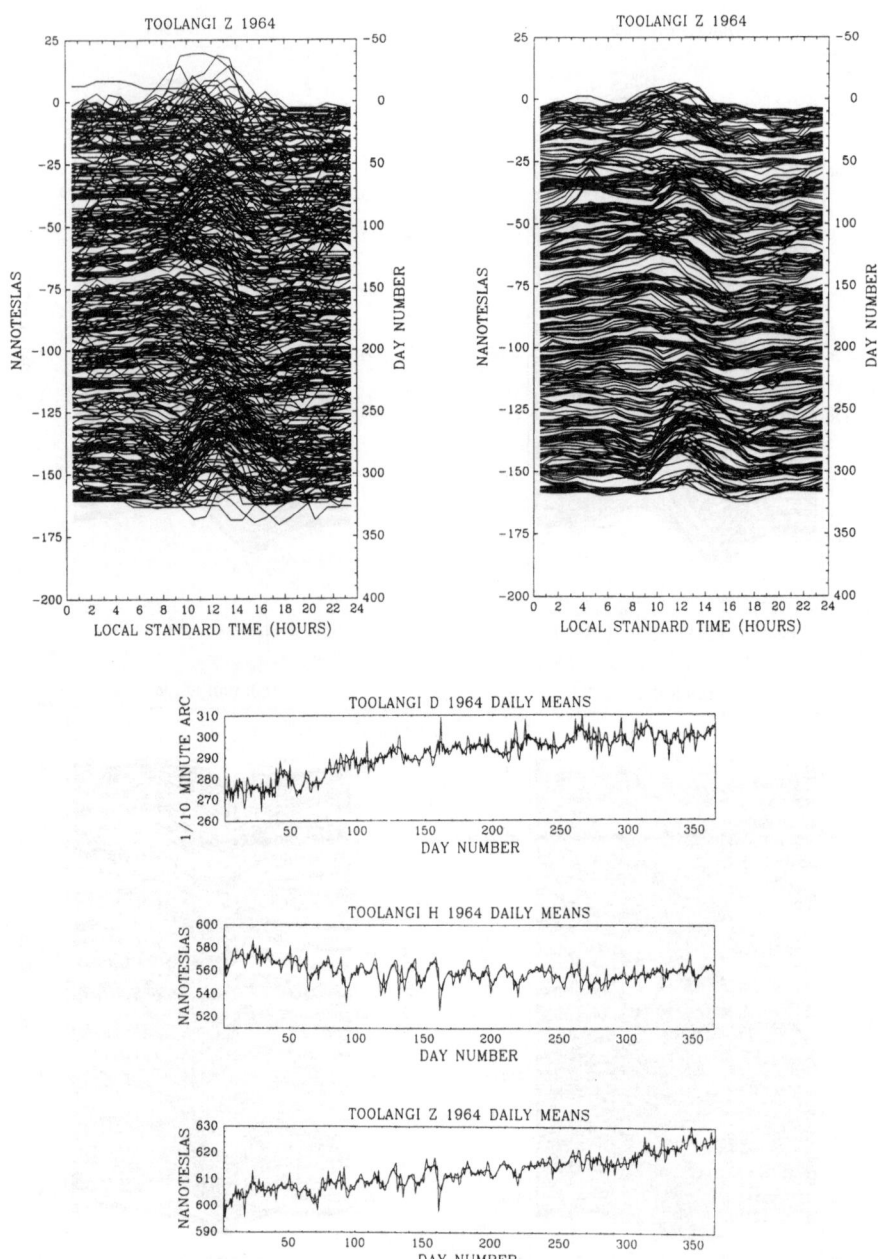

Figure 3b

Displaced daygraph for Vertical Intensity at Toolangi, Australia (TOO); daily displacement is 0.5 nT per day. Daily means for Declination, Horizontal Intensity and Vertical Intensity at TOO are also given; running means of five values are indicated by the smooth line. Banding of daygraphs occurs when the daily mean values are increasing.

The annual change in the Declination daygraphs at AAE is interesting as it appears to be fundamentally different from the much simpler forms obtained for the nonequatorial PMG and TOO; it is associated with a diurnal change in the ionospheric current system component flowing normal to the equator. The graphs for Declination for PMG and TOO appear to be very similar to each other, but a smaller daily displacement has been used to enhance the smaller daily range of Sq in Declination at TOO, which is the observatory of these three which is closest to the Sq overhead current system focus.

There seems to be very little change throughout the year in Horizontal Intensity daygraphs for AAE, whilst the daygraphs for PMG and TOO are typical of those which are to the south or to the north of the Sq overhead current system focus. The banding together of daygraphs is very evident in all the Horizontal Intensity graphs and is clearly present at all three observatories at local midnight, indicating an association with long-period terms or ring-current disturbance effects rather than some diurnally modulated term.

The annual change in vertical intensity appears to be quite different at the three observatories. At AAE, there is a sharp change in the form of the variation at the equinoxes, whilst at PMG there is the expected decrease in the amplitude of the daily variation during local winter months and from the appearance of the daygraphs, it would appear that an afternoon effect in the first 100 days experiences a transition to become a morning effect after the March equinox. At TOO, both the raw data and smoothed daygraphs given in Figure 3b indicate that the annual change in vertical intensity seems to be rather indefinite.

The Lunar Daily Variation

On forming the product of a Fourier series expansion for conductivity of the ionosphere:

$$K = K_0 \sum_{n=0}^{4} a_n \sin[n(t + \phi) + \alpha_n]$$

in which t is UT and ϕ is east longitude, with a term representing the principal lunar semi-diurnal tide in the lunar tide generating potential:

$$M_2 = 0.90812 \quad G_2 \cos(2t - 2s + 2h)$$

one obtains a model for the time-dependence of lunar magnetic tides in the form:

$$L(M_2) = \sum_{n=-2}^{6} A_n \sin(nt - 2s + 2h + \beta_n) \tag{1}$$

in which the summands $n = 1, 2, 3, 4$ are the phase-law tides, $n = 0$ is the long-period term and summands $n = -1, -2$, are the partial tides. The long-period term and partial tide terms are ignored in the traditional Chapman-Miller method of analysis, although the long-period term and partial tides for $n = -1, -2, -3, -4$ are included in the extension of the method presented by WINCH and CUNNINGHAM (1972). It is convenient to denote the lunar magnetic tides collectively in Eq. (1) as $L(2s - 2h)$ and then changes with the season can be considered as sum and difference frequencies and denoted $L(2s - 3h)$ and $L(2s - h)$ for one cycle per year effects. The magnetic variation associated with the lunar elliptic tide is denoted $L(3s - 2h - p)$ with this convention and seasonal changes as $L(3s - 3h - p)$ and $L(3s - h - p)$.

Another source for the lunar magnetic variations is the ocean-dynamo, in which the M_2 tidal movement of the ocean interacts with the main magnetic field to produce a lunar semi-diurnal term whose time dependence is that corresponding to $n = 2$ only in $L(2s - 2h)$. Similarly the lunar elliptic ocean tide contributes to the $n = 2$ term in $L(3s - 2h - p)$.

Spherical harmonic analyses of all these magnetic tides, phase-law, long-period and partial tides, have already been published, e.g., WINCH (1981) and indeed, another purpose of this paper is to illustrate the reconstructed lunar magnetic variations at individual magnetic observatories rather than give the usual world-wide equivalent overhead ionospheric electric current systems. Graphs on the left of Figures 4, 5, 6, present reconstructed lunar magnetic variations in the magnetic elements, X, Y, Z at AAE, PMG and TOO, based on the spherical harmonic coefficients of WINCH (1981) for phase-law tides only of the lunar tides $L(2s - 2h)$, the seasonal variation terms $L(2s - 4h)$, $L(2s - 3h)$, $L(2s - h)$, $L(2s)$, the lunar elliptic tide $L(3s - 2h - p)$ and the seasonal variation terms $L(3s - 3h - p)$, $L(3s - h - p)$.

Graphs on the right of Figures 4, 5, 6, have been computed from both phase-law and partial tides together using the lunar magnetic tides and seasonal changes as for those on the left. There is not a substantial difference between graphs on the left and the corresponding graph on the right in Figures 4, 5, 6, indicating the relatively minor nature of the partial tides. The very different nature of the seasonal change of the lunar magnetic variations from that of the solar daily variation which has a substantial annual mean daygraph is however very evident upon comparing Figures 1, 2, 3 and 4, 5, 6, respectively.

Long-period tides reconstructed from $n = 0$ terms in the lunar magnetic tides used in Figures 4, 5, 6, are given in Figures 4b, 5b, 6b, and can be compared with the graphs of daily mean values over the same epoch given in Figures 1b, 2b, 3b. It does seem likely that these long-period terms contain a very substantial component arising from the effects of the 27-day recurrence tendency in magnetic activity on the magnetic field.

The method presented by MALIN (1970) for the calculation of ocean effects, uses harmonics $n = 1, 3, 4$, evaluated at local midnight to determine $n = 2$ term. Calculations by WINCH (1981) showed that the method was very satisfactory from the point of view of phase-angle differences that were expected for external $n = 2$ terms of $L(2s - 2h)$ on the basis of the Hough function structure of atmospheric tides. That is to say, the method of Malin produced significant changes in both the

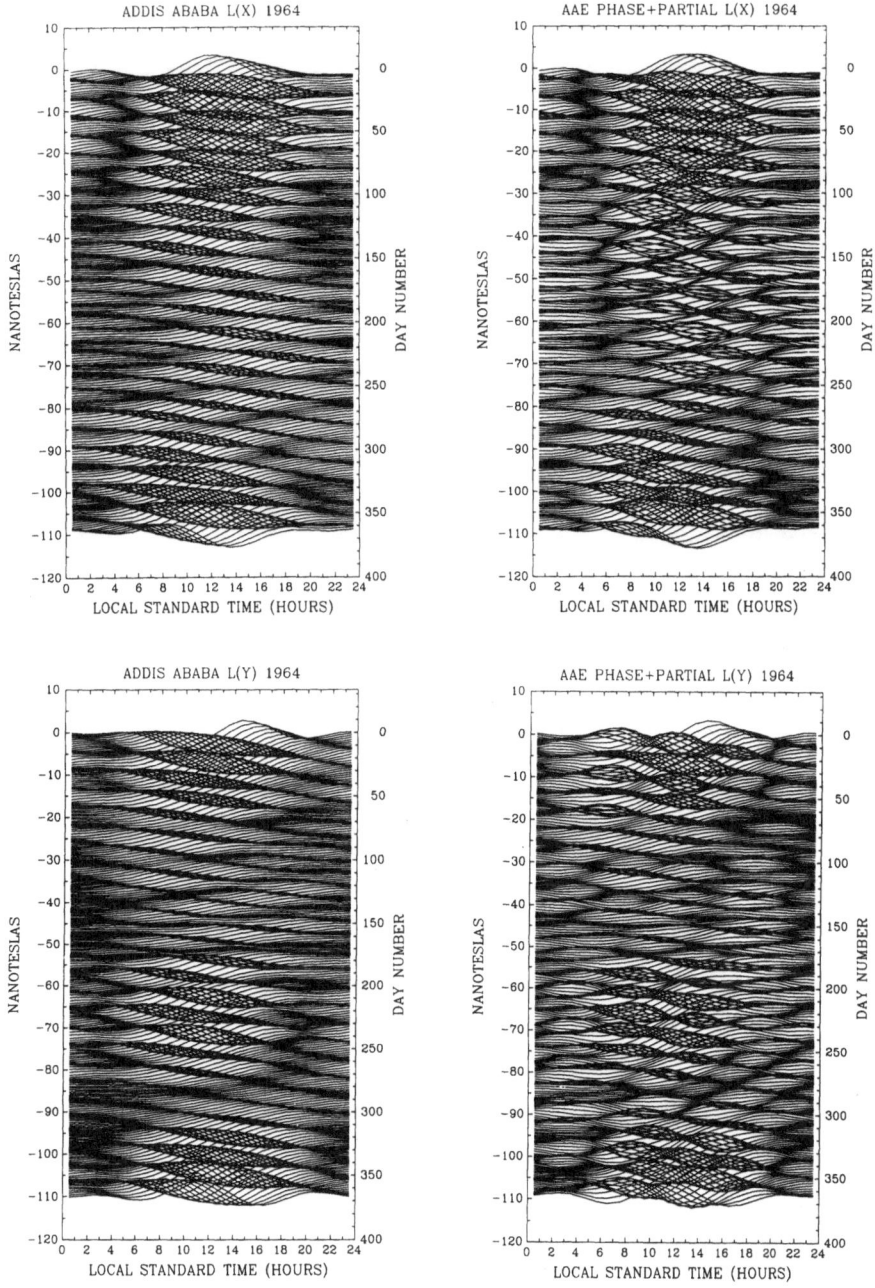

Figure 4a

Displaced daygraphs for lunar magnetic variations in northward (X) and eastward (Y) components at
Addis Ababa, Ethiopia (AAE). Graphs on the left have been reconstructed from spherical harmonic
coefficients for phase-law tides only, whilst graphs on the right have been reconstructed from spherical
harmonic coefficients for both phase-law and partial tides. Daily displacement is 0.3 nT per day.

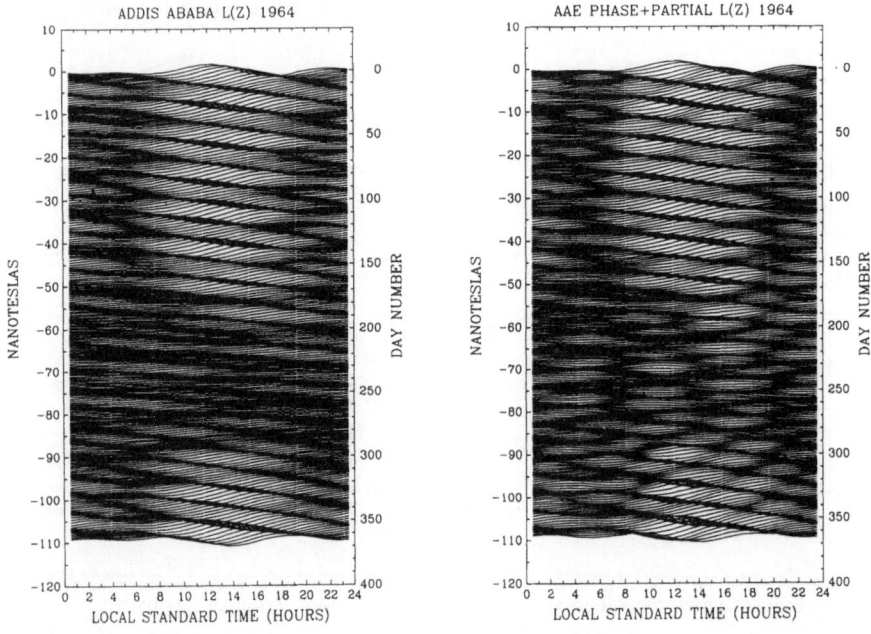

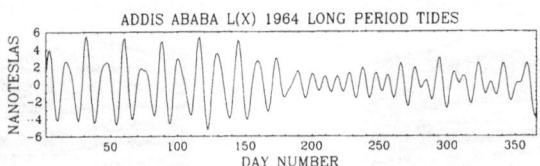

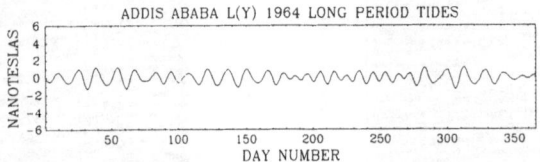

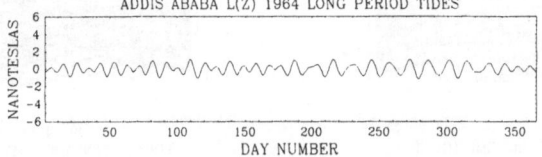

Figure 4b

Displaced daygraphs for lunar magnetic variations in Vertical Intensity at Addis Ababa Ethiopia (AAE);
daily displacement is 0.3 nT per day. Daily mean values for long-period tides at AAE, reconstructed
from spherical harmonic coefficients.

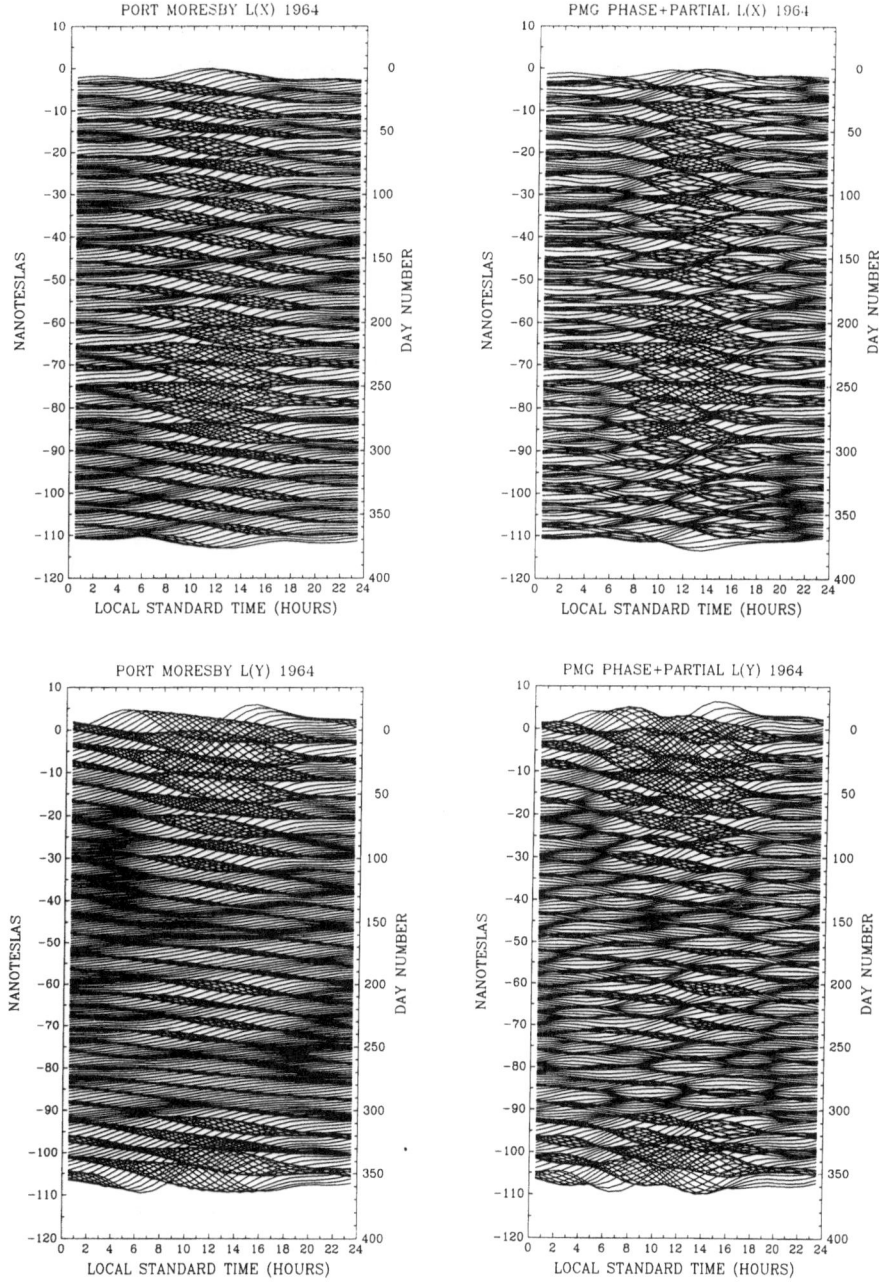

Figure 5a

Displaced daygraphs for lunar magnetic variations in northward (X) and eastward (Y) components at Port Moresby, Papua New Guinea (PMG). Graphs on the left have been reconstructed from spherical harmonic coefficients for phase-law tides only, whilst graphs on the right have been reconstructed from spherical harmonic coefficients for both a phase-law and partial tides. Daily displacement is 0.3 nT per day.

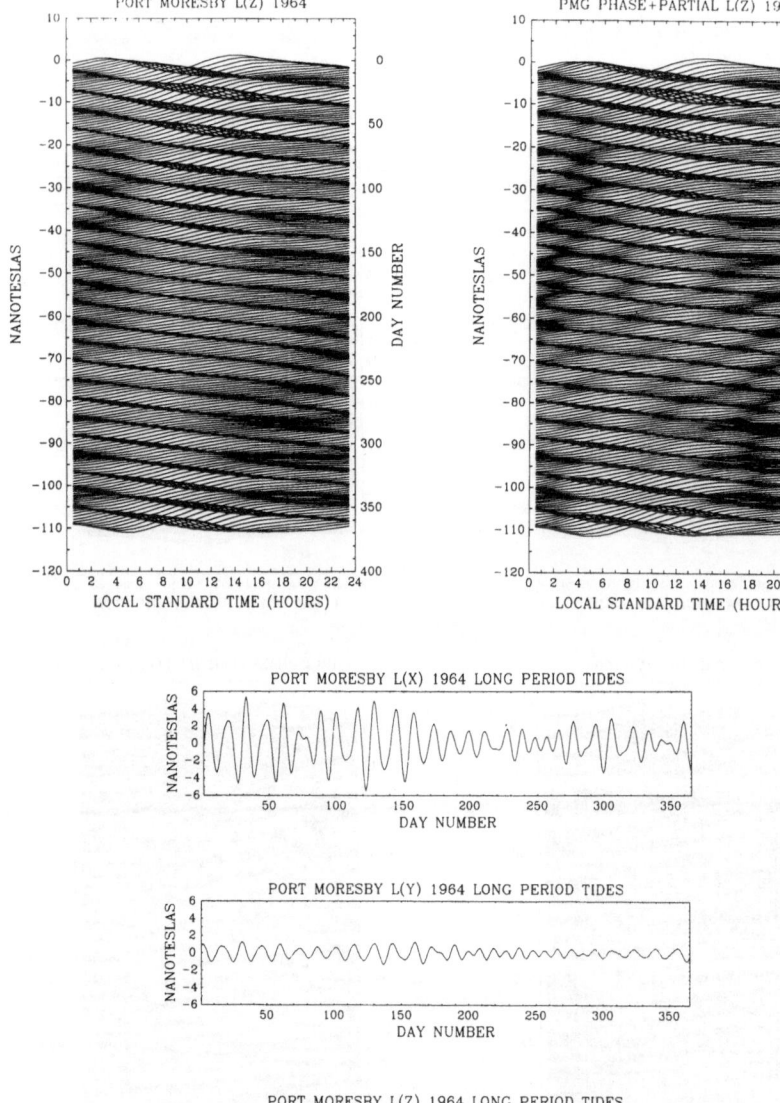

Figure 5b
Displaced daygraphs for lunar magnetic variations in Vertical Intensity at Port Moresby, Papua New Guinea (PMG); daily displacement is 0.3 nT per day. Daily mean values for long-period tides at PMG, reconstructed from spherical harmonic coefficients.

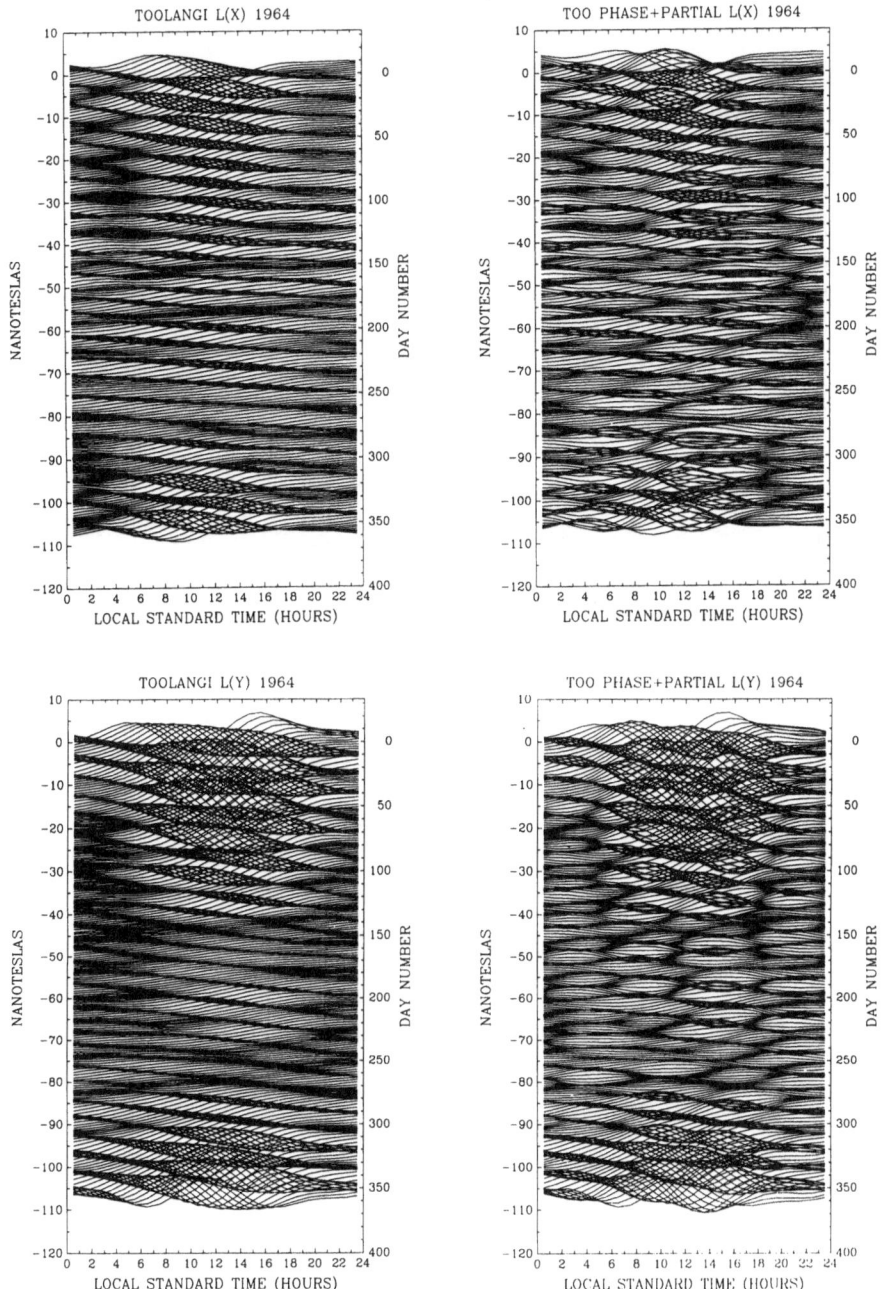

Figure 6a
Displaced daygraphs for lunar magnetic variations in northward (X) and eastward (Y) components at Toolangi, Australia (TOO). Graphs on the right have been reconstructed from spherical harmonic coefficients for phase-law tides only, whilst graphs on the right have been reconstructed from spherical harmonic coefficients for both phase-law and partial tides. Daily displacement is 0.3 nT per day.

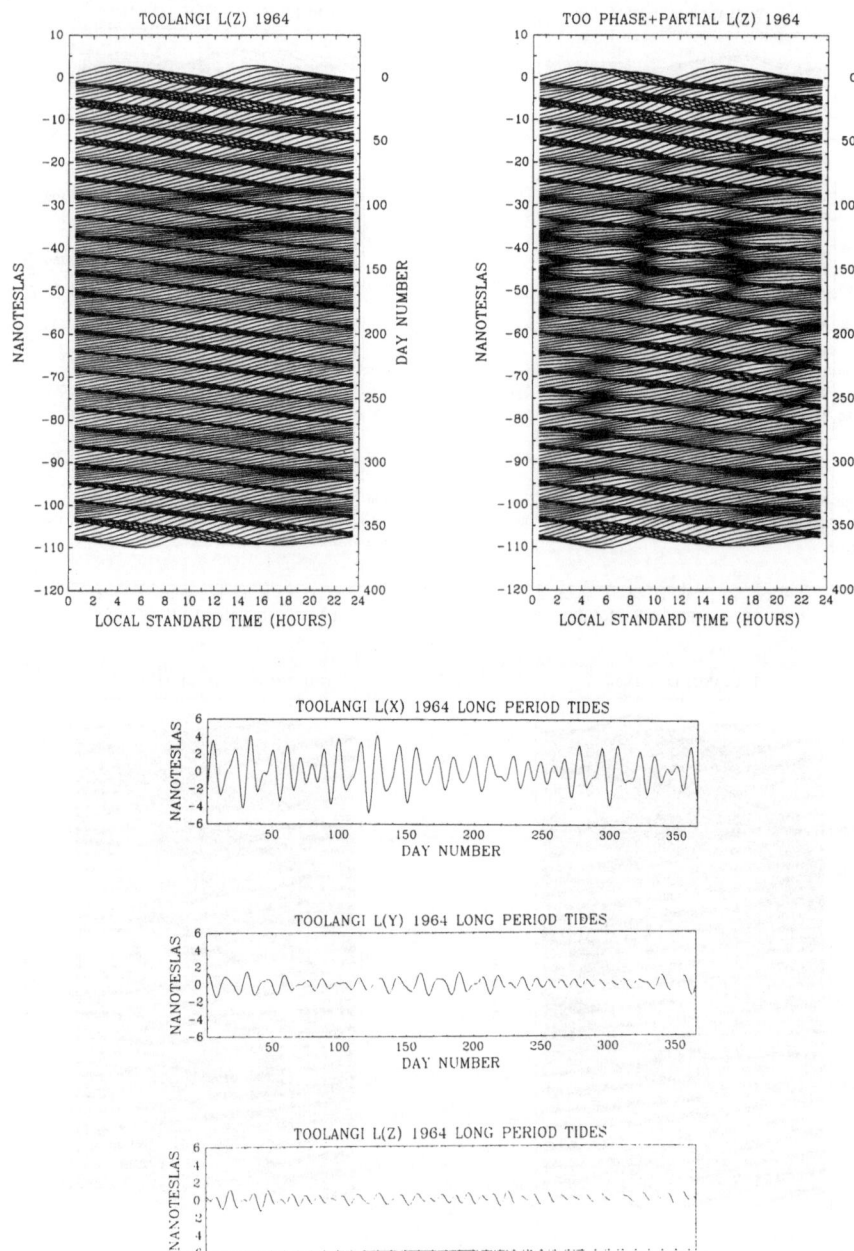

Figure 6b

Displaced daygraphs for lunar magnetic variations in Vertical Intensity at Toolangi, Australia (T); daily displacement is 0.3 nT per day. Daily mean values for long-period tides at TOO, reconstructed from spherical harmonic coefficients.

internal and external components of the $n = 2$ term of $L(2s - 2h)$, whereas if it were only a pure ocean effect, then only the internal term would be involved. The graphs given here, especially for horizontal intensity, show very clearly ring-current effects which occur at all hours of the day including local midnight, and that any analysis of local midnight values for whatever purpose will require a study of the contribution from these ring current effects. Use of data from years with different relative sunspot numbers can be used for this purpose, e.g., STENING and WINCH (1987) found the Malin method gave oceanic components in good agreement between sunspot maximum and minimum years.

Summary

Displaced daygraphs of magnetic observatory data and reconstructed lunar magnetic variation data show very clearly the nature of the annual change on quiet days and the very substantial difference in the annual change of solar and lunar magnetic variations. Displaced daygraphs of horizontal intensity show a banding effect associated with the 27-day recurrence tendency whilst the declination and vertical intensity show banding effects which are more likely to be associated with the lunar magnetic daily variation.

Acknowledgements

Data for Toolangi and Port Moresby were originally provided in yearbook form by the Australian Bureau of Mineral Resources and are now on a magnetic tape containing all available hourly mean value magnetic data for 1964–65. The data have been rewritten with a separate file for each of the 130 magnetic observatories and using three character observatory mnemonics. Data for Addis Ababa were obtained from the compact disc collection of selected magnetic data, distributed by the NOAA/NESDIS National Geophysical Data Center. I am grateful to Dr. Joe Allen for making the compact disc so readily available.

REFERENCES

MALIN, S. R. C. (1970), *Separation of Lunar Daily Geomagnetic Variations into Parts of Ionospheric and Oceanic Origin*, Geophys. J. Roy. Astr. Soc. *21*, 447–455.

STENING, R. J., and WINCH, D. E. (1987), *The Lunar Geomagnetic Tide at Night*, Geophys. J. R. Astr. Soc. *88*, 461–476.

WINCH, D. E. (1981), *Spherical Harmonic Analysis of Geomagnetic Tides, 1964–1965*, Phil. Trans. R. Soc. Lond. A*303*, 1–104.

WINCH, D. E., and CUNNINGHAM, R. A. (1972), *Lunar Magnetic Tides at Wath-eroo, Seasonal, Elliptic, Evectional, Variational and Nodal Components*, J. Geomag. Geoelect. *24*, 381–414.

(Received/accepted July 1, 1988)

PAGEOPH

Reprints from *Pure and Applied Geophysics*

New in the series:

Subduction Zones

Edited by

Larry J. Ruff
Department of Geological
Sciences, University of
Michigan, Ann Arbor, MI, USA

Hiroo Kanamori
Seismological Laboratory,
California Institute of
Technology, Pasadena, CA, USA

Part I:
1988. 352 pages. Paperback
ISBN 3-7643-1928-3

Part II:
1989. 288 pages. Paperback
ISBN 3-7643-2272-1

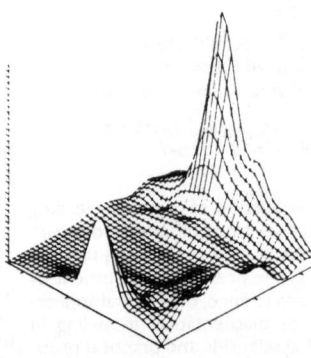

Subduction Zones is a diverse collection of contributions that portray many different facets of the subduction process. Accretionary prism dynamics, subduction initiation, deep mantle penetration of subducted lithosphere, continental crust recycling, and seismicity are some of the topics discussed in the two volumes of *Subduction Zones*. Volume I consists of two sections, *Global-scale reviews* and *Accretionary prism processes*, while Volume II contains the two sections *Tectonics & subduction initiation* and *Earthquake occurrence.*

Seismicity in Mines

Edited by

S. J. Gibowicz
Institute of Geophysics,
Polish Academy of Sciences,
Warsaw, Poland

1989. 398 pages,
Paperback
ISBN 3-7643-2273-X

In this collection of 22 papers, new observations describing seismicity patterns and source mechanism of seismic events induced by mining are reported. The methods describing location of mine tremors with accuracy of the order of a few tens of meters are also included, as well as seismic tomography and its applications in mines. The research on rockbursts undertaken recently in Canada, and recent progress in research on seismicity in mines in South Africa are described in some detail.

Middle Atmosphere

Edited by

Alan R. Plumb
MIT, Cambridge, MA, USA

Robert A. Vincent
University of Adelaide,
Australia

1989. 472 pages.
Paperback
ISBN 3-7643-2290-X

This book presents an up-to-date summary of current research into the dynamics of the middle atmosphere, addressing an ever-increasing need to understand the complex interplay of processes which control the distribution of atmospheric ozone and the impact of stratospheric pollutants on ozone. Special emphasis is given to studies of the southern hemisphere.

Most of the papers contained here were presented at workshops held in Adelaide, South Australia, in May 1987.

Please order through
your bookseller
or Birkhäuser Verlag, P. O. Box 133,
CH-4010 Basel / Switzerland
or, for orders originating from
the USA and Canada, through
Birkhäuser Boston, Inc.,
c/o Springer-Verlag New York, Inc.,
44 Hartz Way, Secaucus,
NJ 07096–2491 / USA

**Birkhäuser
Verlag**
Basel · Boston · Berlin